SIDNEY COLEMAN'S LECTURES ON RELATIVITY

Sidney Coleman (1937–2007) earned his doctorate at Caltech under Murray Gell-Mann. Before completing his thesis, he was hired by Harvard and remained there his entire career. A celebrated particle theorist, he is perhaps best known for his brilliant lectures, given at Harvard and in a series of summer school courses at Erice, Sicily, several of which are collected in *Aspects of Symmetry* (Cambridge, 1984). Three times in the 1960s he taught a graduate course on special and general relativity; this book is based on lecture notes taken by three of his students and compiled by the editors.

DAVID J. GRIFFITHS received his BA (1964) and PhD (1970) from Harvard University, and was Professor of Physics at Reed College from 1978 until 2009. He is author of *Introduction to Electrodynamics* (4th ed. Cambridge University Press, 2017), *Introduction to Quantum Mechanics* (3rd ed. with Darrell Schroeter, Cambridge University Press, 2018), *Introduction to Elementary Particles* (2nd ed. Wiley, 2008), and *Revolutions in Twentieth-Century Physics* (Cambridge, 2013). He is also coeditor of *Quantum Field Theory Lectures of Sidney Coleman* (World Scientific, 2018). He was a PhD student of Sidney Coleman, and this book is based in part on his lecture notes from Coleman's course on relativity.

DAVID DERBES received his BA (1974) from Princeton University, Part III of the Mathematical Tripos at the University of Cambridge (1975), and his PhD (1979) from the University of Edinburgh. He was a high school teacher for forty years, winning a Golden Apple in 2007. He has edited Freeman Dyson's *Advanced Quantum Mechanics* (World Scientific, 2007, 2011), and is coeditor of Lillian Lieber's *The Einstein Theory of Relativity* (Paul Dry Books, Philadelphia, 2008) and *Quantum Field Theory Lectures of Sidney Coleman* (World Scientific, 2018).

RICHARD B. SOHN received his BA (1969) and PhD (1976) in physics from the University of Connecticut, and a master's in physics (1971) from the University of Maryland. He has held roles as a professor in physics, a research scientist, and a software engineer. He is coeditor of *Quantum Field Theory Lectures of Sidney Coleman* (World Scientific, 2018).

SIDNEY COLEMAN'S LECTURES ON RELATIVITY

Compiled and Edited by
David J. Griffiths, David Derbes, Richard B. Sohn

CAMBRIDGE
UNIVERSITY PRESS

CAMBRIDGE
UNIVERSITY PRESS

University Printing House, Cambridge CB2 8BS, United Kingdom

One Liberty Plaza, 20th Floor, New York, NY 10006, USA

477 Williamstown Road, Port Melbourne, VIC 3207, Australia

314–321, 3rd Floor, Plot 3, Splendor Forum, Jasola District Centre, New Delhi – 110025, India

103 Penang Road, #05–06/07, Visioncrest Commercial, Singapore 238467

Cambridge University Press is part of the University of Cambridge.

It furthers the University's mission by disseminating knowledge in the pursuit of education, learning, and research at the highest international levels of excellence.

www.cambridge.org
Information on this title: www.cambridge.org/9781316511725
DOI: 10.1017/9781009053716

© Cambridge University Press 2022

First published 2022

Printed in the United Kingdom by TJ Books Limited, Padstow Cornwall

A catalogue record for this publication is available from the British Library.

Library of Congress Cataloging-in-Publication Data
Names: Coleman, Sidney, 1937–2007, author. | Griffiths, David J. (David Jeffery),
1942– compiler, editor. | Derbes, David, 1952– compiler, editor. | Sohn, Richard B., 1947– compiler, editor.
Title: Sidney Coleman's lectures on relativity / compiled and edited by
David J. Griffiths, David Derbes, Richard B. Sohn.
Other titles: Lectures. Selections
Description: Cambridge ; New York, NY : Cambridge University Press, 2022. |
Includes bibliographical references and index.
Identifiers: LCCN 2021044736 | ISBN 9781316511725 (hardback)
Subjects: LCSH: Relativity (Physics) | BISAC: SCIENCE / Space Science / Cosmology
Classification: LCC QC173.58 .C65 2022 | DDC 530.11–dc23/eng/20211006
LC record available at https://lccn.loc.gov/2021044736

ISBN 978-1-316-51172-5 Hardback

Additional resources for this publication at www.cambridge.org/lecturesonrelativity

Contents

Preface   *page* ix

Part I   Special Relativity

1   The Geometry of Special Relativity   3
     1.1   Introduction   3
              1.1.1   Classical Physical Systems   3
              1.1.2   Symmetries   4
     1.2   Poincaré Invariance   6
              1.2.1   Geometrical Symmetries of Classical Physics   6
              1.2.2   Active and Passive Transformations   11
              1.2.3   Minkowski Space   12
              1.2.4   Topological Structure of the Lorentz Group   13
              1.2.5   Rotations and Boosts   17
              1.2.6   Simultaneous Dilations and Lorentz Transformations   19
     1.3   Time Dilation and Lorentz Contraction   20
              1.3.1   Arc Length and Proper Time   20
              1.3.2   Time Dilation   21
              1.3.3   Lorentz Contraction   22
     1.4   Examples and Paradoxes   23
              1.4.1   The Time Dilation Paradox   23
              1.4.2   The Twin Paradox   25
              1.4.3   Doppler Shift   27
              1.4.4   The Bandits and the Train   27
              1.4.5   The Prisoner's Escape   28
              1.4.6   The Moving Cube   29
              1.4.7   Tachyons   32

2   Relativistic Mechanics   34
     2.1   Tensor Formalism   34
     2.2   Conservation Laws   39
            2.2.1   Conservation Laws Depending Only on Velocity   40
            2.2.2   Conservation Laws including Position   45
     2.3   Lagrangian Particle Mechanics   48
     2.4   Lagrangian Field Theory   51
            2.4.1   Internal Symmetries and Conservation Laws   52
            2.4.2   Invariance under the Poincaré Group   55
            2.4.3   Symmetrization of the Stress Tensor   58

3   Relativistic Electrodynamics   63
     3.1   Lagrangian Formulation   63
            3.1.1   The Free Maxwell Field   63
            3.1.2   Maxwell Field with Source   66
     3.2   Potentials and Fields of a Point Charge   68
            3.2.1   The Action for a Point Charge   68
            3.2.2   Green's Function for the Wave Equation   70
            3.2.3   "In" and "Out" Fields   75
     3.3   Radiation from a Point Charge   76
            3.3.1   The Liénard–Wiechert Potential   76
            3.3.2   The Fields of a Point Charge   78
     3.4   Regularization and Renormalization   80
            3.4.1   Particle Motion with Radiation Reaction   84
            3.4.2   Conservation of Energy   89
            3.4.3   Hyperbolic Motion   90

Part II   General Relativity

4   The Principle of Equivalence   95
     4.1   Gravitational and Inertial Mass   95
     4.2   The Eötvös Experiment   96
     4.3   Gravitation and Geometry   97
     4.4   The Equivalence Principle Revisited   98

5   Differential Geometry   100
     5.1   Manifolds   100
            5.1.1   Vectors   102
            5.1.2   Exterior Calculus   104
            5.1.3   Tensor Densities   109
     5.2   Affine Spaces   115
            5.2.1   Affine Connections   115

5.2.2   How Γ Transforms   116

5.2.3   Parallel Transport of Tensors and Tensor Densities   118

5.2.4   Covariant Derivatives   119

5.3   Riemannian Manifolds   121

     5.3.1   Relation between Affine Connection and Metric   123

     5.3.2   Symmetries of the Riemann Tensor   124

     5.3.3   Flatness and Curvature   127

6   Gravity   132

     6.1   Motion in Curved Spacetime   132

         6.1.1   Program for a Theory of Gravity   132

         6.1.2   Classical Equations in Covariant Form   132

         6.1.3   Tidal Forces   137

     6.2   The Gravitational Field   138

         6.2.1   Einstein's Equation in Empty Space   138

         6.2.2   Alternative Theories   142

         6.2.3   The Source of Gravity   143

         6.2.4   Action Principle Formulation   144

     6.3   Linearized Gravity   150

         6.3.1   Simplifying the Field Equation   150

         6.3.2   Recovering Newton's Law   151

         6.3.3   Gravity Waves   154

7   The Schwarzschild Solution   162

     7.1   Isometries   162

     7.2   The Exterior Solution   166

     7.3   Classic Tests of General Relativity   170

         7.3.1   Precession of the Perihelion of Mercury   171

         7.3.2   Bending of Starlight   176

         7.3.3   Gravitational Redshift   179

         7.3.4   What Do They Really Test?   180

     7.4   The Interior Solution   184

     7.5   The Schwarzschild Singularity   191

         7.5.1   Kruskal Coordinates   193

         7.5.2   Geometry of the Equatorial Surface   195

         7.5.3   Tidal Stress near $r = 0$ 197

8   Conservation and Cosmology   200

     8.1   Conservation Laws   200

         8.1.1   Scalar Conservation Laws   200

         8.1.2   The Energy–Momentum Pseudotensor   201

8.2	The Universe at Large	203
	8.2.1 General Principles	204
	8.2.2 The Robertson–Walker Metric	208
	8.2.3 Redshift and Luminosity	209
8.3	General Relativity and Cosmology	210
	8.3.1 The Friedman Universe	210
	8.3.2 The Cosmological Constant	214
	8.3.3 Singularities in the Robertson–Walker Metric	215

Afterword   220

Appendix A   Compendium of Formulas   223

Appendix B   Final Exams   230
| B.1 | Final Exam, 1966 | 230 |
| B.2 | Final Exam, 1969 | 231 |

Index   234

Preface

In the spring of 1966, the fall of 1967, and the fall of 1969, Sidney Coleman taught Physics 210 at Harvard. The course title was Relativity, and it was divided into two parts: Special Relativity and General Relativity. Einstein's special theory of relativity (1905) is a metatheory in the sense that it lays down restrictions (Lorentz invariance) that *any* physical theory must obey; it is not the story of any *particular* phenomenon, but rather a description of the spacetime arena in which *all* phenomena take place. If you were to propose some new physical theory, the first question would be, "Is it consistent with special relativity?" General relativity, which emerged in pieces from 1909 to 1917, is Einstein's theory of gravity. It is a generalization of the special theory only in the sense that whereas the special theory allows one to do physics in reference frames moving at uniform velocity, the general theory shows how to do physics in *arbitrary* coordinates, with the inclusion of gravity. But it's a misnomer, really: it should be called the theory of gravity.

This book is not verbatim Coleman—we have no such record—it is based on lecture notes taken by three graduate students enrolled in the class: David Griffiths and David Levin in 1966, and David Politzer in 1969, and we thank them for allowing us to use their work. (We also thank Diana Coleman for approving the project.) The two courses were substantially the same, but where they differed we have followed the treatment that seemed cleaner or more complete. In places we have supplied details, clarified explanations, or systematized the notation, but all the arguments are Coleman's. We have not introduced new material or brought the treatment up to date.[1]

In the first iteration of the course Coleman assigned very few problems, though he sometimes suggested examples, and he expected the students to be exploring

---

[1] As David Kaiser notes in "A ψ Is Just a ψ? Pedagogy, Practice, and the Reconstitution of general relativity, 1942-1975" (*Stud. Hist. Phil. Phys.* **29**, 321 (1998)), Coleman's approach, with its emphasis on action principles, was already distinctly "modern."

questions of their own devising. Some of the problems in this book are taken from supplementary notes by the three Davids; others are officially assigned problems from the 1969 course.

Physics 210 was intended for graduate students, and the first part presupposes a fairly sophisticated understanding of special relativity; it focuses on foundational issues and advanced topics not often addressed in the classroom. By contrast, the second part is an *introduction* to general relativity ending with gravitational waves, the Schwarzschild metric, and relativistic cosmology. But throughout the book Coleman's uncanny nose for the subtle and the profound is very much in evidence.

Coleman recommended the following references:

- R. Adler, M. Bazin, and M. Schiffer, *Introduction to General Relativity*, McGraw-Hill (1965).
- V. Fock, *The Theory of Space, Time, and Gravitation*, Pergamon Press (1959). For general relativity.
- C. Møller, *The Theory of Relativity*, 2nd ed., Oxford U. P. (1972). Contains most of the material for this class—the nearest thing to a course textbook.
- W. Pauli, *Theory of Relativity*, Pergamon Press (1958) (reprinted by Dover Publications, 1981). A translation of Pauli's famous article written (at the age of 21) for the *Encyklopädie der mathematischen Wissenschaften*.
- E. Schrödinger, *Space-Time Structure*, Cambridge U. P. (1963). For general relativity.
- J. L. Synge, *Relativity: The Special Theory*, 2nd ed., North Holland (1965) and *Relativity: The General Theory*, North Holland (1960). Exhaustive—look here when all else fails.
- H. Weyl, *Space, Time, Matter*, Dover (1952).
- E. H. Wichmann, "Theory of General Relativity," lecture notes for Berkeley Physics 231 (winter 1965).

David J. Griffiths
David Derbes
Richard B. Sohn

Part I

Special Relativity

1

The Geometry of Special Relativity

1.1 Introduction

1.1.1 Classical Physical Systems

A classical[1] physical system consists of three parts:

1. **Four-dimensional spacetime:** the *arena* of classical physics. We label a point in spacetime (an "event") by its coordinates:

$$x^\mu = (x^0, x^i) = (ct, \mathbf{x}), \tag{1.1}$$

where x^0 represents the time (we'll use units such that $c = 1$)[2] and $\mathbf{x}$ the position. Greek indices near the middle of the alphabet $(\lambda, \mu, \nu, \ldots)$ run from 0 to 3; Roman indices near the middle $(i, j, k, \ldots)$ run from 1 to 3.

2. **Particles and fields:** the *entities* of classical physics.

   (a) **Particles:** A particle is a structureless point object. Its *location*, $\mathbf{x}(t)$, as a function of time, tells you everything there is to say about it (beyond fixed properties such as mass and charge).[3] In 4-vector notation we represent the particle's trajectory (its **world line**) by $x^\mu(s)$, where s is any parameter used to denote points along the curve ($f(s)$ would do just as well, for any monotonic function f):

---

[1] In this book "classical" means "pre-quantum"; it *includes* special relativity.
[2] It's easy to reinsert the c's, when necessary, by dimensional analysis.
[3] We could treat point objects with spin, but let's keep things simple; in this course "particle" means spin 0.

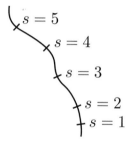

(b) **Fields:** A field is a function of position and time:

$$\varphi^\alpha(x). \tag{1.2}$$

Here α labels the components: one of them, if the field is temperature; six of them, in the case of electromagnetism. (In expressions like this x stands for the four components of x^μ.)

3. **Dynamics:** the *laws of motion*.

1.1.2 Symmetries

A **symmetry** is an operation that leaves an object or a system unchanged (**invariant**). A square, for example, is invariant under rotations (about a perpendicular axis through its center) by 90°, or 180°, or 270°, or reflections (in either diagonal, or a bisector of two opposite sides). Of particular importance to us are invariances of the laws of motion,[4] transformations that carry one possible motion into another. We stipulate that an invariance must have a well-defined inverse.[5]

Mathematically, the invariances of a system form a **group**.

> **Definition:** A group, G, is a set of elements $(a, b, c, \ldots)$ and a law of "multiplication," with the following properties:
>
> 1. It is **closed**: if a and b are in G, so is their product, ab.
> 2. It is **associative**: $a(bc) = (ab)c$.
> 3. It contains a (unique) unit element, 1, such that $1a = a1 = a$ for every element a.
> 4. Each element a has a (unique) inverse, a^{-1}, such that $a^{-1}a = aa^{-1} = 1$.

---

[4] The ancient Greeks thought symmetries pertain to the actual motion: celestial objects ought to move on circles, because a circle is the most perfect (symmetrical) shape. But since the time of Newton we have understood that the more significant invariances apply to the *equations of motion*, and hence to the collection of all *possible* motions—the set of *solutions* to the equations of motion. The sun's gravitational field is spherically symmetric, but planetary orbits don't directly exhibit that symmetry—they're *elliptical*.

[5] This restriction eliminates trivial possibilities such as mapping all points on a particle trajectory onto some fixed point (sitting still at one point being—usually—a solution to the equations of motion). It is necessary in order to ensure that the invariances form a group.

For example, the real numbers (except 0), with multiplication defined in the usual way, constitute an **Abelian** (commutative: $ab = ba$) group. Another group is the set of permutations of three objects (this group is *not* Abelian). We are interested here in the group of invariances of classical physics; "multiplication" in this context means application of two transformations in succession.

Example 1.1

Imagine a quantum mechanical system with nondegenerate energy levels. The state of the system at time $t = 0$ can be expanded in terms of the energy eigenstates:

$$|\psi(0)\rangle = \sum a_n |n\rangle, \tag{1.3}$$

and at any later time

$$|\psi(t)\rangle = \sum a_n e^{-iE_n t/\hbar} |n\rangle. \tag{1.4}$$

But the *phase* of $|n\rangle$ is arbitrary; physical predictions are unaffected by the transformation

$$|n\rangle \rightarrow e^{i\theta_n} |n\rangle, \tag{1.5}$$

for any collection of real numbers θ_n (independent of position and time). This is a huge invariance group, with an infinite number of parameters (if there are infinitely many eigenstates). But for the most part it is a *useless* invariance, which does not help us to solve the equations of motion.

So there exist trivial, accidental, or otherwise inconsequential invariances. One particularly *useful* class consists of the *geometrical* invariances of space and time: translations, rotations, dilations[6] (stretching), and so on. *Question:* What is the group of geometrical invariances of classical physics—the geometrical transformations that leave the laws of classical physics unchanged? A geometrical transformation is a change of coordinates:

$$x^\mu \rightarrow x'^\mu = y^\mu(x). \tag{1.6}$$

In the case of a particle trajectory,

$$x^\mu(s) \rightarrow y^\mu(x(s)). \tag{1.7}$$

Fields are more complicated, because not only do the *components* mix (if it's a vector field, and we're performing a rotation, the $\hat{\mathbf{x}}$ component will pick up $\hat{\mathbf{y}}$ and $\hat{\mathbf{z}}$ terms), but the *argument* (x) must be expressed in terms of the new coordinates (y): schematically,

---

[6] Eds. Coleman calls them "dilatations." Presumably permute:dilate::permutation:dilatation. But most modern authors use "dilation," and "dilatation" seems unnecessarily awkward.

$$\varphi^\alpha(x) \to [\varphi^\alpha(x)]' = F[\varphi^\beta(y^{-1}(x))], \tag{1.8}$$

where F is some function denoting the transformation (mixing) of the components (φ^β), and y^{-1} is the inverse of Eq. 1.6. In words, the new fields at point y are some functions of the old fields at the point x that got mapped into y.

1.2 Poincaré Invariance

1.2.1 Geometrical Symmetries of Classical Physics

We'll focus for the moment on the case of particles. If there were *no* laws of motion (i.e. if *every* particle motion were possible), then *any* geometrical transformation would be an invariance. We'll whittle down this (huge) group by invoking some *actual* laws of motion:

1. **Newton's first law.** The allowed motions for a *free* particle are straight lines in spacetime, so the invariance group must (at a minimum) *take straight lines into straight lines*. One way to characterize a straight line is

$$x^\mu(s) = v^\mu s + b^\mu, \text{ where } v^\mu = \frac{dx^\mu}{ds} \text{ and } b^\mu \text{ are constants,} \tag{1.9}$$

which is the general solution to the differential equation

$$\frac{d^2 x^\mu}{ds^2} = 0. \tag{1.10}$$

But wait: we *could* have used a different parameterization (say, s^3 instead of s); then

$$x^\mu(s) = v^\mu s^3 + b^\mu. \tag{1.11}$$

So Eq. 1.10 is not a reliable way to characterize straight lines—it's *sufficient*, but not *necessary*. Maybe a straight line satisfying 1.10 is transformed into a straight line that *doesn't* satisfy 1.10. In point of fact this worry is misguided: an invariance that carries straight lines into straight lines *automatically* takes linearly parameterized straight lines 1.9 into linearly parameterized straight lines.

   Proof: For transformations that carry straight lines into straight lines:

   (a) **Intersecting (or nonintersecting) straight lines go into intersecting (nonintersecting) straight lines.** If intersecting lines transformed into nonintersecting lines, the transformation for the point of intersection would be ill defined, since it would have to go to two different points—one on each line. And because we have stipulated that invariances have well-defined inverses, the same goes for nonintersecting to intersecting.

(b) **Planes go into planes.** Let P be a point in the plane defined by intersecting lines A and B (but not *on* either line), and draw a line from P intersecting A and B:

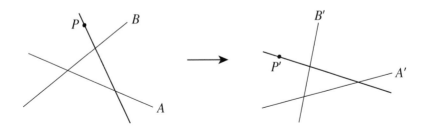

This line transforms into a line intersecting A' and B', so P' lies in the plane defined by A' and B'.

(c) **Parallel lines go into parallel lines.** This follows from (a) and (b).

(d) **Equidistant coplanar parallel lines go into equidistant coplanar parallel lines.** We know that coplanar parallel lines go into coplanar parallel lines, but could it be that equidistant ones (a, b, c) go into *non*equidistant ones (a', b', c')?

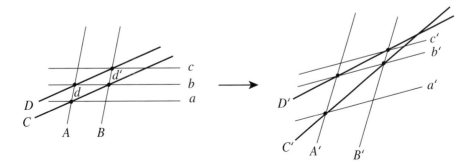

No: draw line A, and let the distance between its intersections with a and b be d. Now draw line B, parallel to A, and construct lines C and D, passing through the four intersections. By simple geometry, C and D are parallel (because a, b, and c are equidistant), and $d' = d$. However, unless a', b', and c' are *also* equidistant, C' and D' will *not* be parallel, violating (c).

So the transformation $x(s) \to y(x(s))$ takes equal intervals $(x(s_3) - x(s_2) = x(s_2) - x(s_1))$ into equal intervals $(y(s_3) - y(s_2) = y(s_2) - y(s_1))$, preserving the linear parameterization. QED

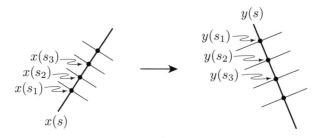

Under the transformation 1.6,

$$x^\mu \to y^\mu(x^\nu),$$

derivatives transform (by the chain rule)[7] as

$$\frac{dx^\mu}{ds} \to \frac{dy^\mu}{ds} = \frac{\partial y^\mu}{\partial x^\nu}\frac{dx^\nu}{ds}, \tag{1.12}$$

$$\frac{d^2 x^\mu}{ds^2} \to \frac{d^2 y^\mu}{ds^2} = \frac{\partial y^\mu}{\partial x^\nu}\frac{d^2 x^\nu}{ds^2} + \frac{\partial^2 y^\mu}{\partial x^\nu \partial x^\lambda}\frac{dx^\nu}{ds}\frac{dx^\lambda}{ds}. \tag{1.13}$$

Because all straight lines $(d^2 x^\mu/ds^2 = 0)$ must transform into straight lines $(d^2 y^\mu/ds^2 = 0)$, it follows that invariances consistent with Newton's first law satisfy

$$\frac{\partial^2 y^\mu}{\partial x^\nu \partial x^\lambda} = 0 \tag{1.14}$$

(for all μ, ν, λ). The general solution is a *linear* function of x:

$$y^\mu = M^\mu_{\ \nu} x^\nu + b^\mu, \tag{1.15}$$

where the 16 elements of $M^\mu_{\ \nu}$ and the 4 components of b^μ are constants. (As a 4×4 matrix, $\det M \neq 0$, since $y(x)$ must have an inverse.) Newton's first law has reduced the geometrical invariances to a 20-parameter group, the **inhomogeneous affine group** (in four dimensions); with $b^\mu = 0$ it becomes the **homogeneous affine group**.

2. **Constancy of the velocity of light.** In empty space, light travels in straight lines, and according to special relativity the speed of light (in vacuum) is a universal constant, independent of the velocity of the source or the observer. If a light signal travels from point **x** to point **x**′, departing at time t and arriving at time t', then

---

[7] We use the **Einstein summation convention**, whereby repeated indices are summed. Thus the third term in Eq. 1.12 carries an implicit summation sign, $\sum_{\nu=0}^{3}$.

$$c(t' - t) = |\mathbf{x}' - \mathbf{x}|, \tag{1.16}$$

or (setting $c = 1$)

$$(t' - t)^2 = (\mathbf{x}' - \mathbf{x})^2 = \sum_{i=1}^{3}[(x^i)' - x^i]^2. \tag{1.17}$$

Introducing the **metric tensor**[8]

$$g_{\mu\nu} \equiv \begin{pmatrix} 1 & 0 & 0 & 0 \\ 0 & -1 & 0 & 0 \\ 0 & 0 & -1 & 0 \\ 0 & 0 & 0 & -1 \end{pmatrix}, \tag{1.18}$$

we have

$$(x' - x)^\mu g_{\mu\nu}(x' - x)^\nu = 0. \tag{1.19}$$

The constancy of the speed of light means that if x and x' satisfy Eq. 1.19, then so too must the transformed coordinates y and y'. What does this tell us about M and b? Well,

$$x^\mu \rightarrow y^\mu = M^\mu_{\ \nu} x^\nu + a^\mu \implies (y' - y)^\mu = M^\mu_{\ \nu}(x' - x)^\nu, \tag{1.20}$$

so

$$M^\mu_{\ \kappa}(x' - x)^\kappa g_{\mu\nu} M^\nu_{\ \sigma}(x' - x)^\sigma = 0, \tag{1.21}$$

or

$$(x' - x)^\kappa \left[M^\mu_{\ \kappa} g_{\mu\nu} M^\nu_{\ \sigma} \right](x' - x)^\sigma = 0. \tag{1.22}$$

This must hold for any x and x' satisfying Eq. 1.19. It follows that[9]

$$M^\mu_{\ \kappa} g_{\mu\nu} M^\nu_{\ \sigma} = \lambda g_{\kappa\sigma} \tag{1.23}$$

for some constant λ; or, in matrix notation,[10]

$$M^T g M = \lambda g. \tag{1.24}$$

---

[8] Some authors use the other **signature** $(-, +, +, +)$; it doesn't matter, as long as you are consistent.

[9] Although 1.23 obviously *guarantees* 1.22, it is not so clear that it is *required* by 1.22. But remember that this must hold for *any* x and x' satisfying 1.19, and from this it is not hard to show that 1.23 is in fact *necessary*.

[10] Reading left to right, the first index (whether up or down) is the *row*, and the second (up or down) is the *column*. The significance of upness and downness will appear in due course. The superscript T denotes the transpose: $(M^T)^\mu_{\ \kappa} = M_\kappa^{\ \mu}$.

What sorts of transformations remain, after invoking Newton's first law and the constancy of the speed of light? We can factor the matrix M as follows:

$$M = M_1 M_2, \text{ where } M_1 = |\det M|^{1/4} I \text{ and } M_2 = \frac{M}{|\det M|^{1/4}} \qquad (1.25)$$

(I is the unit matrix). Thus any M is the product of a pure **dilation** M_1,

$$M_1 = \alpha I, \text{ so } M_1^T g M_1 = \alpha^2 g \text{ and hence } \lambda_1 = \alpha^2, \qquad (1.26)$$

and a dilation-free term M_2 with determinant ± 1, for which

$$(\det M_2)(\det g)(\det M_2) = \lambda_2^4 (\det g) \Rightarrow \lambda_2^4 = 1 \Rightarrow \lambda_2 = \pm 1. \qquad (1.27)$$

Actually, the negative sign is impossible,[11] so (in view of Eq. 1.26) $\lambda = \lambda_1 \lambda_2$ is in fact always positive.

3. **Eliminating dilations.** *Question:* Is our universe invariant under dilations? Imagine performing the Cavendish experiment to measure the gravitational force between two point masses:

$$F = G \frac{m_1 m_2}{r^2}, \qquad (1.28)$$

giving an acceleration to m_1 in the amount

$$a_1 = G \frac{m_2}{r^2}. \qquad (1.29)$$

Under a dilation (change of scale),

$$r \to \lambda r, \quad t \to \lambda t, \quad a \to \lambda^{-1} a. \qquad (1.30)$$

So if dilation doesn't affect G or m_2, then a_1 goes like λ^{-1} but Gm_2/r^2 goes like λ^{-2}. Since G is a universal constant, it can't depend on λ, and since there is no mass continuum (no electron, for example, with slightly larger or smaller mass), mass cannot depend continuously on λ. *Conclusion:* Our universe is *not* invariant under dilations.[12]

---

[11] This follows from **Sylvester's law of inertia**; see, for instance, S. MacLane and G. Birkhoff, *A Survey of Modern Algebra*, 3rd ed., Macmillan (1965) p. 254. In essence, if $M^T g M = -g$ then $Q \equiv (x^0)^2 - (x^1)^2 - (x^2)^2 - (x^3)^2 = -(y^0)^2 + (y^1)^2 + (y^2)^2 + (y^3)^2$, so there is a 3-dimensional subspace ($x^0 = 0$) in which $Q < 0$, and another 3-dimensional subspace ($y^0 = 0$) in which $Q > 0$. But the entire space has only four dimensions, so this is impossible.

[12] This still leaves open the possibility of invariance under *combined* dilations and Lorentz transformations. We'll eliminate that option in Section 1.2.6.

From these three overall constraints ((1) Newton's first law, (2) the constancy of the speed of light, and (3) the absence of scale invariance)[13] it follows that the maximum[14] possible geometrical invariance of classical physics is[15]

$$x^\mu \to y^\mu = \Lambda^\mu{}_\nu x^\nu + b^\mu \quad \text{with} \quad \Lambda^T g \Lambda = g. \qquad (1.31)$$

The group of all such transformations (all possible Λ's and all b's) is called the **Poincaré group** (or the **inhomogeneous Lorentz group**). The subgroup $b^\mu = 0$ is the (homogeneous) **Lorentz group**.

1.2.2 Active and Passive Transformations

I have been thinking of the geometrical invariances as *active* transformations, in which the system is *physically moved* to a new location or orientation (or, in the case of dilations, shrunk or expanded). But one can achieve the same effect (formally) by a *passive* transformation, changing the *coordinates* in the reverse sense, while leaving the system itself fixed. (In criminal circles these are known as the "alibi" and "alias" strategies.) There is a kind of duality here, summarized in the following table:

	Active $(x \to \Lambda x + b)$ "alibi"	**Passive** $(x \to \Lambda^{-1}x - b)$ "alias"
Acts on	position/motion	coordinates
Invariance	solutions $\to$ solutions	form invariance of the equations of motion

---

[13] Caveats: (1) assumes that Newton's first law holds even for inaccessibly high velocities, (2) assumes that the speed of light is independent of the motion of the source even for unattainably high source speeds, and (3) assumes that the law of universal gravitation holds (at least approximately) for all speeds and separation distances. These assumptions are all, of course, open to potential experimental falsification.

[14] I have shown that the geometrical symmetry group is no *bigger* than the Poincaré group, but I have not proved that it couldn't be *smaller*; in principle, some new physical law might reduce it even further. Einstein postulated that there is in fact *no* further reduction; we call this assumption **Lorentz invariance**.

[15] Eds. We have used M for the *generic* linear transformation (1.15); Λ denotes a **Lorentz transformation**, satisfying $\Lambda^T g \Lambda = g$.

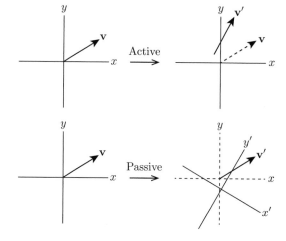

Einstein preferred the alias viewpoint, but the two perspectives are, for the most part, equivalent.

1.2.3 Minkowski Space

Notice that Lorentz transformations

$$\Lambda^T g \Lambda = g \tag{1.32}$$

are *not* (in general) **rotations** in four dimensions. Those would be generated by **orthogonal** matrices,

$$R^T R = I, \tag{1.33}$$

and preserve the (positive definite) quadratic form

$$x^T x = x^T I x = (x^0)^2 + (x^1)^2 + (x^2)^2 + (x^3)^2 \tag{1.34}$$

(that is, if $y = Rx$, then $y^T y = x^T x$). By contrast, Lorentz transformations preserve the indefinite quadratic form

$$x^T g x = (x^0)^2 - (x^1)^2 - (x^2)^2 - (x^3)^2 \tag{1.35}$$

(i.e. if $y = \Lambda x$ then $y^T g y = x^T g x$). If this quantity is positive, then x is said to be **time-like**; if it is negative, x is **space-like**; if it is zero, x is **light-like**. More generally, if $y = \Lambda x$ and $z = \Lambda w$ then $y^T g z = (\Lambda x)^T g (\Lambda w) = x^T \Lambda^T g \Lambda w = x^T g w$. This invariant quantity is the **scalar** (or **dot**) **product** of x and w; if $x^T g w = 0$, x and w are **orthogonal**.[16]

---

[16] Don't confuse orthogonal *vectors* with orthogonal *matrices* (1.33)—same word, different meanings.

Minkowski space (the 4-dimensional spacetime of special relativity) separates into distinct regions, as illustrated in the following figure (which you must imagine includes an undrawable z-coordinate, so the cones are really hyper-cones):

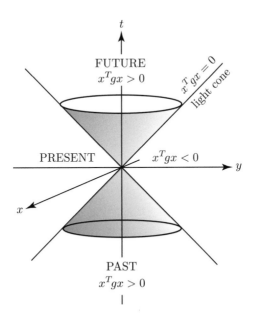

If you are sitting at the origin ($x = y = z = t = 0$), your **future** is the locus of all points in spacetime that you can influence; your **past** is the locus of all points that can have influenced you. As for the **present**, you cannot affect anything there, and it cannot affect you—to do so would require a signal propagating faster than the speed of light.

1.2.4 Topological Structure of the Lorentz Group

Orthogonal transformations in three dimensions fall into two disjoint "components":

(1) **Rotations:** determinant $+1$ (3 parameters; e.g. the Euler angles),
(2) **Reflections:** determinant -1 (3 parameters; e.g. the Euler angles).

A 3-dimensional rotation can be represented by a point within a sphere of radius π (with antipodal surface points identified): the axis of rotation defines the direction to the point, and the angle of rotation tells you its radial coordinate. Thus the north pole would specify a rotation by $180°$ about the north–south axis (which has the same effect as a rotation by $180°$ about the south–north axis). All rotations are continuously connected (to one another, and to the identity), but reflections are *not* continuously connected to rotations—they are represented by a separate sphere.

Question: How many parameters characterize a (homogeneous) Lorentz transformation, $\Lambda^T g \Lambda = g$ (at most 16, of course, since Λ is a 4×4 matrix, but presumably fewer than that)? And how many disjoint components does the Lorentz group possess? If we know how four basis vectors (for instance, $(1,0,0,0)$, $(0,1,0,0)$, $(0,0,1,0)$, and $(0,0,0,1)$) transform under Λ, then we know how *any* vector transforms. Let's start with $(1,0,0,0)$; it gets transformed (under Λ) to some vector (x^0, x^1, x^2, x^3), with

$$x^T g x = (x^0)^2 - (x^1)^2 - (x^2)^2 - (x^3)^2 = 1 - 0 - 0 - 0 = 1, \qquad (1.36)$$

so

$$x^0 = \pm\sqrt{1 + (x^1)^2 + (x^2)^2 + (x^3)^2}. \qquad (1.37)$$

This defines a hyperboloid of two sheets:

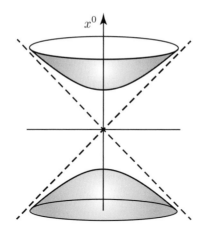

Every x^1, x^2, and x^3 is possible, but once they are specified, x^0 is determined, up to an overall sign, so there are three continuous parameters, and one discrete choice.

Similarly, $(0, 1, 0, 0)$ transforms to (y^0, y^1, y^2, y^3), with

$$y^T g y = (y^0)^2 - (y^1)^2 - (y^2)^2 - (y^3)^2 = 0 - 1 - 0 - 0 = -1, \qquad (1.38)$$

or

$$(y^0)^2 + 1 = (y^1)^2 + (y^2)^2 + (y^3)^2. \qquad (1.39)$$

This time it's a hyperboloid of *one* sheet:

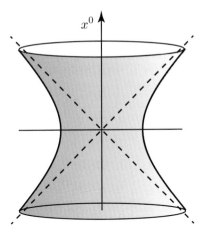

where y^0 determines the radius of the (y^1, y^2, y^3) sphere, leaving two angular variables; again, three parameters. However,

$$y^T gx = y^0 x^0 - y^1 x^1 - y^2 x^2 - y^3 x^3 = 0 \cdot 1 - 1 \cdot 0 - 0 \cdot 0 - 0 \cdot 0 = 0 \quad (1.40)$$

eliminates one variable (we could solve, for instance, for y^3), so there remain two (continuous) parameters, and no new discrete choices.

The same goes for $(0,0,1,0) \rightarrow (z^0, z^1, z^2, z^3)$, except that there are now *two* orthogonality constraints $(z^T gx = z^T gy = 0)$, leaving just one additional free parameter. Finally, for $(0,0,0,1) \rightarrow (w^0, w^1, w^2, w^3)$ we have

$$(w^0)^2 - (w^1)^2 - (w^2)^2 - (w^3)^2 = -1, \quad (1.41)$$
$$w^0 x^0 - w^1 x^1 - w^2 x^2 - w^3 x^3 = 0, \quad (1.42)$$
$$w^0 y^0 - w^1 y^1 - w^2 y^2 - w^3 y^3 = 0, \quad (1.43)$$
$$w^0 z^0 - w^1 z^1 - w^2 z^2 - w^3 z^3 = 0. \quad (1.44)$$

From the last three we can solve for w^1, w^2, and w^3 (in terms of w^0 and the other parameters); then 1.41 determines w^0 up to a sign. So there are no new (continuous) parameters, but one discrete sign choice.

Conclusion:

> The Lorentz group has four disjoint components, and six continuous parameters.

We classify the four disjoint components of the Lorentz group according to whether the sign of the time stays the same (**orthochronous**) or changes (**nonorthochronous**), and whether the spatial part turns a right-handed coordinate system into another right-handed system or (by including a reflection) into a left-handed system. An example of a right-handed orthochronous transformation is the identity:

$$\Lambda = I \equiv \begin{pmatrix} 1 & 0 & 0 & 0 \\ 0 & 1 & 0 & 0 \\ 0 & 0 & 1 & 0 \\ 0 & 0 & 0 & 1 \end{pmatrix}, \tag{1.45}$$

an example of a left-handed orthochronous transformation is **parity**:

$$\Lambda = P \equiv \begin{pmatrix} 1 & 0 & 0 & 0 \\ 0 & -1 & 0 & 0 \\ 0 & 0 & -1 & 0 \\ 0 & 0 & 0 & -1 \end{pmatrix}, \tag{1.46}$$

an example of a right-handed nonorthochronous transformation is **time reversal**:

$$\Lambda = T \equiv \begin{pmatrix} -1 & 0 & 0 & 0 \\ 0 & 1 & 0 & 0 \\ 0 & 0 & 1 & 0 \\ 0 & 0 & 0 & 1 \end{pmatrix}, \tag{1.47}$$

and an example of a left-handed nonorthochronous transformation is the product PT:

$$\Lambda = PT = \begin{pmatrix} -1 & 0 & 0 & 0 \\ 0 & -1 & 0 & 0 \\ 0 & 0 & -1 & 0 \\ 0 & 0 & 0 & -1 \end{pmatrix}. \tag{1.48}$$

Notice (from Eq. 1.32) that

$$\det(\Lambda) = +1 \text{ (\textbf{proper}) or } -1 \text{ (\textbf{improper}).} \tag{1.49}$$

I'll use the symbol $\mathcal{L}_+^\uparrow$ to denote the proper (subscript $+$) orthochronous (superscript $\uparrow$) sector; $-$ for improper and $\downarrow$ for nonorthochronous:

Symbol	$\mathcal{L}_+^\uparrow$	$\mathcal{L}_-^\uparrow$	$\mathcal{L}_-^\downarrow$	$\mathcal{L}_+^\downarrow$
Name	right-handed orthochronous	left-handed orthochronous	right-handed nonorthochronous	left-handed nonorthochronous
Example	I	P	T	PT
$\det(\Lambda)$	1 (proper)	-1 (improper)	-1 (improper)	1 (proper)

Notice that $\mathcal{L}_+^\uparrow$ contains the identity (and all transformations continuously connected to the identity); I'll call it $\mathcal{L}_0$. Suppose Λ is in $\mathcal{L}_-^\uparrow$. Since $P^2 = 1$, I can write $\Lambda = P(P\Lambda)$, and $P\Lambda$ is in $\mathcal{L}_0$. So any element of $\mathcal{L}_-^\uparrow$ can be expressed as the product of P with an element of $\mathcal{L}_0$. The same goes for the other two sectors:

$$\mathcal{L}_-^\uparrow = P\mathcal{L}_0, \quad \mathcal{L}_-^\downarrow = T\mathcal{L}_0, \quad \mathcal{L}_+^\downarrow = PT\mathcal{L}_0. \tag{1.50}$$

Thus the discrete invariances P and T, together with $\mathcal{L}_0$, generate the entire group. For the most part we will confine our attention to $\mathcal{L}_0$, the "connected part" of the (homogeneous) Lorentz group.

1.2.5 Rotations and Boosts

Within $\mathcal{L}_0$, two kinds of transformations are of special interest: spatial **rotations**, and **boosts**. The former act only on the spatial coordinates:

$$R = \begin{pmatrix} 1 & 0 & 0 & 0 \\ 0 & & & \\ 0 & & R_3 & \\ 0 & & & \end{pmatrix}, \tag{1.51}$$

where R_3 is a 3×3 orthogonal matrix. The rotations constitute a 3-parameter subgroup (the ordinary 3-dimensional rotation group, SO(3)). Boosts comprise another 3-parameter set. A boost (or "pure" Lorentz transformation)[17] acts in a plane that includes the time axis; a boost in the xt plane transforms a unit vector in the t direction as follows:

$$B : \begin{pmatrix} 1 \\ 0 \\ 0 \\ 0 \end{pmatrix} \rightarrow \begin{pmatrix} a \\ b \\ 0 \\ 0 \end{pmatrix}, \tag{1.52}$$

with $a^2 - b^2 = 1$ and $a > 0$ (orthochronous), so we can write

$$a = \cosh\phi, \quad b = \sinh\phi \tag{1.53}$$

for some (real) number ϕ. Meanwhile, it takes a unit vector in the x direction,

$$B : \begin{pmatrix} 0 \\ 1 \\ 0 \\ 0 \end{pmatrix} \rightarrow \begin{pmatrix} c \\ d \\ 0 \\ 0 \end{pmatrix}. \tag{1.54}$$

---

[17] Eds. Coleman calls them "accelerations," but "boost" is the standard term.

The transformed vectors must be orthogonal, because the original vectors were:

$$c \cosh\phi - d \sinh\phi = 0 \implies \frac{c}{\sinh\phi} = \frac{d}{\cosh\phi} \equiv \alpha, \tag{1.55}$$

and their "lengths" are preserved ($c^2 - d^2 = -1$), so $\alpha = \pm 1$. Which sign do we want? Evidently

$$B = \begin{pmatrix} \cosh\phi & \pm\sinh\phi & 0 & 0 \\ \sinh\phi & \pm\cosh\phi & 0 & 0 \\ 0 & 0 & 1 & 0 \\ 0 & 0 & 0 & 1 \end{pmatrix}, \tag{1.56}$$

but $\det(B) = +1$, so we need the plus sign.

Under the boost B, a particle at rest, whose trajectory in spacetime (using t as the parameter) is $x = 0$, $t = s$, is transformed to $x' = \sinh\phi\, s$, $t' = \cosh\phi\, s$, and its velocity is

$$v = \frac{x'}{t'} = \tanh\phi. \tag{1.57}$$

Notice that whereas ϕ (the **rapidity**) can be any (real) number, $\tanh\phi$ is always between -1 and $+1$: the limiting speed achievable by a boost is c (which is 1 in our units). Note also that a boost looks very much like a rotation, except that (and this is crucial) the circular functions (sine and cosine) are replaced by hyperbolics (sinh and cosh). Equation 1.56 can be cast in a more familiar form by solving 1.57 for $\cosh\phi$ and $\sinh\phi$ in terms of v:

$$B = \begin{pmatrix} \gamma & \gamma v & 0 & 0 \\ \gamma v & \gamma & 0 & 0 \\ 0 & 0 & 1 & 0 \\ 0 & 0 & 0 & 1 \end{pmatrix}, \quad \text{where} \quad \gamma \equiv \frac{1}{\sqrt{1 - v^2}}. \tag{1.58}$$

Every transformation in $\mathcal{L}_0$ can be expressed as the product (in either order) of a rotation and a boost:

$$\Lambda = BR = R'B'. \tag{1.59}$$

Proof: Since Λ carries the vector $(1,0,0,0)$ into a vector of "length" 1,

$$\Lambda \begin{pmatrix} 1 \\ 0 \\ 0 \\ 0 \end{pmatrix} = \begin{pmatrix} \cosh\phi \\ \hat{e}_x \sinh\phi \\ \hat{e}_y \sinh\phi \\ \hat{e}_z \sinh\phi \end{pmatrix} \tag{1.60}$$

(for some real ϕ), where $\hat{\mathbf{e}}$ is a unit 3-vector. We might as well choose our axes so $\hat{\mathbf{e}}$ is in the x direction. Then

$$\Lambda \begin{pmatrix} 1 \\ 0 \\ 0 \\ 0 \end{pmatrix} = \begin{pmatrix} \cosh\phi \\ \sinh\phi \\ 0 \\ 0 \end{pmatrix} \quad\Rightarrow\quad \Lambda = \begin{pmatrix} \cosh\phi & ? & ? & ? \\ \sinh\phi & ? & ? & ? \\ 0 & ? & ? & ? \\ 0 & ? & ? & ? \end{pmatrix}. \tag{1.61}$$

Let B be the *boost* (1.56) that has the same effect on $(1,0,0,0)$:

$$B = \begin{pmatrix} \cosh\phi & \sinh\phi & 0 & 0 \\ \sinh\phi & \cosh\phi & 0 & 0 \\ 0 & 0 & 1 & 0 \\ 0 & 0 & 0 & 1 \end{pmatrix}. \tag{1.62}$$

Define

$$L \equiv B^{-1}\Lambda = \begin{pmatrix} 1 & a & b & c \\ 0 & ? & ? & ? \\ 0 & ? & ? & ? \\ 0 & ? & ? & ? \end{pmatrix}. \tag{1.63}$$

Actually, the top row has to be $(1,0,0,0)$, because L is itself a Lorentz transformation, and hence satisfies $L^T g L = g$:

$$\begin{pmatrix} 1 & 0 & 0 & 0 \\ a & ? & ? & ? \\ b & ? & ? & ? \\ c & ? & ? & ? \end{pmatrix} \begin{pmatrix} 1 & 0 & 0 & 0 \\ 0 & -1 & 0 & 0 \\ 0 & 0 & -1 & 0 \\ 0 & 0 & 0 & -1 \end{pmatrix} \begin{pmatrix} 1 & a & b & c \\ 0 & ? & ? & ? \\ 0 & ? & ? & ? \\ 0 & ? & ? & ? \end{pmatrix} \tag{1.64}$$

$$= \begin{pmatrix} 1 & a & b & c \\ a & ? & ? & ? \\ b & ? & ? & ? \\ c & ? & ? & ? \end{pmatrix} = \begin{pmatrix} 1 & 0 & 0 & 0 \\ 0 & -1 & 0 & 0 \\ 0 & 0 & -1 & 0 \\ 0 & 0 & 0 & -1 \end{pmatrix} \quad\Rightarrow a = b = c = 0.$$

So L is in fact a rotation:

$$R = B^{-1}\Lambda \quad\Rightarrow\quad \Lambda = BR. \quad \text{QED} \tag{1.65}$$

1.2.6 Simultaneous Dilations and Lorentz Transformations

In Section 1.2.1 we eliminated pure dilations (λI) as elements of the geometrical invariance group of classical physics, but what about dilations combined with

Lorentz transformations: $\lambda\Lambda$? Suppose the group contained both $\lambda_1\Lambda$ and $\lambda_2\Lambda$ for one and the same Λ; in that case it would also contain

$$(\lambda_1\Lambda)(\lambda_2\Lambda)^{-1} = (\lambda_1\lambda_2^{-1})I, \tag{1.66}$$

which would be a pure dilation unless $\lambda_1 = \lambda_2$. So there cannot be two *different* λ's for a given Λ; λ must be uniquely determined by Λ—every Lorentz transformation Λ carries a particular dilation λ. Is that possible? Because every Lorentz transformation is the product of a rotation and a boost ($\Lambda = RB$), and rotations are certainly in the invariance group, so too is

$$R^{-1}(\lambda RB) = \lambda B, \tag{1.67}$$

so (by the same argument as before) λ depends only on the boost, not on the rotation: $\lambda(\mathbf{v})$. In fact, by rotational invariance, it can only depend on the *magnitude* of $\mathbf{v}$: $\lambda(v^2)$. Now $B(\mathbf{v})B(-\mathbf{v}) = I$, so the invariance group must also contain

$$I = (\lambda(\mathbf{v})B(\mathbf{v}))\,(\lambda(\mathbf{v})B(\mathbf{v}))^{-1} = \lambda(v^2)B(\mathbf{v})\lambda(v^2)B(-\mathbf{v}) = \left(\lambda(v^2)\right)^2, \tag{1.68}$$

and hence $\lambda(v^2) = \pm 1$. But $\lambda(0) = 1$ (the identity), and we assume continuity, so $\lambda(v^2) = 1$: dilations are not allowed *even if* they are tied to Lorentz transformations.

1.3 Time Dilation and Lorentz Contraction

1.3.1 Arc Length and Proper Time

In Minkowski space the analog to arc length along a world line $x^\mu(s)$ is

$$\tau(b) - \tau(a) = \int_a^b ds \left(g_{\mu\nu} \frac{dx^\mu}{ds} \frac{dx^\nu}{ds} \right)^{1/2} = \int_a^b \sqrt{dt^2 - dx^2 - dy^2 - dz^2}. \tag{1.69}$$

It is independent of parameterization (unchanged if we use $s' = f(s)$ in place of s), and it is Lorentz invariant. (The square root is OK, since world lines are everywhere time-like.) If the particle is at rest, the integral is just the elapsed time. If the particle is moving at constant velocity, it can be brought to rest by a Lorentz transformation, and since arc length is invariant, it is still the time elapsed on the particle's own watch—its **proper time**, τ. But what about a particle that speeds up or slows down? The curved path in spacetime can be approximated by short line segments, each corresponding to a different **comoving** observer traveling at constant velocity. When it is necessary to change directions (to "pass the baton" to the next comoving observer), the two observers are at the same spacetime point, and can synchronize their clocks unambiguously. Thus the sum of the observers' time intervals (which is to say, the total arc length) still corresponds to the elapsed time on the particle's own clock. *Conclusion:* Arc length is proper time, *even when the particle accelerates.* Proper time is the "natural" parameterization for a world line; it is defined (up to an additive constant) by

$$\boxed{g_{\mu\nu} \frac{dx^{\mu}}{d\tau} \frac{dx^{\nu}}{d\tau} = 1.} \tag{1.70}$$

Problem 1.1

Alice and Bob both travel from spacetime point a to spacetime point b. Alice goes by the straight line path (in Minkowski space); Bob wanders around—his world line is curved. *Question:* Which trip takes longer, according to each traveler's own watch?

1.3.2 Time Dilation

Suppose observer $\mathcal{O}$ is at rest at the origin. Observer $\mathcal{O}'$ starts out at the origin, at $t = t' = 0$, and moves away (say, in the x direction) at constant speed $v = dx/dt$. When $\mathcal{O}$'s clock reads time t, what time t' does $\mathcal{O}$ observe on the (moving) clock carried by $\mathcal{O}'$? That is, what is the elapsed (proper) time on the $\mathcal{O}'$ clock? From Eq. 1.69,

$$t' = \tau = \int \sqrt{1 - \left(\frac{dx}{dt}\right)^2 - \left(\frac{dy}{dt}\right)^2 - \left(\frac{dz}{dt}\right)^2} \, dt = \sqrt{1 - v^2}\, t = \frac{1}{\gamma} t. \tag{1.71}$$

Notice that t' is *smaller* than t; the *moving clock runs slow*. This is known as **time dilation**.

Incidentally, if $\mathcal{O}$ observes $\mathcal{O}'$ going at velocity $\mathbf{v}$, then $\mathcal{O}'$ observes $\mathcal{O}$ going at velocity $-\mathbf{v}$. That is to say,[18]

$$(B(\mathbf{v}))^{-1} = B(-\mathbf{v}). \tag{1.72}$$

---

[18] Eds. Actually, Coleman already used this in Eq. 1.68.

Proof: Choose axes such that the motion is along the x direction. From Eq. 1.56,

$$B(v) = \begin{pmatrix} \cosh\phi & \sinh\phi & 0 & 0 \\ \sinh\phi & \cosh\phi & 0 & 0 \\ 0 & 0 & 1 & 0 \\ 0 & 0 & 0 & 1 \end{pmatrix}, \tag{1.73}$$

where $v = \tanh\phi$, and hence $-v = -\tanh\phi = \tanh(-\phi)$. Then

$$B(-v) = B(-\phi) = \begin{pmatrix} \cosh\phi & -\sinh\phi & 0 & 0 \\ -\sinh\phi & \cosh\phi & 0 & 0 \\ 0 & 0 & 1 & 0 \\ 0 & 0 & 0 & 1 \end{pmatrix}. \tag{1.74}$$

Therefore

$$B(-v)B(v) = \begin{pmatrix} \cosh\phi & -\sinh\phi & 0 & 0 \\ -\sinh\phi & \cosh\phi & 0 & 0 \\ 0 & 0 & 1 & 0 \\ 0 & 0 & 0 & 1 \end{pmatrix} \begin{pmatrix} \cosh\phi & \sinh\phi & 0 & 0 \\ \sinh\phi & \cosh\phi & 0 & 0 \\ 0 & 0 & 1 & 0 \\ 0 & 0 & 0 & 1 \end{pmatrix}$$

$$= \begin{pmatrix} (\cosh^2\phi - \sinh^2\phi) & 0 & 0 & 0 \\ 0 & (\cosh^2\phi - \sinh^2\phi) & 0 & 0 \\ 0 & 0 & 1 & 0 \\ 0 & 0 & 0 & 1 \end{pmatrix} = I. \quad \text{QED}$$

1.3.3 Lorentz Contraction

Two stripes (A and B) are painted across a road, a distance d apart. A car, going at constant speed v, takes time t to go from A to B:

$$d = vt. \tag{1.75}$$

Now examine the same process from the perspective of an observer in the car (using clocks and meter sticks moving with the car):

$$d' = v't'. \tag{1.76}$$

The speeds are the same (as we saw in Section 1.3.2), but the time interval on the moving clock is reduced (it's running slow, Eq. 1.71), so

$$d' = \sqrt{1 - v^2}\, d = \frac{1}{\gamma} d. \tag{1.77}$$

Lengths shrink, for a moving observer, by just the factor necessary to compensate for time dilation. This is called Lorentz–Fitzgerald contraction, or **Lorentz contraction**, for short.

Lorentz contraction only affects dimensions *parallel to the direction of motion*; lengths *perpendicular* to the motion are not contracted.[19] *Beware:* An **observation** is what you get *after* correcting for the time the signal took to get to you; what you **observe**, therefore, is not at all the same as what you *see* (or hear). Time dilation and Lorentz contraction pertain to what you *observe*, and relativity is almost always talking about *observations*.

1.4 Examples and Paradoxes

1.4.1 The Time Dilation Paradox

Time dilation raises an apparent paradox: If $\mathcal{O}$ says the $\mathcal{O}'$ clock is running slow, $\mathcal{O}'$ can say with equal justice that the $\mathcal{O}$ clock is running slow (and by the same factor). Who's right? They *both* are! It's a matter of **simultaneity**, which is different for the two observers. On a Minkowski diagram, *lines of simultaneity make the same angle with the light cone as does the time axis*. Thus, in the following figure, lines of simultaneity for $\mathcal{O}$ (which are, of course, horizontal) all make the same angle with the light cone (to wit, $45°$) as the t-axis does, while lines of simultaneity for $\mathcal{O}'$ all make the same angle with the light cone as the t'-axis does (to wit, α).

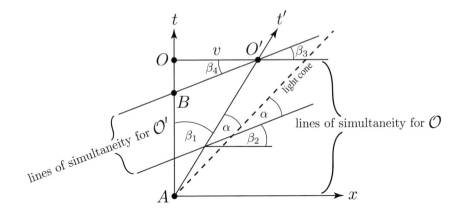

First I'll resolve the paradox, then I'll confirm the rule. Suppose $AO = 1$, the time shown on the $\mathcal{O}$ clock. Then $\sqrt{1 - v^2}$ is the time shown simultaneously (according to $\mathcal{O}$) on the $\mathcal{O}'$ clock; $\mathcal{O}$ reports that the $\mathcal{O}'$ clock is running slow. By simple geometry, $\beta_1 = \beta_2 = \beta_3 = \beta_4 \equiv \beta$ (note that $\alpha + \beta = 45°$), and so

$$\frac{OB}{v} = \frac{OO'}{AO} = \frac{v}{1} \Rightarrow OB = v^2 \Rightarrow AB = 1 - v^2. \qquad (1.78)$$

---

[19] Eds. For a nice proof, see E. F. Taylor and J. A. Wheeler, *Spacetime Physics*, Freeman, San Francisco (1966), page 21.

Therefore $\mathcal{O}'$ says that when the $\mathcal{O}'$ clock reads $\sqrt{1-v^2}$, the $\mathcal{O}$ clock reads $(1-v^2)$, and hence that the $\mathcal{O}$ clock is running slow (by the same factor, $\sqrt{1-v^2}$). Paradox resolved.

Now let's justify the rule for constructing lines of simultaneity. We want to show that S' is a line of simultaneity for $\mathcal{O}'$ if $\alpha = \beta$:

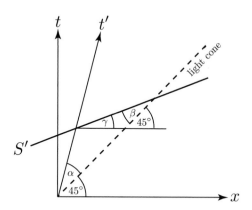

Proof: The equation for the t'-axis is $x' = 0$, or (using 1.73)

$$x' = \Lambda^1{}_0 t + \Lambda^1{}_1 x = (\sinh\phi)t + (\cosh\phi)x = 0, \quad t = -(\coth\phi)\,x.$$

The equation for S' is $t' = $ constant (if it is to be a line of simultaneity in $\mathcal{O}'$):

$$t' = \Lambda^0{}_0 t + \Lambda^0{}_1 x = (\cosh\phi)t + (\sinh\phi)x = \text{const}, \quad t = -(\tanh\phi)\,x + \text{const}.$$

So the slope of the t'-axis is

$$\tan(\alpha + 45°) = -\coth\phi,$$

and the slope of S' is

$$\tan\gamma = -\tanh\phi.$$

Now, $\gamma + \beta + (180° - 45°) = 180°$, so $\gamma = 45° - \beta$, and hence

$$\tan\gamma = \tan(45° - \beta) = \frac{\tan 45° - \tan\beta}{1 + \tan 45° \tan\beta} = \frac{1 - \tan\beta}{1 + \tan\beta} = -\tanh\phi$$

and

$$\tan(\alpha + 45°) = \frac{\tan\alpha + 1}{1 - \tan\alpha} = -\coth\phi \implies \frac{1 - \tan\alpha}{1 + \tan\alpha} = -\tanh\phi.$$

So $\alpha = \beta$.   QED

1.4.2 The Twin Paradox

Alice boards a rocket ship, which accelerates uniformly in the x direction. After a certain time, it decelerates (at the same rate) back to $v = 0$. Then it reverses direction, and returns to earth in the same way. Because moving clocks run slow, not as much time will have elapsed on the rocket clock as on a stationary earth clock, so at their reunion Alice will have aged less than her twin brother Bob, who stays at home. Suppose the trip takes 40 years by her watch, and the acceleration is g. How many years has Bob aged in the process?

We need to determine the rocket's position (x) as a function of time (t). By "uniform acceleration" we mean that a passenger *on the rocket* experiences an unchanging acceleration, g. From *her* perspective she is at a fixed position (seat 23B, or (x_0, y_0, z_0)), and her watch reads proper time, τ. Thus her coordinates are $y^\mu = (\tau, x_0, y_0, z_0)$, and her velocity is

$$\dot{y}^\mu = (1,0,0,0), \tag{1.79}$$

where the dot denotes differentiation with respect to proper time. Of course, she is not in an inertial reference frame; these coordinates refer to her **instantaneously comoving inertial frame**. You cannot get her acceleration by differentiating again, because it involves transfer to a new comoving frame. We'll get it instead by an indirect route.

It follows from 1.70 that

$$g_{\mu\nu}\ddot{x}^\mu\dot{x}^\nu + g_{\mu\nu}\dot{x}^\mu\ddot{x}^\nu = 0, \quad \text{so} \quad g_{\mu\nu}\dot{x}^\mu\ddot{x}^\nu = 0. \tag{1.80}$$

Thus **proper acceleration** ($\ddot{x}$) is always orthogonal to **proper velocity** ($\dot{x}$). So her acceleration must have the form

$$\ddot{y}^\mu = (0, g, 0, 0). \tag{1.81}$$

Thus

$$g_{\mu\nu}\ddot{x}^\mu\ddot{x}^\nu = g_{\mu\nu}\ddot{y}^\mu\ddot{y}^\nu = -g^2. \tag{1.82}$$

We'll take this as the (Lorentz-invariant) characterization of "uniform acceleration." Now let's examine her motion from the earth's perspective. Proper velocity always has "length" 1 (Eq. 1.70),

$$g_{\mu\nu}\dot{x}^\mu\dot{x}^\nu = 1; \tag{1.83}$$

for motion in the x direction we have

$$\dot{x}^\mu = (\cosh\varphi, \sinh\varphi, 0, 0) \quad \text{so} \quad \ddot{x}^\mu = \dot{\varphi}(\sinh\varphi, \cosh\varphi, 0, 0). \tag{1.84}$$

Hence

$$g_{\mu\nu}\ddot{x}^{\mu}\ddot{x}^{\nu} = (\dot{\varphi})^2(\sinh^2\varphi - \cosh^2\varphi) = -(\dot{\varphi})^2. \tag{1.85}$$

For uniform acceleration, therefore, Eq. 1.82 says $\dot{\varphi} = g$, or $\varphi = g\tau$. Setting $x = t = 0$ at $\tau = 0$,

$$\dot{x}^0 = \frac{dt}{d\tau} = \cosh\varphi = \cosh(g\tau) \quad \Rightarrow \quad t = \frac{1}{g}\sinh(g\tau), \tag{1.86}$$

$$\dot{x}^1 = \frac{dx}{d\tau} = \sinh\varphi = \sinh(g\tau) \quad \Rightarrow \quad x = \frac{1}{g}(\cosh(g\tau) - 1). \tag{1.87}$$

These are parametric equations for a hyperbola; $gt = \sinh(g\tau)$ and $(gx + 1) = \cosh(g\tau)$, so

$$(gx + 1)^2 - (gt)^2 = 1. \tag{1.88}$$

For this reason, uniform acceleration is known as **hyperbolic motion**.

In the figure below, each of the four segments is a hyperbolic arc (the scale on the x-axis is not the same as on the t-axis; with equal scales the slope would never be less than 1).

If we measure time in years and distance in light-years, the unit of acceleration would be

$$1\,\frac{\text{light-yr}}{\text{yr}^2} = \frac{c}{\text{yr}} \approx \frac{3 \times 10^8\,\text{m/s}}{\pi \times 10^7\,\text{s}} \approx 9.5\,\text{m/s}^2, \tag{1.89}$$

which is pretty close to the acceleration of gravity on earth. That means we can take $g = 1$ in our units, for a reasonably comfortable ride. The acceleration segment lasts 10 years, by Alice's watch (the whole trip takes her 40 years). But her brother has aged

$$t = 4\sinh(10) \approx 2\,e^{10} = 44,000\,\text{yr} \tag{1.90}$$

(note that earth time goes up exponentially with rocket time).

1.4.3 Doppler Shift

Consider a spaceship approaching you (at rest on the earth) with constant speed v. The spaceship sends out a light pulse every second (by the *spaceship* clock). At time $t = t' = 0$, when the rocket is a distance d away, the first signal is sent, so it gets to you at $t = d$. The second signal is sent when the *spaceship* clock reads $t' = 1 = t\sqrt{1 - v^2}$, so yours reads $t = 1/\sqrt{1 - v^2}$. At that time (in your system) the spaceship is at

$$d - \frac{v}{\sqrt{1 - v^2}} \tag{1.91}$$

(which is also the time it takes for the pulse to reach you). Thus the second signal leaves at $1/\sqrt{1 - v^2}$ (your time), and arrives at

$$t = \frac{1}{\sqrt{1 - v^2}} + d - \frac{v}{\sqrt{1 - v^2}}. \tag{1.92}$$

The interval between signals received is

$$(1) \quad \Delta t = \frac{1 - v}{\sqrt{1 - v^2}} = \sqrt{\frac{1 - v}{1 + v}}. \tag{1.93}$$

If the rocket is moving *away* from you, v has the opposite sign, and

$$(2) \quad \Delta t = \sqrt{\frac{1 + v}{1 - v}}. \tag{1.94}$$

In case (1) you *see* the spaceship clock running *faster* than yours (the Doppler effect swamps time dilation); in case (2) you *see* greater time dilation than expected (Doppler and dilation conspire). But always you *observe* time dilation: $t = t'/\sqrt{1 - v^2}$.

1.4.4 The Bandits and the Train

A row of bandits regularly fires on the daily train as it passes by at a speed close to c. One day a bandit calls in sick, so there is a gap in the line. The bandit chief points out that the train will be Lorentz contracted, and there will come an interval (when the foreshortened train is directly opposite the missing bandit) when no bullets hit the train. The engineer notes that from the perspective of a person on the train, it is the row of bandits that will be contracted, and the fact that one is missing will hardly be noticeable; the train will be fired upon without interruption.

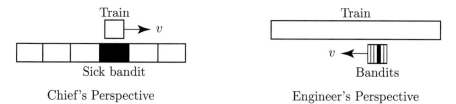

| Chief's Perspective | Engineer's Perspective |

Who is right? Does there occur a moment when no bullets hit the train? *Answer:* They *both* are—for their respective reference systems. The problem is conflicting notions of simultaneity:

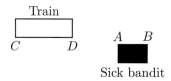

When D meets B,

- the chief says C is already past A (C meets A *before* D meets B),
- the engineer says C has not yet reached A (C meets A *after* D meets B).

There is no contradiction, just a different perspective concerning the sequence of two events.

1.4.5 The Prisoner's Escape

A (spherical) prisoner proposes to escape by running so fast that Lorentz contraction will permit him to slip between the bars:

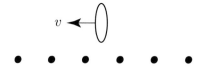

His cellmate retorts that in the escaping prisoner's reference frame it is the distance between the bars that will be Lorentz contracted, and it will be *even more* difficult for him to get out. Who is right? (This is a better problem than the train and bandit paradox, because they *cannot* both be correct: at the end of the day either he's a free man, or he's *not*.)

Answer: He does *not* escape. The trouble arises when he changes direction to slip through the bars:

In the nonrelativistic case (left) there is no Lorentz contraction, the spacing between the bars is (presumably!) less than his diameter (a), and he cannot get through; in the relativistic case (right) he is indeed Lorentz contracted *along the direction of motion*—he becomes an ellipsoid (or a spheroid, or something), but the critical dimension a is unchanged, and it is no easier (nor more difficult) for him to slip through. The essential point is that *dimensions perpendicular to the direction of motion are not contracted*.

1.4.6 The Moving Cube

Imagine looking at a cube, of side 1, very far away (so there is no parallax—rays reaching your eye from all parts of the cube are parallel). The cube is at rest (so are you), and oriented so that you can see two faces. The view from above is shown in the first figure; what you see is shown in the second figure:

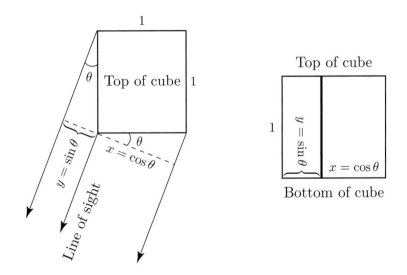

Question: What do you see if the cube is *moving*, to the left, at speed v? This is a rare case in which we deliberately ask what you *see*, not what you *observe*; what you *observe* is just a Lorentz-contracted version of the figure on the right, but what you *see* must take into account the fact that light from more distant parts of the cube takes longer to reach your eye. In the figure below I have drawn the top face of the cube, in your (stationary) reference frame, where it is Lorentz contracted along the direction of motion. At the same instant you receive the light from the left corner (O), you also get light from the right corner (A), but since the latter had to travel a greater distance (AB) it must have left somewhat earlier, when the cube was in the "old" position, as indicated.

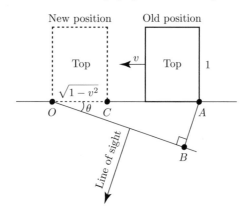

By simple trigonometry,

$$OA \equiv L, \quad AB = L\sin\theta, \quad AC = vt, \tag{1.95}$$

where $t = AB$ is the time it takes light to travel the "extra" distance. So

$$L = OC + AC = \sqrt{1 - v^2} + v(L\sin\theta), \tag{1.96}$$

and hence

$$L = \frac{\sqrt{1-v^2}}{1 - v\sin\theta}, \quad OB = L\cos\theta = \frac{\sqrt{1-v^2}\,\cos\theta}{1 - v\sin\theta} = x. \tag{1.97}$$

This replaces x in the previous figure (notice that it reduces to $\cos\theta$ when $v \to 0$).

What about y? Again, the light from O arrives at the same time as the light from the back corner (C), which had to travel an extra distance CD, and therefore must have left earlier, when the cube was in the "old" position:

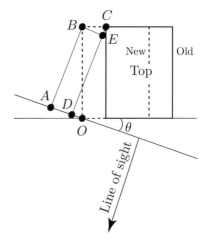

We're looking for $y = OD$. All the acute angles are equal to θ, so

$$BC \equiv \ell, \quad CE = \ell \sin\theta, \quad BE = \ell \cos\theta, \quad AB = \cos\theta, \quad AO = \sin\theta. \quad (1.98)$$

The delay time is

$$CD = AB + CE = \ell \sin\theta + \cos\theta, \quad (1.99)$$

so ℓ (the distance the cube moves as light goes from C to D) is

$$\ell = v(\ell \sin\theta + \cos\theta) \quad \Rightarrow \quad \ell = \frac{v\cos\theta}{1 - v\sin\theta}. \quad (1.100)$$

Then (since $AD = BE$)

$$y = OD = AO - AD = \sin\theta - \frac{v\cos^2\theta}{1 - v\sin\theta} = \frac{\sin\theta - v}{1 - v\sin\theta}. \quad (1.101)$$

What you *see*, then, is

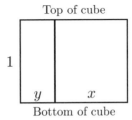

Top of cube

1

y　x

Bottom of cube

where x and y are given by Eqs. 1.97 and 1.101. Now comes a small miracle:

$$x^2 + y^2 = \frac{\sin^2\theta - 2v\sin\theta + v^2 + \cos^2\theta - v^2\cos^2\theta}{(1 - v\sin\theta)^2} = 1, \quad (1.102)$$

so we might as well define a new angle θ', such that $x = \cos\theta'$ and $y = \sin\theta'$, and what we *see* looks exactly like what we saw for a cube at rest, only rotated at a different angle. It doesn't look like a *contracted* cube at all, but rather like a *rotated* cube! We *observe* a contracted cube, but we *see* a rotated cube. What is more, we could construct any *other* object out of cubical "Lego," so the same conclusion holds quite generally.[20]

---

[20] Eds. In 1939 George Gamow published a famous book, *Mr. Tompkins in Wonderland,* in which he imagined a world where the speed of light was just a few mph, and he described seeing a Lorentz-contracted bicycle with a skinny rider, as it passed by. His blunder was corrected by many authors: he had failed to distinguish what you see from what you observe. R. Penrose, *Proc. Cambridge Phil. Soc.* **55**, 137 (1959); J. Terrell, *Phys. Rev.* **116**, 1041 (1959); V. F. Weisskopf, *Physics Today* **13**, 24 (September 1960); M. L. Boas, *Am. J. Phys.* **29**, 283 (1961).

1.4.7 Tachyons

Could there exist particles that travel faster than light (**tachyons**)? They are perfectly consistent with Lorentz invariance. You could not convert an ordinary particle into a tachyon by Lorentz transformation, but one tachyon is connectable to another tachyon by Lorentz transformation (though it may be going backward in time). A tachyon's velocity would be space-like, and by Lorentz transformation the world line of a tachyon moving at constant velocity could be made to coincide with (say) the x-axis—it would take no time at all to get from point a to point b: the particle would be in two places at the same time (and everywhere in between). We would presumably like to exclude such outlandish behavior, but this is an *independent* assumption; it does not follow from Lorentz invariance alone.

Problem 1.2

Aberration. A spaceship moves with velocity **v** along its axis of symmetry. A star is at rest; the vector from the spaceship to the star makes an angle θ with this axis. What is the angle at which an observer on the spaceship sees the star?

Problem 1.3

Milne's model of an expanding universe. Imagine a collection of noninteracting particles (galaxies, if you like). They all start out ($t = 0$) at the origin, but with "randomly" distributed velocities. If the distribution of velocities is Lorentz invariant, then we can take any particle to be the one "at rest." At a later time t, the faster-moving particles will be farther away. This is a primitive model for an expanding universe.

   To construct a Lorentz-invariant velocity distribution, note that

$$g_{\mu\nu}v^{\mu}v^{\nu} = v^2 = 1 \tag{1.103}$$

(Eq. 1.70), where $v^{\mu} \equiv dx^{\mu}/d\tau$ is the proper velocity, τ is the proper time (for the particle in question), and v^0 is positive. Thus the velocities all lie on the forward hyperboloid:

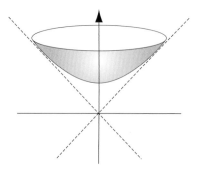

Let's express the number of particles in the volume d^4v (of velocity space) as[21]

$$N\,\delta(v^2 - 1)\,d^4v, \tag{1.104}$$

for some constant N. The delta function is manifestly Lorentz invariant, as is d^4v, so this defines a Lorentz-invariant distribution of velocities on the hyperboloid. What is the resulting distribution in the 3-velocity[22] $\mathbf{v} = d\mathbf{x}/d\tau$? Using the standard formula for the delta function of a function,

$$\delta(f(z)) = \sum_i \frac{1}{|f'(z_i)|}\delta(z - z_i), \tag{1.105}$$

where the sum is over the zeros of f ($f(z_i) = 0$),[23]

$$
\begin{aligned}
\delta((v^0)^2 - \mathbf{v}^2 - 1) &= \frac{1}{2|\sqrt{\mathbf{v}^2+1}|}\delta(v^0 - \sqrt{\mathbf{v}^2+1}) \\
&\quad + \frac{1}{2|-\sqrt{\mathbf{v}^2+1}|}\delta(v^0 + \sqrt{\mathbf{v}^2+1}) \\
&= \frac{1}{2v^0}\delta(v^0 - \sqrt{\mathbf{v}^2+1})
\end{aligned}
\tag{1.106}
$$

(the other term vanishes because $v^0 > 0$). Thus the number of particles in the volume d^4v is

$$N\,\delta(v^2-1)\,d^4v = \frac{N}{2v^0}\delta\left(v^0 - \sqrt{\mathbf{v}^2+1}\right)dv^0 d^3\mathbf{v}. \tag{1.107}$$

Integrating over v^0 we get the number of particles with velocities in the range $d^3\mathbf{v}$:

$$\frac{N}{2\sqrt{\mathbf{v}^2+1}}d^3\mathbf{v}. \tag{1.108}$$

(a) Find the density of galaxies (the number per unit volume) as a function of position and time: $\rho(\mathbf{x},t)$.

(b) What is "Hubble's law" (speed $|\mathbf{v}|$ as a function of distance $|\mathbf{x}|$) in this universe?

(c) What would someone (at rest at the origin) actually *see* (as opposed to *observe*) at time t—what density of galaxies, as a function of distance, and what speed law?

---

[21] Note that the Lorentz-invariant density is unique (up to the overall constant N). Given the density at one point, A, I can, by Lorentz transformation, carry that point to any other point B on the hyperboloid, and thus determine the density at B. There cannot exist two *different* Lorentz-invariant distributions.

[22] Note that this is *not* the ordinary 3-velocity $d\mathbf{x}/dt$, but the *proper* 3-velocity $\mathbf{v} = d\mathbf{x}/d\tau$.

[23] Ordinarily I would write the square of a 3-vector $(\mathbf{v}\cdot\mathbf{v} = |\mathbf{v}|^2)$ as v^2, but this notation has been preempted for the 4-vector $(v^2 = (v^0)^2 - \mathbf{v}\cdot\mathbf{v})$, so I'll write the square of the 3-vector using boldface: $\mathbf{v}^2$ (even though it's a scalar).

2

Relativistic Mechanics

2.1 Tensor Formalism

Under a homogeneous Lorentz transformation, position and time transform as

$$x^\mu \to y^\mu = \Lambda^\mu_{\ \nu} x^\nu; \quad \Lambda^\mu_{\ \nu} = \frac{\partial y^\mu}{\partial x^\nu}. \tag{2.1}$$

A **4-vector** is any set of four "components" that transform the same way:

$$a^\mu \to \Lambda^\mu_{\ \nu} a^\nu. \tag{2.2}$$

For example, proper velocity ($v^\mu = dx^\mu/d\tau$) is a 4-vector, as is proper acceleration ($d^2x^\mu/d\tau^2$); τ is a scalar (invariant).[1] A scalar *field* transforms as

$$\varphi(x) \to \varphi(\Lambda^{-1}x). \tag{2.3}$$

For a *vector* field, both the components and the arguments transform (Eq. 1.8):

$$A^\mu(x) \to \Lambda^\mu_{\ \nu} A^\nu(\Lambda^{-1}x). \tag{2.4}$$

What about *derivatives* of a scalar field (with respect to the coordinates x^μ)? Do they constitute a vector field (like the gradient in three dimensions)? That is, do they transform according to Eq. 2.2? They do *not*. By the chain rule,

$$\frac{\partial}{\partial x^\mu} \to \frac{\partial}{\partial y^\mu} = \frac{\partial x^\nu}{\partial y^\mu} \frac{\partial}{\partial x^\nu}, \tag{2.5}$$

so, adopting the convenient shorthand

$$\partial_\mu \equiv \frac{\partial}{\partial x^\mu}, \tag{2.6}$$

---

[1] As we shall see, $\Lambda^\mu_{\ \nu}$ is itself a tensor, but for now we shall think of it as just a collection of 16 coefficients, which may be displayed as a 4×4 matrix.

we have

$$\partial_\mu \varphi(x) \to (\Lambda^{-1})_\mu{}^\nu \partial_\nu \varphi(\Lambda^{-1}x), \text{ where } (\Lambda^{-1})_\mu{}^\nu = \frac{\partial x^\nu}{\partial y^\mu}. \tag{2.7}$$

The coefficients here are *different*: comparing Eqs. 2.1 and 2.7, $x \leftrightarrow y$ (so $\Lambda \to \Lambda^{-1}$) and $\mu \leftrightarrow \nu$ (so $\Lambda \to \Lambda^T$). This notation makes sense:

$$\Lambda^\mu{}_\nu (\Lambda^{-1})_\lambda{}^\nu = \frac{\partial y^\mu}{\partial x^\nu} \frac{\partial x^\nu}{\partial y^\lambda} = \delta_\lambda^\mu = I_\lambda^\mu, \tag{2.8}$$

where I is the 4×4 unit matrix. We are led to distinguish, then, two different kinds of "4-vectors": **contravariant** (archetype: position/time, x^μ, index *up*) and **covariant** (archetype: gradient, ∂_μ, index *down*).[2]

Contravariant 4-vector	upper index	$a^\mu \to \Lambda^\mu{}_\nu a^\nu$
Covariant 4-vector	lower index	$b_\mu \to (\Lambda^{-1})_\mu{}^\nu b_\nu$

A simple example of a **tensor** is the product of components of two 4-vectors. It transforms with *two* factors of Λ, one for each index:

$$a^\mu b^\nu \to \Lambda^\mu{}_\sigma \Lambda^\nu{}_\tau a^\sigma b^\tau. \tag{2.10}$$

More generally, a (contravariant second-rank) tensor $t^{\mu\nu}$ is a collection of 16 "components" that transform the same way:

$$t^{\mu\nu} \to \Lambda^\mu{}_\sigma \Lambda^\nu{}_\tau t^{\sigma\tau}. \tag{2.11}$$

An *n*th rank tensor has n indices (4^n components), and transforms with n factors of Λ (or, for covariant indices, Λ^{-1T}).

We have not proved (yet) that the metric (1.18) is a tensor, but *assuming* it is,

$$g_{\mu\nu} \to (\Lambda^{-1})_\mu{}^\sigma (\Lambda^{-1})_\nu{}^\tau g_{\sigma\tau}, \tag{2.12}$$

or, in matrix notation,

$$g \to (\Lambda^{-1})^T g (\Lambda^{-1}). \tag{2.13}$$

But Λ^{-1} is itself a Lorentz transformation, so Eq. 1.31 says

$$g_{\mu\nu} \to g_{\mu\nu}. \tag{2.14}$$

---

[2] In Euclidean space the analogous invariance under rotations is

$$x \to Rx \text{ (contravariant)}, \quad \partial \to (R^{-1})^T \partial \text{ (covariant)}. \tag{2.9}$$

But rotation matrices are orthogonal: $R^{-1} = R^T$, so there is no difference between covariant and contravariant, and the gradient is an ordinary vector.

Thus the metric is a **numerically invariant** tensor, whose elements have the same values in any inertial system.[3] We define the contravariant metric tensor $g^{\mu\nu}$ as the *matrix inverse* of $g_{\mu\nu}$, but transforming contravariantly:[4]

$$g_{\mu\nu} = \begin{pmatrix} 1 & 0 & 0 & 0 \\ 0 & -1 & 0 & 0 \\ 0 & 0 & -1 & 0 \\ 0 & 0 & 0 & -1 \end{pmatrix}; \quad g^{\mu\nu} = \begin{pmatrix} 1 & 0 & 0 & 0 \\ 0 & -1 & 0 & 0 \\ 0 & 0 & -1 & 0 \\ 0 & 0 & 0 & -1 \end{pmatrix}. \tag{2.15}$$

You can use the metrics to raise and lower indices, creating an associated covariant vector from a contravariant one, or a mixed tensor from a fully covariant one, and so on:

$$a_\mu = g_{\mu\nu}a^\nu, \quad b^\mu{}_\nu = g^{\mu\lambda}b_{\lambda\nu}, \quad c^{\mu\kappa\tau} = g^{\mu\nu}c_\nu{}^{\kappa\tau}, \quad \text{etc.} \tag{2.16}$$

We use the same letter (a, b, c) for the different—covariant, contravariant, mixed— versions. Thus we have the covariant position–time 4-vector

$$x_\mu = g_{\mu\nu}x^\nu \tag{2.17}$$

and the contravariant gradient

$$\partial^\mu = g^{\mu\nu}\partial_\nu = g^{\mu\nu}\frac{\partial}{\partial x^\nu} = \frac{\partial}{\partial x_\mu} \tag{2.18}$$

(notice how a covariant index in the denominator becomes a contravariant index in the numerator). In Minkowski space, raising or lowering a spatial index costs you a minus sign (raising or lowering a temporal index is free).[5] Notice that

$$g_{\mu\sigma}g^{\sigma\nu} = g_\mu{}^\nu = \delta_\mu^\nu. \tag{2.19}$$

The scalar product of two 4-vectors,

$$g_{\mu\nu}a^\mu b^\nu = a^\mu b_\mu = a_\nu b^\nu = g^{\mu\nu}a_\mu b_\nu = a^0 b^0 - \mathbf{a}\cdot\mathbf{b} \equiv a \cdot b \tag{2.20}$$

is a tensor of rank 0 (invariant). More generally, if you sum over like indices, one covariant and one contravariant, you get a new tensor of rank reduced by 2; this is called **contraction**. Schematically,[6]

---

[3] Eds. There is precisely one other invariant tensor in Minkowski space: the (fourth rank) Levi-Civita tensor density $\epsilon^{\mu\nu\lambda\sigma}$.

[4] Eds. The fact that they are numerically identical, and both are invariant under Lorentz transformations, is an invitation to sloppiness, such as writing $g_{\mu\nu} = g^{\mu\nu}$ (a notational abomination). Please respect the positions of the indices; in general relativity, where they are *not* the same, you will regret any bad habits you develop now.

[5] For authors who use the metric $(-1, 1, 1, 1)$ (footnote 8 in Chapter 1) raising/lowering a spatial index is free, and a temporal index costs the minus sign.

[6] For reasons that will become clear in a moment, each index (whether covariant or contravariant) occupies an exclusive "slot" (horizontal position). So T_ν^μ is an illegal expression—I don't know whether μ occupies the first slot or the second. That's why I have been careful to write $\Lambda^\mu{}_\nu$, and it's one reason why I call 2.21

$$\sum_{\lambda=0}^{3} T^{\mu_1 \cdots \mu_n \lambda}_{\nu_1 \cdots \nu_m \lambda} = S^{\mu_1 \cdots \mu_n}_{\nu_1 \cdots \nu_m}. \tag{2.21}$$

A second-rank tensor is **symmetric** if it doesn't change when you switch the indices: $S_{\nu\mu} = S_{\mu\nu}$; it is **antisymmetric** if it changes sign: $A_{\nu\mu} = -A_{\mu\nu}$. Any tensor can be decomposed into a symmetric part and an antisymmetric part:

$$T_{\mu\nu} = S_{\mu\nu} + A_{\mu\nu}, \tag{2.22}$$

where

$$S_{\mu\nu} \equiv \tfrac{1}{2}(T_{\mu\nu} + T_{\nu\mu}), \quad A_{\mu\nu} \equiv \tfrac{1}{2}(T_{\mu\nu} - T_{\nu\mu}). \tag{2.23}$$

In general, $T_{\mu\nu}$ has 16 independent elements ($T_{00}, T_{01}, \ldots, T_{33}$), 10 in the symmetric piece, and 6 in the antisymmetric piece. Lorentz transformations do not mix the symmetric and antisymmetric parts. The **trace** of a second-rank tensor is

$$\mathrm{Tr}(T) \equiv T^{\lambda}_{\lambda} = T_{\lambda}{}^{\lambda} = T^{0}{}_{0} + T^{1}{}_{1} + T^{2}{}_{2} + T^{3}{}_{3} = T_{00} - T_{11} - T_{22} - T_{33}. \tag{2.24}$$

A symmetric tensor (including the symmetric *part* of *any* tensor) can be further decomposed into a (symmetric) **traceless** piece, $\hat{T}_{\mu\nu}$, and a multiple of the metric tensor:

$$T_{\mu\nu} = \hat{T}_{\mu\nu} + \tfrac{1}{4} \mathrm{Tr}(T)\, g_{\mu\nu}. \tag{2.25}$$

Thus the 10 independent components in $S_{\mu\nu}$ consist of 9 elements in $\hat{T}$ plus the trace. Once again, Lorentz transformations do not mix the traceless and metric parts.

Comment: In the language of group theory, what we just did was decompose the 16-dimensional representation of the Lorentz group (by generic second-rank tensors) into the direct sum of a 9-dimensional representation (by symmetric traceless second-rank tensors) plus a 6-dimensional representation (by antisymmetric second-rank tensors) and a 1-dimensional representation (the trace). Symbolically,

$$16 = 9 \oplus 6 \oplus 1. \tag{2.26}$$

A **representation** of a group G is a set of (square) matrices (D), one for each group element, whose multiplication mimics that of the group elements themselves:

$$D(g_1)D(g_2) = D(g_1 g_2). \tag{2.27}$$

"schematic"; really, there should be a gap below every superscript, and a gap above every subscript. The other reason is that it doesn't matter *where* among the upper indices the λ occurs (or where among the lower indices λ appears); once the matching indices are summed, you get a tensor of reduced rank (though it may be a *different* tensor, depending on where the λ's are situated). Finally, in view of the Einstein convention, the summation sign is unnecessary. (But these notational niceties are irrelevant in Eq. 2.21.)

"Reducing" a representation means separating the "carrier space" on which the representation acts (e.g. second-rank tensors) into subspaces that do not intermix under the action of the group elements. A representation that cannot be reduced is "irreducible," and the theory of group representations consists mainly in identifying all the **irreducible representations** of a given group. (We did the reduction here by the "seat of our pants," but there are, of course, more systematic methods.) Irreducible representations of the pertinent invariance group determine the independent building blocks for a physical theory. That's why scalars, vectors, antisymmetric tensors, and so on, play such an important role in classical mechanics.[7]

Example 2.1

Hooke's law. Consider the elasticity **stress tensor**, T_{ij} (in three dimensions). It tells you the force per unit area (or "stress") acting in the i direction on a patch of area oriented in the j direction; T_{ij} is a symmetric tensor, but otherwise unrestricted. Let S_{ij} be the **strain tensor** (likewise symmetric); it measures the resulting deformation of the medium. Hooke's law, in its general form, says that *stress is proportional to strain* (for small departures from equilibrium), though the constant of proportionality need not be the same for all components.

Let us assume that the medium is homogeneous and isotropic, so the system is invariant under rotations. *Question:* How many independent constants (**elastic moduli**) are there? We want to decompose T and S into irreducible representations of the rotation group. As before

$$T_{ij} = \hat{T}_{ij} + \frac{1}{3}\delta_{ij}\sum_k T_{kk}, \quad S_{ij} = \hat{S}_{ij} + \frac{1}{3}\delta_{ij}\sum_k S_{kk}. \tag{2.28}$$

Thus the (6-dimensional) representation by symmetric tensors decomposes into a 5-dimensional representation by *traceless* symmetric tensors plus a 1-dimensional representation (the trace), and we are left with two independent laws:

$$\sum_k T_{kk} = \mu \sum_k S_{kk} \quad \text{and} \quad \hat{T}_{ij} = \lambda \hat{S}_{ij}. \tag{2.29}$$

But, for instance, the trace of S cannot be a linear function of $\hat{T}_{ij}$, because they belong to different irreducible representations of the rotation group. Evidently there are just *two* constants here: **Young's modulus** and the **shear modulus**.[8]

Problem 2.1

Prove the contraction law (Eq. 2.21). That is, show that the contraction of a tensor with n indices (n_u up and n_d down) is a tensor with $n - 2$ indices ($n_u - 1$ up and $n_d - 1$ down).

---

[7] Similarly, the proton and neutron belong (together with six less familiar particles) to an 8-dimensional representation of the symmetry group SU(3) of the strong interactions, the **baryon octet**.

[8] Remember, this all assumes rotational invariance; for solids with crystal structure the symmetry group is smaller, the tensor representations decompose further, and there may be different moduli in different directions.

Problem 2.2

(a) Show that the Lorentz transform of a symmetric tensor is symmetric, and the Lorentz transform of an antisymmetric tensor is antisymmetric.

(b) Show that the Lorentz transform of a traceless tensor is traceless.

(c) Show that if $s_{\mu\nu}$ is symmetric, so too is $s^{\mu\nu}$.

(d) In d dimensions, how many independent elements are there in a symmetric second-rank tensor? How many in an antisymmetric tensor?

(e) In view of footnote 6, why do I keep writing δ^{μ}_{ν}, instead of $\delta_{\nu}{}^{\mu}$ or $\delta^{\mu}{}_{\nu}$?

2.2 Conservation Laws

We begin by enumerating the possible kinematic conservation laws in a Lorentz-invariant particle theory.[9] Consider the Minkowski (spacetime) diagram for a generic scattering process (there could be more particles involved, of course):

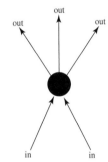

In the distant past,[10] when the particles are far apart, they move at constant velocity (that's Newton's first law); the position/time 4-vector for particle a has the form

$$x^{(a)\text{in}} = b^{(a)\text{in}} + v^{(a)\text{in}}\tau. \tag{2.30}$$

Does this mean there are eight independent parameters for each particle? No, because $v^2 = 1$, and changing the origin for proper time is trivial, so really just six parameters—the initial (spatial) position and velocity—determine each incoming trajectory, as in nonrelativistic mechanics. Similarly, the outgoing trajectories are

$$x^{(a)\text{out}} = b^{(a)\text{out}} + v^{(a)\text{out}}\tau. \tag{2.31}$$

---

[9] There may also be conserved "internal" properties, such as charge and lepton number, but we are concerned here with *kinematic* conservation laws for classical, featureless particles.

[10] I am talking about *asymptotic* conservation laws: things that are the same in the faraway future as they were in the long-ago past. Of course, these quantities may well be conserved moment-by-moment throughout the process.

We're looking for asymptotic conservation laws of the form

$$F\left(\{b^{(a)\,\text{out}}, v^{(a)\,\text{out}}\}\right) = F\left(\{b^{(a)\,\text{in}}, v^{(a)\,\text{in}}\}\right), \tag{2.32}$$

for some function F (I put in the curly brackets to indicate that all the different particles are included). What are the possibilities?

2.2.1 Conservation Laws Depending Only on Velocity

Consider first laws depending only on the velocities:

$$F\left(\{v^{(a)\,\text{out}}\}\right) = F\left(\{v^{(a)\,\text{in}}\}\right). \tag{2.33}$$

Imagine expanding F in a power series; we'd get a term linear in the velocities, and a term quadratic in the velocities, and so on. For relativistic invariance we must equate tensors belonging to the same representation of the Lorentz group; we can without loss of generality consider only irreducible tensor conservation laws. Moreover, we're only interested in *additive* functions of the various velocities:

$$F\left(\{v^{(a)}\}\right) = \sum_a f_a\left(v^{(a)}\right). \tag{2.34}$$

That is, there should be no "cross terms" involving particles that could, after all, be very distant:

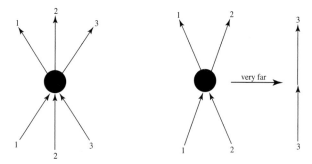

If, for example, the conservation law for the three-body scattering on the left included a term of the form $v^{(2)} \cdot v^{(3)}$, it would also be present for the scattering on the right, even though particle 3 cannot possibly affect the scattering of 1 and 2.

- **Scalar.** The only scalar you can construct from a single velocity is $v^2 = 1$, and we're not allowed to use cross terms such as $v^1 \cdot v^2$. So there are no kinematic scalar conservation laws.[11]

---

[11] But remember (footnote 9), there can be any number of scalar conservation laws that do not involve the kinematic quantities v and a at all—including, in principle, conservation of *mass* (though this is in fact *not* conserved, relativistically).

- **Vector.** The only vector law you can form is

$$\sum_{\text{out}} \lambda_a \, v^{\mu(a)} = \sum_{\text{in}} \lambda_a \, v^{\mu(a)} \tag{2.35}$$

for some coefficients λ_a.

- **Tensor.** For second-rank tensors the candidates are traceless/symmetric (the trace itself is a scalar, and we already ruled that out) or antisymmetric. You can't make an antisymmetric form ($v^{\mu(a)} \, v^{\nu(b)} - v^{\nu(a)} \, v^{\mu(b)}$ would do it, but we're not allowed to use cross terms). This leaves only the traceless symmetric form

$$\sum_{\text{out}} \mu_a \left(v^{\mu(a)} \, v^{\nu(a)} - \tfrac{1}{4} g_{\mu\nu} \right) = \sum_{\text{in}} \mu_a \left(v^{\mu(a)} \, v^{\nu(a)} - \tfrac{1}{4} g_{\mu\nu} \right). \tag{2.36}$$

This gives us *nine* conservation laws ... but there are only *six* parameters! No scattering would be possible if there existed such a conservation law; it conserves too many things. In fact, there can't be a conservation law belonging to any irreducible representation of the Lorentz group with dimension greater than six, and presumably this rules out all tensors of rank higher than 2.[12]

So the only possibility is a *vector* conservation law (and there can only be *one* of them, since two would mean eight conserved quantities):

$$\sum_{a} \lambda_a \, v^{\mu(a)}. \tag{2.37}$$

But *is* this quantity in fact conserved, and if so what are the coefficients λ_a? At low velocities

$$v = \frac{1}{\sqrt{1 - \beta^2}} (1, \boldsymbol{\beta}) \approx (1, \boldsymbol{\beta}), \tag{2.38}$$

(where $\boldsymbol{\beta} \equiv d\mathbf{x}/dt$ is the *ordinary* velocity) and, picking λ_a to be the *mass* of particle a, the spatial terms reduce to the nonrelativistic conservation of *momentum*. So this looks promising. We define the **relativistic momentum** 4-vector

$$p^{\mu} \equiv m v^{\mu}. \tag{2.39}$$

Its temporal component is

$$p^0 = \frac{m}{\sqrt{1 - \beta^2}} = m \left(1 + \tfrac{1}{2}\beta^2 + \cdots \right) = m + \tfrac{1}{2} m \beta^2 + \cdots \tag{2.40}$$

where we recognize the nonrelativistic kinetic energy as the second term, and identify the total as **relativistic energy**. Notice that if p^0 is conserved then mass

---

[12] Well ... I haven't really proved that the decomposition of a tensor of rank 27 (say) could not include a representation of rank 5 (or 3) ... but take it from me.

itself (the **rest energy** m) is *not*; in general, the sum of the incoming masses is not the same as the sum of the outgoing masses. (If mass *is* conserved, we call it an **elastic** collision.) From 2.39 it follows that

$$p^2 = m^2. \tag{2.41}$$

Example 2.2

Photon rocket. For particles of matter, $p^\mu = mv^\mu$. But what about photons? Their mass is zero, and their (proper) velocity is infinite (because of the square root factor in 2.38), so the product is indeterminate. Luckily, quantum mechanics steps in: Planck says $E = \hbar\omega$ (which suggests $p^0 = \hbar k^0$), and de Broglie says $\mathbf{p} = \hbar\mathbf{k}$, so this invites

$$p^\mu = \hbar k^\mu, \tag{2.42}$$

where ω is the (angular) frequency and $\mathbf{k}$ is the wave number ($\mathbf{k}^2 = \omega^2$). In rocket propulsion, higher exhaust velocity means greater efficiency. The most efficient rocket possible would be ... a *flashlight*, pointing backward!

How much thrust do you get from the emission of a single photon? In its rest frame, the velocity 4-vector of the rocket is $v^\mu = (1, \mathbf{0})$, and its momentum is $p^\mu = (m, \mathbf{0})$. It emits a photon in the $-x$ direction: $k^\mu = (\omega, -\omega, 0, 0)$, with momentum $p^\mu = \hbar(\omega, -\omega, 0, 0) = (p, -p, 0, 0)$, where $p \equiv \hbar\omega$. By conservation of momentum,

$$(m, 0, 0, 0) = p'^\mu + (p, -p, 0, 0), \tag{2.43}$$

so the momentum of the rocket after the emission is

$$p'^\mu = (m - p, p, 0, 0). \tag{2.44}$$

In particular,

$$p'^\mu p'_\mu = (m - p)^2 - p^2 = m^2 - 2mp, \tag{2.45}$$

so its mass is now

$$m' = m\sqrt{1 - \frac{2p}{m}} \approx m - p \tag{2.46}$$

(assuming $p \ll m$, which is to say that the energy of the photon is much less than the rest energy of the rocket). The *change* in mass of the rocket is $\Delta m = -p$. Its 4-velocity (in the old rocket rest frame) is now

$$v'^\mu \approx \frac{p'^\mu}{m - p} = \left(1, \frac{p}{m - p}, 0, 0\right) \approx \left(1, \frac{p}{m}, 0, 0\right). \tag{2.47}$$

The *change* in velocity is

$$\Delta v^\mu = v'^\mu - v^\mu \approx \left(0, \frac{p}{m}, 0, 0\right), \tag{2.48}$$

so

$$\frac{(\Delta v^\mu)(\Delta v_\mu)}{(\Delta\tau)^2} \approx -\frac{p^2}{m^2(\Delta\tau)^2} \approx -\frac{(\Delta m)^2}{m^2(\Delta\tau)^2}, \tag{2.49}$$

where $\Delta\tau$ is the interval of proper time between one photon emission and the next. Evidently the proper acceleration $(d^2x^\mu/d\tau^2)$ is given by

$$a^\mu a_\mu \approx -\frac{1}{m^2}\left(\frac{dm}{d\tau}\right)^2. \tag{2.50}$$

But the acceleration is in the x direction $(a^\mu = (b, a, 0, 0))$, and the acceleration 4-vector is always orthogonal to the velocity 4-vector (here $v^\mu = (1, 0, 0, 0)$), so $b = 0$, and $a^\mu a_\mu = -a^2$. Therefore the magnitude of the spatial acceleration is

$$a = -\frac{1}{m}\frac{dm}{d\tau} \tag{2.51}$$

(the minus sign reflects the fact that a *positive* acceleration means a *decreasing* mass). In particular, if a is constant then

$$m = m_0 e^{-a\tau}. \tag{2.52}$$

Suppose (as in Section 1.4.2) the trip takes 40 years (of rocket time), and the acceleration $a = 1$. How much fuel should you bring?

$$\frac{m_0}{m} = e^{40} = 2 \times 10^{17}. \tag{2.53}$$

Even with the most efficient rocket possible, the space capsule will only account for a miniscule fraction 5×10^{-18} of the initial load—the rest is all fuel.

Example 2.3

Here's a different approach to the Doppler effect for light (Section 1.4.3). The electromagnetic potential is a 4-vector, and a plane wave is one possible case:

$$A_\mu(x) = e_\mu e^{i(\mathbf{k}\cdot\mathbf{x}-\omega t)}, \tag{2.54}$$

where e_μ is related to the polarization, and (to satisfy Maxwell's equations) $\omega = |\mathbf{k}|$. We might as well let $k^0 \equiv \omega$, so

$$k^2 \equiv k^\mu k_\mu = \omega^2 - \mathbf{k}^2 = 0 \quad \text{and} \quad k \cdot x = \omega t - \mathbf{k}\cdot\mathbf{x}. \tag{2.55}$$

Thus

$$e^{i(\mathbf{k}\cdot\mathbf{x}-\omega t)} = e^{-ik\cdot x} \tag{2.56}$$

is invariant.

A star emits light at frequency ω_0 (by its own clock) in the direction of the unit vector $\mathbf{n}$. Passing by at (ordinary) velocity $-\boldsymbol{\beta}$ (according to the star), we receive the signal at frequency ω (by our clock). *Question:* What is ω, as a function of $\omega_0, \boldsymbol{\beta}$ (the velocity of the star relative to us), and $\mathbf{n}$?

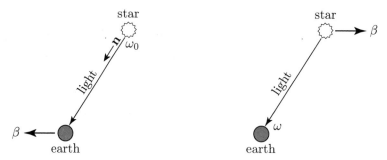

Solution: In the star's reference frame, the wave vector of the light and the earth's proper velocity are

$$k^\mu = (1, \mathbf{n})\,\omega_0, \quad v_e^\mu = \frac{1}{\sqrt{1 - \beta^2}}(1, -\boldsymbol{\beta}). \tag{2.57}$$

In the earth's reference frame, the same two quantities are

$$\bar{k}^\mu = (\omega, \ldots), \quad \bar{v}_e^\mu = (1, 0, 0, 0). \tag{2.58}$$

From the latter two it follows that $\bar{k} \cdot \bar{v}_e = \omega$. But the scalar product of two 4-vectors is invariant, so we can just as well calculate it in the star frame:

$$\omega = k \cdot v_e = \frac{1 + \mathbf{n} \cdot \boldsymbol{\beta}}{\sqrt{1 - \beta^2}}\,\omega_0. \tag{2.59}$$

This is the relativistic Doppler shift, for light; it differs from the classical result by the time-dilation factor in the denominator. (If $\mathbf{n}$ is parallel or opposite to $\boldsymbol{\beta}$, we recover 1.93 and 1.94.)

Problem 2.3

(a) A spaceship has the shape of a cube of side s. Its propulsion mechanism is such that its velocity is parallel to an edge of the cube. It moves with speed v directly toward a star at rest. The star emits energy, in the form of light, isotropically in its rest frame, with power P. The distance from the spaceship to the star is r. If the front face of the spaceship is black, so it absorbs all light striking it, what is the rate at which the spaceship receives energy, as measured in the rest frame of the ship? (Assume $r \gg s$.)
(b) If you have the right formula, you will find that you can bake a cake by sending it rapidly enough toward a burning kitchen match one light-year away. Is this consistent with the conservation of energy? If so, how?

Problem 2.4

A spaceship, with specifications as in the previous problem, and of mass M, is moving through empty space at fixed velocity. Suddenly it enters a cloud of dust particles, at rest, each of mass m; the cloud itself has a mass density ρ. Collisions with the dust act to slow the spaceship. (For simplicity, assume that such collisions are elastic head-on collisions; that is to say that the final masses of both the ship and the dust particles are the same as their initial masses, and that the final velocities are aligned along the initial velocity of the ship. Also assume that $M \gg m$.) The spaceship captain orders the jets turned on to maintain the ship's initial velocity. If the captain has the perfectly efficient "flashlight drive" (Example 2.2) at what rate (per unit proper time) does the spaceship lose mass?

Problem 2.5

Technology has advanced since the last problem, and a new propulsion mechanism has supplemented the flashlight drive. It is called the "vacuum-cleaner drive," and it works in the following manner. The spaceship absorbs any dust particles of mass m which it encounters, and shoots them out its back end with a lesser mass, m'. (For example, this could be done by collecting pairs of hydrogen nuclei and turning them into helium nuclei.) The mass of the ship itself does not change as a result of this process. If the parameters of ship and cloud are as in the preceding problem, what proper acceleration does this drive produce?

2.2.2 Conservation Laws including Position

Now let's look for possible conservation laws involving *position* (as well as velocity). The restriction to additive quantities (2.34) is not absolutely forced, this time, since there could be contributions depending (say) on inverse powers of the distance between two particles (potential energy, for instance).[13] Nevertheless, we shall consider only additive quantities:

$$F\left(\{v^{(a)}, b^{(a)}\}\right) = \sum_a f_a\left(v^{(a)}, b^{(a)}\right). \tag{2.60}$$

Any conserved quantity must be invariant under a trivial change in the origin for proper time:

$$\tau \to \tau + \lambda, \quad \text{which entails (from (2.30))} \quad b \to b + \lambda v. \tag{2.61}$$

One can construct three scalars from the two vectors b and v: b^2, v^2, and $b \cdot v$. But $v^2 = 1$, and the only vector independent of reparameterization (2.61) is $b - (b \cdot v)v$ (the component of b perpendicular to v), so in fact there is only *one* scalar left:

$$w \equiv (b - (b \cdot v)v)^2 = b^2 - 2(b \cdot v)^2 + (b \cdot v)^2 v^2 = b^2 - (b \cdot v)^2. \tag{2.62}$$

Again, we enumerate the possible conservation laws according to their representation of the Lorentz group:

---

[13] In the asymptotic limit the particles are infinitely far apart, and any such terms would vanish, but we might contemplate, for example, a dependence on the "distance of closest approach if they were to continue on their original trajectories." However, this does seem rather contrived.

- **Scalar.** The most general additive scalar structure is

$$F = \sum_a f_a \left(w^{(a)} \right) \tag{2.63}$$

where f_a denotes some function of the argument $w^{(a)}$. This is invariant under Lorentz transformations and changes of parameterization. But how about invariance under *translations* (and thus the full Poincaré group)? We didn't have to worry about that in the case of conservation laws dependent only on v, because velocities are unchanged by translation (and the same goes for reparameterization). But any viable conservation law should also be invariant under

$$b^{(a)} \to b^{(a)} + d, \tag{2.64}$$

which means that all *derivatives* of the conserved quantity (with respect to d) must *also* be conserved.[14]

- **Vector.** The first derivative yields

$$\frac{\partial F}{\partial d_\mu} = \sum_a \frac{df_a}{dw^{(a)}} \frac{\partial w^{(a)}}{\partial b_\mu} = \sum_a f_a' \left[2b_v g^{\mu v} - 2(b \cdot v)v_v g^{\mu v} \right], \tag{2.65}$$

so

$$F^\mu = \sum_a f_a' \left[b^{\mu(a)} - (b^{(a)} \cdot v^{(a)})v^{\mu(a)} \right] \tag{2.66}$$

is *also* conserved—the derivative of the *scalar* conservation law yields a *vector* conservation law.

- **Symmetric tensor.** The *second* derivative of F produces a conserved second-rank tensor. It's a *symmetric* tensor, because of the equality of the cross derivatives: $\partial^2/\partial d_\mu \partial d_v = \partial^2/\partial d_v \partial d_\mu$. But a symmetric tensor has 10 components—or 9, if it's traceless—and together with momentum this means at least 13 conserved quantities. If there are just two outgoing particles there are only 12 variables to be determined (6 for each one). To abort this cascade of conservation laws it must be that $f_a' = 0$: f_a is a constant, and F doesn't depend on the kinematic quantities

---

[14] An analogous argument applies in nonrelativistic mechanics, where the asymptotic kinetic energy

$$T = \sum_a \tfrac{1}{2} m^{(a)} \mathbf{v}^{(a)} \cdot \mathbf{v}^{(a)}$$

is conserved in a scattering process. If this is to be consistent with Galilean relativity,

$$t \to t, \ \mathbf{x} \to \mathbf{x} + \mathbf{v}_0 t \quad \Rightarrow \quad \mathbf{v}^{(a)} = \frac{d\mathbf{x}^{(a)}}{dt} \to \mathbf{v}^{(a)} + \mathbf{v}_0,$$

then the first derivative of T (with respect to $\mathbf{v}^{(a)}$) must also be conserved ($\sum m^{(a)}\mathbf{v}^{(a)}$) and likewise the second derivative ($\sum m^{(a)}$)—conservation of momentum and conservation of mass, respectively.

b and v at *all*. Similar considerations rule out starting with a vector conservation law involving b:

$$\sum_a \left\{ g_a \left(w^{(a)} \right) v^\mu + h_a \left(w^{(a)} \right) \left[b^{\mu(a)} - (b^{(a)} \cdot v^{(a)}) v^{\mu(a)} \right] \right\}. \tag{2.67}$$

- **Antisymmetric tensor.** However, we *can* construct a (potentially conserved) *antisymmetric* tensor:

$$J^{\mu\nu} = \sum_a f_a \left(w^{(a)} \right) \left[b^{\mu(a)} v^{\nu(a)} - b^{\nu(a)} v^{\mu(a)} \right]. \tag{2.68}$$

This carries only 6 independent components, so we are OK even with momentum also conserved ($10 < 12$). Again, translation invariance requires f_a to be a constant,

$$J^{\mu\nu} = \sum_a \mu_a \left[b^{\mu(a)} v^{\nu(a)} - b^{\nu(a)} v^{\mu(a)} \right], \tag{2.69}$$

and the derivative (with respect to d_λ) yields another conserved quantity:

$$J^{\mu\nu\lambda} = \sum_a \mu_a \left[g^{\mu\lambda} v^{\nu(a)} - g^{\nu\lambda} v^{\mu(a)} \right]. \tag{2.70}$$

But this is automatic (given conservation of momentum), provided $\mu_a = m_a$. Thus (adopting a more suggestive notation)

$$J^{\mu\nu} = \sum_a m_a \left[b^{\mu(a)} v^{\nu(a)} - b^{\nu(a)} v^{\mu(a)} \right] = \sum_a \left[x^\mu p^\nu - x^\nu p^\mu \right]^{(a)} \tag{2.71}$$

is a possible conserved quantity.

What would it represent physically? The spatial components are

$$J_{ij} = \sum_a \left[x_i p_j - x_j p_i \right]^{(a)}, \tag{2.72}$$

which (in the nonrelativistic limit) would be angular momentum (now revealed to be part of an antisymmetric tensor, not—as one might have guessed—a 4-vector). How about the other 3 components? Evidently

$$J^{0i} = \sum_a \left[x^0 p^i - x^i p^0 \right]^{(a)}, \quad \text{making a vector} \quad \sum_a \left[t\mathbf{p} - \mathbf{x} p^0 \right]^{(a)}, \tag{2.73}$$

where t is the time, and p^0 is the relativistic energy. The sum of the particle momenta is the total momentum, $\mathbf{P}$, so $\sum t\mathbf{p} = t\mathbf{P}$, and

$$\sum_a \mathbf{x}^{(a)} p^{0(a)} = \mathbf{X}E, \tag{2.74}$$

where $\mathbf{X}$ is the **center of energy** (relativistic generalization of center of mass), and $E = \sum p^{0(a)}$ is the total energy. Thus the conservation of J^{0i} says

$$\mathbf{X} = \mathbf{V}t + \text{constant}, \quad \text{where} \quad \mathbf{V} \equiv \frac{\mathbf{P}}{E}. \tag{2.75}$$

In words, the center of energy moves at constant velocity—no surprise. But it is a curious fact that *non*relativistically this is a consequence of conservation of *momentum*, whereas relativistically it is related to the conservation of *angular momentum*.

Conclusion: There are just two possible additive asymptotic kinematic conservation laws: conservation of momentum (a 4-vector) and conservation of angular momentum (an antisymmetric second-rank tensor). For a particle of mass m, the momentum is

$$\boxed{p^{\mu} \equiv m v^{\mu},} \tag{2.76}$$

and the angular momentum is

$$\boxed{J^{\mu\nu} \equiv x^{\mu} p^{\nu} - x^{\nu} p^{\mu}.} \tag{2.77}$$

Together these conservation laws impose 10 constraints on the process (still leaving a little room to maneuver, even if there are only two particles—12 parameters—in the final state).[15]

2.3 Lagrangian Particle Mechanics

In nonrelativistic particle mechanics the **action** (I) is the time integral of the **Lagrangian** L, which is a function of the generalized coordinates q and their time derivatives $\dot{q}$:

$$I = \int_{t_1}^{t_2} L(q, \dot{q}) \, dt. \tag{2.78}$$

The action depends on the trajectory, $q(t)$, and **Hamilton's principle** says that it is an extremum for the path nature actually takes, from q_1 at t_1 to q_2 at t_2:

$$\delta I = \int_{t_1}^{t_2} \left(\frac{\partial L}{\partial q} \delta q + \frac{\partial L}{\partial \dot{q}} \delta \dot{q} \right) dt = 0. \tag{2.79}$$

---

[15] Asked about multiplicative conservation laws, such as parity, Coleman remarked that in classical mechanics there are no kinematical conservation laws corresponding to discrete symmetries. He added that the distinction between additive and multiplicative conservation is to some extent a matter of notation: if $a + b + c + \cdots$ is conserved, then so too is $e^{a+b+c+\cdots} = e^a e^b e^c \cdots$.

But

$$\delta\dot{q} = \frac{\partial}{\partial t}\delta q, \tag{2.80}$$

so, integrating by parts and using the stipulation that $\delta q = 0$ at the endpoints,

$$\delta I = \int_{t_1}^{t_2}\left(\frac{\partial L}{\partial q} - \frac{\partial}{\partial t}\frac{\partial L}{\partial\dot{q}}\right)\delta q\, dt = 0. \tag{2.81}$$

And this holds for *arbitrary* $\delta q(t)$, so

$$\frac{\partial L}{\partial q} - \frac{d}{dt}\left(\frac{\partial L}{\partial\dot{q}}\right) = 0. \tag{2.82}$$

This is the **Euler–Lagrange equation**[16]—the equation of motion in Lagrangian mechanics (though in practice it is often simpler to work directly with $\delta I = 0$).

The Lagrangian is not uniquely defined by the motion; anything whose integral is an extremum for the actual motion will do. For example, you can always add a constant to the Lagrangian, or multiply it by a constant, or add a total time derivative—you'll still get the same equation of motion. Action principle formulations are therefore far from unique, and in principle one should never ask "What is *the* Lagrangian for this system?" The appropriate question is rather "What is *a* convenient and viable Lagrangian for this system?" Having said that, I will (as everyone does) speak freely of "the" Lagrangian, and "the" action.

In nonrelativistic particle mechanics there's a canonical recipe for constructing L: kinetic energy minus potential energy. For a single particle in one dimension,

$$L = T - V = \tfrac{1}{2}m\dot{x}^2 - V(x). \tag{2.83}$$

The Euler–Lagrange equation then delivers

$$-\frac{dV}{dx} - \frac{d}{dt}(m\dot{x}) = 0 \quad \Rightarrow \quad m\ddot{x} = -\frac{dV}{dx} = F, \tag{2.84}$$

Newton's second law.

Lagrangian mechanics carries over in a natural way to the relativistic régime, where t is treated on an equal footing with the spatial coordinates and is replaced (as the independent variable) by the world line parameter s. The Lagrangian is now a function of x^μ and $\dot{x}^\mu$ (where the dot now denotes differentiation with respect to s), and the action is an integral over s.[17] For a *free* particle of mass m a tempting choice (based on 2.83) would be

---

[16] If there are several generalized coordinates, there's an Euler–Lagrange equation for each one.

[17] One advantage of the Lagrangian formulation is that it is easy to implement general principles: if L is a scalar, the resulting theory is guaranteed to be Lorentz invariant; if I is independent of parameterization, so too are the equations of motion; if you want your theory to have some invariance, just make sure L shares that symmetry.

$$L = -\frac{m}{2}\,\dot{x}^\mu \dot{x}_\mu. \tag{2.85}$$

Hamilton's principle says

$$\delta I = -m \int_{s_1}^{s_2} \dot{x}^\mu \left(\delta \dot{x}_\mu\right) ds = m \int_{s_1}^{s_2} \ddot{x}^\mu \,\delta x_\mu \, ds = 0 \quad \Rightarrow \quad m\ddot{x}^\mu = 0. \tag{2.86}$$

However, this is not invariant under reparameterization ($s \to f(s)$), and we realize belatedly that neither was I. We need a Lagrangian that goes like s^{-1}. How about[18]

$$L = -m\sqrt{g_{\mu\nu} \frac{dx^\mu}{ds} \frac{dx^\nu}{ds}} \;? \tag{2.87}$$

Then

$$I = -m \int_{s_1}^{s_2} \sqrt{g_{\mu\nu} \frac{dx^\mu}{ds} \frac{dx^\nu}{ds}} \, ds \tag{2.88}$$

and

$$\delta I = -m \int_{s_1}^{s_2} \frac{1}{\sqrt{\dot{x}^\nu \dot{x}_\nu}} \dot{x}^\mu \,\delta \dot{x}_\mu \, ds = m \int_{s_1}^{s_2} \frac{d}{ds}\left(\frac{\dot{x}^\mu}{\sqrt{\dot{x}^\nu \dot{x}_\nu}}\right) \delta x_\mu \, ds = 0. \tag{2.89}$$

The equation of motion, then, is

$$m\frac{d}{ds}\left(\frac{\dot{x}^\mu}{\sqrt{\dot{x}^\nu \dot{x}_\nu}}\right) = 0. \tag{2.90}$$

This *is* independent of parameterization, and it is more illuminating if we now[19] adopt proper time as our parameter, because (Eq. 1.70) the denominator is then 1:

$$m\frac{d^2 x^\mu}{d\tau^2} = 0. \tag{2.91}$$

Adding a force term to the Lagrangian,

$$L = -m\sqrt{g_{\mu\nu} \frac{dx^\mu}{ds} \frac{dx^\nu}{ds}} - F_\mu(s)x^\mu, \tag{2.92}$$

we get

$$m\frac{d}{ds}\left(\frac{\dot{x}^\mu}{\sqrt{\dot{x}^\nu \dot{x}_\nu}}\right) - F^\mu(s) = 0. \tag{2.93}$$

---

[18] I chose the coefficient out front to match the nonrelativistic expression in the appropriate limit: using $s = t$,

$$L = -m\sqrt{1 - \mathbf{v}\cdot\mathbf{v}} \approx -m(1 - \tfrac{1}{2}\mathbf{v}\cdot\mathbf{v}) = \tfrac{1}{2}mv^2 + \text{a constant.}$$

[19] Note that you can't do this in the Lagrangian itself (2.87). It is a classic blunder in Lagrangian mechanics to insert facts about the solution prematurely, before doing the variation. This is all the more dangerous because occasionally you can get away with it.

Or, choosing proper time parameterization,

$$m\frac{d^2x^\mu}{d\tau^2} = F^\mu(\tau). \tag{2.94}$$

This is the relativistic version of Newton's second law, with the 4-vector **Minkowski force** F^μ.

Problem 2.6

If we stipulate right from the start that proper time (τ) is to be used as the world line parameter, then the Lagrangian 2.85 works (with the dots denoting $d/d\tau$). Confirm this, and construct the analog to Eq. 2.92.

2.4 Lagrangian Field Theory

The same strategy works for relativistic *fields*, except that we now start with the **Lagrange density**, $\mathcal{L}$, instead of the Lagrangian itself.[20] It is a function of the fields (ϕ) and all four of their derivatives $\partial_\mu\phi$, and the action is now an integral over some region of spacetime.

Example 2.4

The Klein–Gordon equation. Suppose

$$\mathcal{L}(\phi, \partial_\mu\phi) = \tfrac{1}{2}\left(\partial_\mu\phi\,\partial^\mu\phi - \mu^2\phi^2\right). \tag{2.95}$$

Then

$$I = \int d^4x\, \tfrac{1}{2}\left(\partial_\mu\phi\,\partial^\mu\phi - \mu^2\phi^2\right), \tag{2.96}$$

$$\delta I = \int d^4x\left(\partial_\mu\phi\,\partial^\mu(\delta\phi) - \mu^2\phi\,(\delta\phi)\right) = -\int d^4x\left(\partial^\mu\partial_\mu\phi + \mu^2\phi\right)\delta\phi = 0.$$

But $\delta\phi$ is arbitrary (except that, as before, we stipulate that it vanishes on the boundary), so

$$\partial^\mu\partial_\mu\phi + \mu^2\phi = 0. \tag{2.97}$$

This is the **Klein–Gordon equation** for a free scalar field with mass μ.

---

[20] Informally, we often refer to $\mathcal{L}$ as "the Lagrangian."

Of course, there may be more than one field, and the fields might be *tensors*, with multiple components; in that case the Lagrange density is a function of all the fields and their first derivatives, $\mathcal{L}(\{\phi^a, \partial_\mu \phi^a\})$.[21] Varying the fields,

$$\delta I = \sum_a \int d^4x \left[\frac{\partial \mathcal{L}}{\partial \phi^a} \delta \phi^a + \frac{\partial \mathcal{L}}{\partial (\partial_\mu \phi^a)} \delta(\partial_\mu \phi^a) \right] = 0. \tag{2.98}$$

Assuming (as always) that the fields are specified on the boundary, integrating by parts, and adopting the notation

$$\pi^{\mu a} \equiv \frac{\partial \mathcal{L}}{\partial (\partial_\mu \phi^a)}, \tag{2.99}$$

we find

$$\delta I = \sum_a \int d^4x \left[\frac{\partial \mathcal{L}}{\partial \phi^a} - \partial_\mu \pi^{\mu a} \right] \delta \phi^a = 0. \tag{2.100}$$

The action principle then yields the Euler–Lagrange field equations,

$$\frac{\partial \mathcal{L}}{\partial \phi^a} - \partial^\mu \pi_\mu^a = 0 \quad \text{for all } a. \tag{2.101}$$

2.4.1 Internal Symmetries and Conservation Laws

Suppose we integrate $\mathcal{L}$ over the shaded volume V, bounded by surfaces σ_1 and σ_2, and extending on the sides to spatial infinity (where the fields presumably vanish):

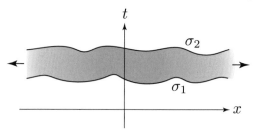

Assume that the normal vector is everywhere time-like, over both surfaces (I'll call such a surface "space-like"). I propose again to study the change in the action when the fields are varied:

$$\delta I_V = \int_V d^4x \, \delta \mathcal{L}, \tag{2.102}$$

---

[21] The inclusion of higher derivatives generally yields an energy that runs negative, making the system unstable. You might contemplate actions of the form $\int d^4x \, d^4y \, \mathcal{L}(\phi^a(x), \phi^b(y))$, but these are not quantizable.

only this time $\delta\mathcal{L}$ does *not* vanish on the boundary (if it *did* we'd just recover the Euler–Lagrange equations). Rather,

$$\delta I_V = \sum_a \int_V d^4x \left[\frac{\partial\mathcal{L}}{\partial\phi^a} \delta\phi^a + \pi_\mu^a \partial^\mu (\delta\phi^a) \right]. \tag{2.103}$$

Again, we integrate the second term by parts, using the product rule

$$\partial^\mu (A_\mu B) = A_\mu(\partial^\mu B) + B(\partial^\mu A_\mu). \tag{2.104}$$

Thus

$$\delta I_V = \sum_a \int_V d^4x \left[\frac{\partial\mathcal{L}}{\partial\phi^a} \delta\phi^a + \partial^\mu \left(\pi_\mu^a \delta\phi^a \right) - \left(\partial^\mu \pi_\mu^a \right) \delta\phi^a \right]. \tag{2.105}$$

The first and third terms in the integrand cancel (by Euler–Lagrange), and the 4-dimensional analog to **Gauss's theorem** converts the middle term into a surface integral:

$$\delta I_V = \sum_a \int_V d^4x\, \partial^\mu \left(\pi_\mu^a \delta\phi^a \right) = \oint_S d\sigma^\mu \sum_a \left(\pi_\mu^a \delta\phi^a \right)$$

$$= \int_{\sigma_2} d\sigma^\mu \sum_a \left(\pi_\mu^a \delta\phi^a \right) - \int_{\sigma_1} d\sigma^\mu \sum_a \left(\pi_\mu^a \delta\phi^a \right). \tag{2.106}$$

(Here S is the surface bounding V, with the element of "area" pointing outward—hence the minus sign on the last term, with σ_1 and σ_2 pointing forward in time. The "sides" at spatial infinity do not contribute, because π_μ^a is zero out there.) If $\delta\phi^a$ vanishes on the boundary we get zero, of course, but that's not what we are interested in now. *If the field variations are such as to leave $\mathcal{L}$ invariant*, the left hand side is zero, and hence

$$\int_\sigma d\sigma^\mu \sum_a \left(\pi_\mu^a \delta\phi^a \right) \tag{2.107}$$

is a conserved quantity, the same for any space-like surface σ.

Example 2.5

Suppose the Lagrangian is the sum of two Klein–Gordon Lagrangians (with the same mass):

$$\mathcal{L} = \sum_{a=1}^{2} \frac{1}{2} \left((\partial_\mu\phi^a)(\partial^\mu\phi^a) - \mu^2\phi^a\phi^a \right). \tag{2.108}$$

It is invariant under

$$\phi^1 \to \phi^1\cos\theta + \phi^2\sin\theta, \quad \phi^2 \to -\phi^1\sin\theta + \phi^2\cos\theta, \tag{2.109}$$

for all θ, but we only need the *infinitesimal* variations:

$$\delta\phi^1 = \phi^2 \, \delta\theta, \quad \delta\phi^2 = -\phi^1 \, \delta\theta. \tag{2.110}$$

For this Lagrangian,

$$\pi_\mu^a = \frac{\partial \mathcal{L}}{\partial(\partial^\mu \phi^a)} = \partial_\mu \phi^a, \tag{2.111}$$

so

$$\sum_a \pi_\mu^a \, \delta\phi^a = \left[(\partial_\mu \phi^1)\phi^2 - (\partial_\mu \phi^2)\phi^1 \right] \delta\theta = J_\mu \, \delta\theta, \tag{2.112}$$

where

$$J^\mu \equiv (\partial^\mu \phi^1)\phi^2 - (\partial^\mu \phi^2)\phi^1. \tag{2.113}$$

Dropping the constant factor $\delta\theta$, Eq. 2.106 says

$$\int_{\sigma_1} d\sigma_\mu J^\mu = \int_{\sigma_2} d\sigma_\mu J^\mu. \tag{2.114}$$

In particular, if we choose σ_1 and σ_2 to be constant-time surfaces, so that $d\sigma_\mu = (d^3\mathbf{x}, \mathbf{0})$,

$$\int_{t_1} d^3\mathbf{x} \, J^0 = \int_{t_2} d^3\mathbf{x} \, J^0. \tag{2.115}$$

Thus

$$Q \equiv \int d^3\mathbf{x} \, J^0(\mathbf{x}, t) \tag{2.116}$$

is conserved (independent of t).

An invariance of this sort—a reshuffling of the field variables that leaves the Lagrangian unchanged—is called an **internal symmetry**:

$$\delta\phi^a = \sum_b S^{ab} \phi^b \, \delta\theta. \tag{2.117}$$

We introduce the **current**[22]

$$J_\mu \equiv \sum_{a,b} \pi_\mu^a S^{ab} \phi^b, \tag{2.118}$$

and obtain a conserved **charge**

$$Q \equiv \int_\sigma d\sigma_\mu J^\mu = \int d^3\mathbf{x} \, J^0(\mathbf{x}, t). \tag{2.119}$$

---

[22] J_μ is not uniquely determined by this prescription, of course—we could just as well use any multiple of θ to parameterize the variation, and this would change S^{ab}, but the resulting Q would still be conserved. In practice, the coefficient is chosen for convenience.

But we can say something much stronger. Suppose σ_1 and σ_2 differ only by a little "bump," with boundary S', enclosing a tiny volume V':

Running the previous argument in reverse,

$$\int_{\sigma_2} d\sigma_\mu J^\mu = \int_{\sigma_1} d\sigma_\mu J^\mu \;\Rightarrow\; \int_{S'} d\sigma_\mu J^\mu = 0 \;\Rightarrow\; \int_{V'} d^4x\,(\partial^\mu J_\mu) = 0. \quad (2.120)$$

But V' is arbitrary, so the current must be divergenceless:

$$\partial_\mu J^\mu = 0. \qquad (2.121)$$

This is the **local conservation** law.[23] It is a stronger condition than **global conservation** (Q independent of time): the latter would allow "charge" to disappear in one place, as long as it immediately reappears somewhere else, whereas the former says that any charge leaving a volume V must flow out through the surface—it tracks the movement of charge from place to place. In view of Eq. 2.119, J^0 is the charge per unit volume, and if Q_V is the charge in V,

$$\frac{dQ_V}{dt} = \int_V d^3\mathbf{x}\,\partial_0 J^0(\mathbf{x},t) = -\int_V d^3\mathbf{x}\,\partial_i J^i = -\int_V d^3\mathbf{x}\,\nabla\cdot\mathbf{J} = -\int_\sigma d\boldsymbol{\sigma}\cdot\mathbf{J}.$$
$$(2.122)$$

Evidently $\mathbf{J}$ represents the charge per unit area, per unit time, flowing out through the surface. In this way, internal symmetries lead to scalar conservation laws (charge, lepton number, and so on) and locally conserved currents, in relativistic field theory.

Problem 2.7

Consider the Lagrangian in Example 2.5:

(a) Use the Euler–Lagrange equations to obtain the equations of motion.
(b) Use the equations of motion to show explicitly that Q (Eq. 2.116) is independent of time.

2.4.2 *Invariance under the Poincaré Group*

We turn now to the *geometric* invariances of the Lagrangian, and the associated conservation laws.[24] Although $\mathcal{L}$ is a scalar, it is a function of x, because the fields

---

[23] In electrodynamics it is called the **continuity equation**.
[24] Eds. The relation between symmetries and conservation laws was elucidated by E. Noether in 1918, and it plays a major role in Coleman's treatment. But as far as we can tell he never mentioned her name.

are. So δI_V is *not* zero under a Poincaré transformation, unless we take care to transform the *volume* as well as the Lagrangian. Doing so, we get two terms:

$$\delta I_V = \sum_a \oint_S d\sigma^\mu \, \pi^a_\mu \, \delta\phi^a + \oint_S d\sigma^\mu \delta x_\mu \, \mathcal{L}. \tag{2.123}$$

The first term is the result of transforming the fields (calculated the same way we did for internal symmetries, Eq. 2.106); the second term is the result of moving the boundaries an amount δx^μ.

We need to determine the forms of $\delta\phi^a$ and δx^μ under the Poincaré group. First consider a *scalar* field, $\phi(x)$:

$$\phi(x) \to \phi(\Lambda^{-1}x - b), \quad \mathcal{L}(x) \to \mathcal{L}(\Lambda^{-1}x - b) \tag{2.124}$$

(Eqs. 1.31 and 2.3). Actually, to avoid cumbersome notation let's work with the inverse transformation,

$$\phi(x) \to \phi(\Lambda x + b), \quad \mathcal{L}(x) \to \mathcal{L}(\Lambda x + b). \tag{2.125}$$

For an *infinitesimal* Lorentz transformation, $\Lambda = (1 + \delta\Lambda)$, Eq. 1.31 says

$$\Lambda^T g \Lambda = g \implies (1 + \delta\Lambda^T)g(1 + \delta\Lambda) \implies (\delta\Lambda^T)g + g(\delta\Lambda) = 0. \tag{2.126}$$

If we write $\delta\Lambda^\mu_{\;\nu} \equiv \epsilon^\mu_{\;\nu}$ (using ϵ to remind ourselves that it is infinitesimal),

$$\left(\epsilon^\mu_{\;\lambda}\right)^T g^\lambda_{\;\nu} + g^\mu_{\;\lambda}\epsilon^\lambda_{\;\nu} = 0 \implies \epsilon_\nu^{\;\mu} + \epsilon^\mu_{\;\nu} = 0 \implies \epsilon_{\nu\mu} + \epsilon_{\mu\nu} = 0. \tag{2.127}$$

Thus $\epsilon_{\mu\nu}$ is antisymmetric, and writing $\delta b^\mu \equiv \epsilon^\mu$, the general infinitesimal Poincaré transformation becomes

$$\delta x^\mu = \epsilon^\mu_{\;\nu} x^\nu + \epsilon^\mu. \tag{2.128}$$

For scalar fields, then,

$$\delta\phi = \phi(\Lambda x + b) - \phi(x) \approx (\partial_\mu\phi)\,\delta x^\mu. \tag{2.129}$$

For *tensor* fields the components mix *and* the argument transforms:

$$\delta\phi^a = \sum_b S^{ab}_{\mu\nu} \epsilon^{\mu\nu}\phi^b + (\partial_\mu\phi^a)\,\delta x^\mu, \tag{2.130}$$

where S is some matrix that depends on the kinds of tensors involved (we might as well make it antisymmetric in $\mu \leftrightarrow \nu$, since any symmetric part would disappear in the contraction with $\epsilon^{\mu\nu}$). The first term is just the most general form linear in the fields and in $\epsilon^{\mu\nu}$ (but not ϵ^μ, since tensor components do not mix under translations).

Returning, then, to Eq. 2.123,

$$\delta I_V = \oint_S d\sigma^\mu \left\{ \mathcal{L}\, \delta x_\mu - \sum_a \pi_\mu^a\, (\partial_\nu \phi^a)\delta x^\nu - \sum_{a,b} \pi_\mu^a\, S_{\lambda\nu}^{ab}\, \phi^b\, \epsilon^{\lambda\nu} \right\} = 0 \quad (2.131)$$

(the minus signs revert to the natural convention, 2.124).

- **The coefficient of ϵ^μ.** Suppose $\epsilon^\mu_\nu = 0$, so (2.128) $\delta x_\mu = \epsilon_\mu = g_{\mu\nu}\epsilon^\nu$. Then

$$\delta I_V = \oint_S d\sigma^\mu \left\{ \mathcal{L}\, g_{\mu\nu} - \sum_a \pi_\mu^a\, (\partial_\nu \phi^a) \right\} \epsilon^\nu = 0. \quad (2.132)$$

Defining the **canonical energy/momentum/stress tensor** (or "stress tensor," for short)

$$T_{\mu\nu}^{\mathrm{CAN}} \equiv -\mathcal{L}\, g_{\mu\nu} + \sum_a \pi_\mu^a\, (\partial_\nu \phi^a), \quad (2.133)$$

we conclude that the momentum 4-vector

$$p_\nu \equiv \int_\sigma d\sigma^\mu\, T_{\mu\nu}^{\mathrm{CAN}} \quad (2.134)$$

is a constant (the same for any space-like surface σ).

This can again be generalized to a *local* conservation law, if we demand invariance under *local* translations, i.e. if we permit ϵ^ν to vary from point to point—otherwise σ_2 is just a displaced (translated) version of σ_1, whereas to obtain the *local* law we require σ_2 to differ from σ_1 only in a small region V'.[25] With **local translation invariance** we have

$$\partial^\mu T_{\mu\nu}^{\mathrm{CAN}} = 0. \quad (2.135)$$

- **The coefficient of $\epsilon^{\mu\nu}$.** Now we take $\epsilon_\mu = 0$, so $\delta x^\mu = \epsilon^{\mu\sigma} x_\sigma$. Then (Eq. 2.131)

$$\delta I_V = \oint_S d\sigma^\mu \left\{ \mathcal{L}\, \epsilon_{\mu\sigma}\, x^\sigma - \sum_a \pi_\mu^a (\partial_\nu \phi^a)\epsilon^{\nu\sigma} x_\sigma - \sum_{a,b} \pi_\mu^a\, S_{\lambda\nu}^{ab}\, \phi^b\, \epsilon^{\lambda\nu} \right\}$$

$$= \oint_S d\sigma^\mu \left\{ \mathcal{L}\, g_{\mu\nu}\, x_\sigma - \sum_a \pi_\mu^a(\partial_\nu \phi^a)\, x_\sigma - \sum_{a,b} \pi_\mu^a\, S_{\nu\sigma}^{ab}\, \phi^b \right\} \epsilon^{\nu\sigma} = 0.$$

$$(2.136)$$

---

[25] In truth, there is some subtlety here, since for ϵ^ν to factor out of the integral, it has to be locally constant (left figure),

but this violates the requirement that the normal to the surface be everywhere time-like. To be rigorous we would have to use a surface like the one on the right, and then show that, in the limit, the regions (a) and (b), where ϵ^ν is not constant, cause no trouble.

Since $\epsilon^{v\sigma}$ is antisymmetric, only the antisymmetric part of the term in curly brackets survives, so we can replace it by its antisymmetric part:

$$\delta I_V = \frac{1}{2} \oint_S d\sigma^\mu \left\{ \left[\mathcal{L} g_{\mu v} - \sum_a \pi_\mu^a (\partial_v \phi^a) \right] x_\sigma - \left[\mathcal{L} g_{\mu\sigma} - \sum_a \pi_\mu^a (\partial_\sigma \phi^a) \right] x_v \right.$$

$$\left. - \sum_{a,b} \pi_\mu^a S_{v\sigma}^{ab} \phi^b + \sum_{a,b} \pi_\mu^a S_{\sigma v}^{ab} \phi^b \right\} \epsilon^{v\sigma} = 0. \tag{2.137}$$

But $S_{\sigma v}^{ab}$ is antisymmetric in its lower indices, so

$$\delta I_V = \frac{1}{2} \oint_S d\sigma^\mu \left\{ -T_{\mu v}^{\text{CAN}} x_\sigma + T_{\mu\sigma}^{\text{CAN}} x_v - 2 \sum_{a,b} \pi_\mu^a S_{v\sigma}^{ab} \phi^b \right\} \epsilon^{v\sigma} = 0. \tag{2.138}$$

Define

$$M_{\mu v\sigma} \equiv -T_{\mu v}^{\text{CAN}} x_\sigma + T_{\mu\sigma}^{\text{CAN}} x_v - 2 \sum_{a,b} \pi_\mu^a S_{v\sigma}^{ab} \phi^b. \tag{2.139}$$

Then

$$\oint_S d\sigma^\mu M_{\mu v\sigma} \epsilon^{v\sigma} = 0, \tag{2.140}$$

and hence

$$J^{v\sigma} \equiv \int_\sigma d\sigma_\mu M^{\mu v\sigma} \tag{2.141}$$

is a conserved quantity (the same for any space-like surface σ). Physically, $M^{\mu v\sigma}$ represents **angular momentum density**; $J^{v\sigma}$ consists of two terms, one of which $(Tx - Tx)$ is readily interpreted as **orbital** angular momentum, and the other is "intrinsic" angular momentum, or **spin**.

Thus invariance under the 10-parameter Poincaré group leads to 10 conserved quantities, a vector (4 components) and an antisymmetric tensor (6 components).

2.4.3 Symmetrization of the Stress Tensor

Notice (Eq. 2.133) that the canonical stress tensor is not symmetric. Nor is it unique, as an energy/momentum density, since the addition of a perfect divergence would yield the same integrated value. In general relativity we will require a *symmetric* $T_{\mu v}$, so I pause now to symmetrize the stress tensor.[26] Returning to Eq. 2.131,

---

[26] Eds. This was first accomplished by F. J. Belinfante in 1940; Coleman uses a method due to J. Schwinger.

$$\delta I_V = \oint_S d\sigma^\mu \left\{ \mathcal{L}\, \delta x_\mu - \sum_a \pi_\mu^a\, (\partial_\nu \phi^a) \delta x^\nu - \sum_{a,b} \pi_\mu^a\, S_{\lambda\nu}^{ab}\, \phi^b\, \epsilon^{\lambda\nu} \right\} = 0, \quad (2.142)$$

we introduce

$$f_{\mu\lambda\nu} \equiv \sum_{a,b} \pi_\mu^a\, S_{\lambda\nu}^{ab}\, \phi^b + B_{\mu\lambda\nu}, \qquad (2.143)$$

$$B_{\mu\lambda\nu} \equiv \sum_{a,b} \left[\pi_\nu^a\, S_{\lambda\mu}^{ab}\, \phi^b + \pi_\lambda^a\, S_{\nu\mu}^{ab}\, \phi^b \right]. \qquad (2.144)$$

Note that f is *antisymmetric* in its first two indices, while B is *symmetric* in its last two indices:

$$f_{\mu\lambda\nu} = -f_{\lambda\mu\nu}, \quad B_{\mu\lambda\nu} = B_{\mu\nu\lambda}. \qquad (2.145)$$

Recall (Eq. 2.128) that $\delta x^\mu = \epsilon^\mu_{\ \nu} x^\nu + \epsilon^\mu$, so $\partial_\nu(\delta x^\mu) = \epsilon^\mu_{\ \nu}$, and hence

$$\epsilon^{\lambda\nu} = \partial^\nu(\delta x^\lambda) = -\partial^\lambda(\delta x^\nu). \qquad (2.146)$$

Because of the symmetry of B and the antisymmetry of ϵ,

$$f_{\mu\lambda\nu}\epsilon^{\lambda\nu} = \sum_{a,b} \pi_\mu^a\, S_{\lambda\nu}^{ab}\, \phi^b\, \epsilon^{\lambda\nu}. \qquad (2.147)$$

So the third term in Eq. 2.142 becomes

$$\oint_S d\sigma^\mu \left(-\sum_{a,b} \pi_\mu^a\, S_{\lambda\nu}^{ab}\, \phi^b\, \epsilon^{\lambda\nu} \right) = \oint_S d\sigma^\mu \left[f_{\mu\lambda\nu}\partial^\lambda(\delta x^\nu) \right]$$

$$= -\oint_S d\sigma^\mu \left[f_{\lambda\mu\nu}\partial^\lambda(\delta x^\nu) \right]$$

$$= -\oint_S d\sigma^\mu \partial^\lambda(f_{\lambda\mu\nu}\delta x^\nu) + \oint_S d\sigma^\mu (\partial^\lambda f_{\lambda\mu\nu})\delta x^\nu. \qquad (2.148)$$

But

$$-\oint_S d\sigma^\mu \partial^\lambda(f_{\lambda\mu\nu}\delta x^\nu) = \oint_S d\sigma^\mu \partial^\lambda(f_{\mu\lambda\nu}\delta x^\nu) = 0. \qquad (2.149)$$

Proof: For a surface at constant time, the second expression becomes

$$\int d^3\mathbf{x}\, \partial^\lambda(f_{0\lambda\nu}\delta x^\nu) = \int d^3\mathbf{x}\, \partial^0(f_{00\nu}\delta x^\nu) + \int d^3\mathbf{x}\, \partial^k(f_{0k\nu}\delta x^\nu), \qquad (2.150)$$

but $f_{00\nu} = 0$ (by antisymmetry), and Gauss's theorem turns the second integral into a surface integral at spatial infinity, where f vanishes, so the whole thing is zero.   QED

Conclusion: The variation in I_V is

$$\delta I_V = \oint_S d\sigma^\mu \left\{ \mathcal{L} g_{\mu\nu} \delta x^\nu - \sum_a \pi_\mu^a (\partial_\nu \phi^a) \delta x^\nu + \left(\partial^\lambda f_{\lambda\mu\nu}\right) \delta x^\nu \right\}$$

$$= -\oint_S d\sigma^\mu T_{\mu\nu}^{\text{SYM}} \delta x^\nu, \tag{2.151}$$

where we define the **symmetrized stress tensor**

$$T_{\mu\nu}^{\text{SYM}} \equiv -\mathcal{L} g_{\mu\nu} + \sum_a \pi_\mu^a (\partial_\nu \phi^a) - \partial^\lambda f_{\lambda\mu\nu} = T_{\mu\nu}^{\text{CAN}} - \partial^\lambda f_{\lambda\mu\nu}. \tag{2.152}$$

It differs from the canonical stress tensor (2.133) by the divergence of f. From now on I will drop the superscript "SYM," and call it simply *the* stress tensor. Physically it represents the energy/momentum density; the **momentum 4-vector** is

$$p^\mu \equiv \int_\sigma d\sigma_\nu T^{\mu\nu}. \tag{2.153}$$

It is conserved (the same for any space-like surface σ), and therefore (see Eqs. 2.120 and 2.121) T itself is *locally* conserved,

$$\partial_\mu T^{\mu\nu} = 0. \tag{2.154}$$

From the $\epsilon^{\mu\nu}$ term in δx^ν we now obtain (Problem 2.8)

$$J^{\mu\nu} \equiv \int_\sigma d\sigma_\lambda \left(-T^{\lambda\mu} x^\nu + T^{\lambda\nu} x^\mu\right) \tag{2.155}$$

(the conserved angular momentum), from which it follows that

$$\partial^\mu \left(T_{\mu\lambda} x_\sigma - T_{\mu\sigma} x_\lambda\right) = T_{\mu\lambda} \delta_\sigma^\mu - T_{\mu\sigma} \delta_\lambda^\mu = T_{\sigma\lambda} - T_{\lambda\sigma} = 0, \tag{2.156}$$

so T is indeed symmetric. Since we have changed T only by a divergence, the integrated quantities (p^μ and $J^{\mu\nu}$) are the same as before, though expressed in somewhat cleaner form (the "spin" term in $J^{\mu\nu}$ has been absorbed into $T^{\mu\nu}$).

Problem 2.8

Show that $M_{\mu\nu\sigma}$ (Eq. 2.139) becomes, in terms of $T_{\mu\nu}$ (Eq. 2.152),

$$M_{\mu\nu\sigma} = (-T_{\mu\nu} x_\sigma + T_{\mu\sigma} x_\nu) + \partial^\lambda(-f_{\lambda\mu\nu} x_\sigma + f_{\lambda\mu\sigma} x_\nu),$$

and thus confirm Eq. 2.155.

Problem 2.9

Find $S^{ab}_{\mu\nu}$ (Eq. 2.130) for a *vector* field, A^μ.

Example 2.6

The stress tensor for a free particle. In Section 2.2 we worked out the possible asymptotically conserved relativistic momentum for *particles* (Eq. 2.76),

$$p^\mu = mv^\mu, \tag{2.157}$$

and relativistic angular momentum (Eq. 2.77),

$$J^{\mu\nu} = x^\mu p^\nu - x^\nu p^\mu. \tag{2.158}$$

In Section 2.4.2 we obtained the analogous conserved quantities for *fields* (Eqs. 2.153 and 2.155), in terms of the stress tensor. It is natural to wonder whether we could adapt the latter method to the particle case.

Using the result of Problem 2.6, we take the action for a free particle to be

$$I = \frac{m}{2} \int_{\tau_1}^{\tau_2} \dot{x}_\mu \dot{x}^\mu \, d\tau \tag{2.159}$$

(with the dots denoting differentiation with respect to proper time). So, under a change of coordinates,

$$\begin{aligned} \delta I &= m \int_{\tau_1}^{\tau_2} \dot{x}_\mu \, \delta\dot{x}^\mu \, d\tau = m \int_{\tau_1}^{\tau_2} \dot{x}_\mu \frac{d}{d\tau}(\delta x^\mu) \, d\tau \\ &= -m \int_{\tau_1}^{\tau_2} \ddot{x}_\mu \, \delta x^\mu \, d\tau + m \, \dot{x}_\mu \, \delta x^\mu \Big|_{\tau_1}^{\tau_2}. \end{aligned} \tag{2.160}$$

If δx^μ vanishes at τ_1 and τ_2, then Hamilton's principle says $\delta I = 0$, and hence

$$\ddot{x}_\mu = 0 \tag{2.161}$$

(the equation of motion for a free particle). If, instead, δx^μ is an infinitesimal Poincaré transformation (2.128),

$$\delta x^\mu = \epsilon^{\mu\nu} x_\nu + \epsilon^\mu, \tag{2.162}$$

then $\delta I = 0$ (because $\mathcal{L}$ is a scalar), so

$$m \, \dot{x}_\mu \left(\epsilon^{\mu\nu} x_\nu + \epsilon^\mu \right) \tag{2.163}$$

is the same at τ_1 and τ_2 for all (antisymmetric) $\epsilon^{\mu\nu}$ and ϵ^μ. Thus (choosing first $\epsilon^{\mu\nu} = 0$ and then $\epsilon^\mu = 0$)

$$p^\mu \equiv m\dot{x}^\mu \quad \text{and} \quad J^{\mu\nu} \equiv m(x^\mu \dot{x}^\nu - x^\nu \dot{x}^\mu) \tag{2.164}$$

are (again) conserved (the same at τ_2 as at τ_1).

Can we construct a stress tensor for the particle case? If the path followed is $y^\mu(\tau)$ it is natural to try

$$T_{\mu\nu}(x) \equiv m \int_{-\infty}^{\infty} \dot{y}_\mu \, \dot{y}_\nu \, \delta^4(x - y(\tau)) \, d\tau. \tag{2.165}$$

It is plainly symmetric. Is it divergenceless? Well,

$$\partial^\mu T_{\mu\nu} = m \int_{-\infty}^{\infty} \dot{y}_\mu \, \dot{y}_\nu \, \partial^\mu \left(\delta^4(x - y(\tau)) \right) d\tau. \tag{2.166}$$

The derivative is with respect to x, but since it acts on a function of $(x - y(\tau))$,

$$\frac{\partial}{\partial x^\mu} = -\frac{\partial}{\partial y^\mu}, \qquad \dot{y}_\mu \partial^\mu = -\frac{dy_\mu}{d\tau} \frac{\partial}{\partial y_\mu} = -\frac{d}{d\tau}. \tag{2.167}$$

Therefore

$$\partial^\mu T_{\mu\nu} = -m \int_{-\infty}^{\infty} \dot{y}_\nu \frac{d}{d\tau} \left(\delta^4(x - y(\tau)) \right) d\tau$$

$$= m \int_{-\infty}^{\infty} \ddot{y}_\nu \, \delta^4(x - y(\tau)) \, d\tau = 0 \tag{2.168}$$

(because for a *free* particle $\ddot{y}_\nu = 0$). Yes, it is divergenceless. For a free particle in its own rest frame, $\mathbf{y}(\tau) = \mathbf{y}$ is a constant, $t(\tau) = \tau$, $v_\mu = \dot{y}_\mu = (1, \mathbf{0})$, so

$$T_{00}(x) = m \, \delta^3(\mathbf{x} - \mathbf{y}) \tag{2.169}$$

and all other elements are zero. So T_{00} is indeed the (rest) energy density. *Conclusion:* Equation 2.165 satisfies all the requirements for a stress tensor. Indeed, a particle is, in a sense, just a highly localized field.

Problem 2.10

Consider the free scalar field, discussed in Example 2.4:

$$\mathcal{L} = \tfrac{1}{2} \left(\partial_\mu \phi \, \partial^\mu \phi - \mu^2 \phi^2 \right).$$

A general solution to the equation of motion can be written in the form

$$\phi(t, \mathbf{x}) = \frac{1}{(2\pi)^{3/2}} \int d^3k \, \frac{a(\mathbf{k})}{\sqrt{2\omega}} \, e^{i(\mathbf{k} \cdot \mathbf{x} - \omega t)} + \text{complex conjugate},$$

where $\omega \equiv \sqrt{\mathbf{k}^2 + \mu^2}$. (The complex conjugate is there to keep ϕ real.) Find the four components of p^μ as integrals over the unknown coefficients (a). [Once you have the answer, you'll see why, in quantum field theory, $a^*a/\hbar^4$ is called the "number density" of particles in momentum space.] *Hint:*

$$\int_{-\infty}^{\infty} dx \, e^{i\kappa x} = 2\pi \, \delta(\kappa). \tag{2.170}$$

3

Relativistic Electrodynamics

3.1 Lagrangian Formulation

3.1.1 The Free Maxwell Field

The Lagrangian for a free (noninteracting) electromagnetic field is

$$\mathcal{L} = -\tfrac{1}{4} \left(\partial_\mu A_\nu - \partial_\nu A_\mu \right) \left(\partial^\mu A^\nu - \partial^\nu A^\mu \right). \tag{3.1}$$

Varying the fields and combining terms (Problem 3.1),

$$\delta I = - \int d^4x \, \partial_\mu \, (\delta A_\nu) \, (\partial^\mu A^\nu - \partial^\nu A^\mu)$$

$$= \int d^4x \, (\delta A_\nu) \, \partial_\mu \, (\partial^\mu A^\nu - \partial^\nu A^\mu) = 0, \tag{3.2}$$

so the equations of motion are

$$\partial_\mu \, (\partial^\mu A^\nu - \partial^\nu A^\mu) = 0. \tag{3.3}$$

It is convenient to introduce the antisymmetric tensor[1]

$$F^{\mu\nu} \equiv \partial^\mu A^\nu - \partial^\nu A^\mu \tag{3.4}$$

in terms of which the equations of motion read, more compactly,

$$\partial_\mu F^{\mu\nu} = 0. \tag{3.5}$$

Because it is antisymmetric, the diagonal elements F^{00} and F^{ii} are trivially zero. Under 3-dimensional rotations the elements F^{0i} transform as the components of a 3-vector (nothing happens to the 0), so we write

$$F^{i0} \equiv E_i. \tag{3.6}$$

---

[1] Note that $F^{\mu\nu}$ is *not* an independent dynamical variable, just shorthand notation. In this formulation the (vector) field is A^μ (which we would ordinarily call the "potential"), and the electromagnetic "fields" **E** and **B** are, as we'll see in a moment, auxiliary constructs.

This defines **E**, the **electric field**. Similarly,

$$F^{ij} \equiv -\epsilon_{ijk} B_k, \tag{3.7}$$

which defines the **magnetic field, B**.[2] Meanwhile, the ordinary scalar and vector potentials (V and **A**) are embedded in A^μ:[3]

$$A^\mu = (V, \mathbf{A}). \tag{3.8}$$

What are the field equations in this 3-dimensional notation? For $\nu = 0$ Eq. 3.5 says

$$\partial_i(E_i) = 0, \quad \text{or} \quad \boxed{\nabla \cdot \mathbf{E} = 0.} \tag{3.9}$$

For $\nu = i$ we get

$$\partial_0 F^{0i} + \partial_j F^{ji} = 0 \quad \Rightarrow \quad \frac{\partial(-E_i)}{\partial t} + \frac{\partial}{\partial x^j}(-\epsilon_{jik} B_k) = 0, \tag{3.10}$$

or

$$\boxed{\frac{\partial \mathbf{E}}{\partial t} - \nabla \times \mathbf{B} = \mathbf{0}.} \tag{3.11}$$

Meanwhile, Eq. 3.4, with $\mu = i$ and $\nu = j$, says $-\epsilon_{ijk} B_k = \partial^i A^j - \partial^j A^i$. Multiplying both sides by ϵ_{ijm} (and summing on i and j),[4]

$$-2B_m = -\epsilon_{ijm}(\partial_i A^j - \partial_j A^i) = -2(\nabla \times \mathbf{A})_m. \tag{3.12}$$

Thus

$$\mathbf{B} = \nabla \times \mathbf{A}, \quad \text{and therefore} \quad \boxed{\nabla \cdot \mathbf{B} = 0.} \tag{3.13}$$

If $\mu = 0$ and $\nu = i$,

$$-E_i = \frac{\partial A^i}{\partial t} + \frac{\partial A^0}{\partial x^i} \quad \Rightarrow \quad \mathbf{E} = -\frac{\partial \mathbf{A}}{\partial t} - \nabla V, \tag{3.14}$$

or, taking the curl,

$$\boxed{\frac{\partial \mathbf{B}}{\partial t} + \nabla \times \mathbf{E} = \mathbf{0}.} \tag{3.15}$$

---

[2] Here ϵ_{ijk} is the 3-dimensional **Levi-Civita symbol**, with $\epsilon_{123} \equiv 1$; **E** and **B** are *not* elements of 4-vectors, and for them there is no covariant/contravariant distinction, so we will always write the indices *down*. We do, however, copy the Einstein summation convention, with the stipulation that repeated Roman letters are summed from 1 to 3.

[3] Eds. Yes, this leaves the expression A_i ambiguous: Is it the ith component of the covariant 4-vector A_μ, or the ith component of the 3-vector **A** (they differ by a sign)? Coleman finesses this sort of problem by being cheerfully inconsistent. We will reserve the boldface **A** for the ordinary 3-vector potential.

[4] Note that $\epsilon_{ijm}\epsilon_{ijk} = 2\delta_{mk}$.

The boxed results are Maxwell's equations in regions free of charge and current. Because they come from a Lorentz-invariant Lagrangian, they are (collectively) Lorentz invariant, by construction.

Although A^μ is the fundamental field, it is not directly measurable—rather, **E** and **B** are the measurable quantities. Indeed, a **gauge transformation**

$$A^\mu \to A^\mu + \partial^\mu \chi \tag{3.16}$$

(for an arbitrary scalar function χ) leaves $F^{\mu\nu}$ unchanged, and the equations of motion unaltered. We consider two solutions *physically equivalent* if they differ only by such a gauge transformation.[5] In the **Lorenz gauge**[6] we choose χ so that A^μ is divergenceless:[7]

$$\partial_\mu A^\mu = 0. \tag{3.17}$$

In the Lorenz gauge the equation of motion simplifies to

$$\Box^2 A^\mu = 0, \tag{3.18}$$

where

$$\Box^2 \equiv \partial_\mu \partial^\mu = \partial_0^2 - \nabla^2 \tag{3.19}$$

is the **d'Alembertian** (4-dimensional generalization of the **Laplacian**, ∇^2). Thus the components of A^μ satisfy the homogeneous wave equation.

The electromagnetic stress tensor (see Problem 3.2) is

$$T^{\mu\nu} = F^{\mu\lambda} F_\lambda{}^\nu + \tfrac{1}{4} g^{\mu\nu} F_{\lambda\sigma} F^{\lambda\sigma}. \tag{3.20}$$

In terms of **E** and **B**,

$$
\begin{aligned}
F_{\lambda\sigma} F^{\lambda\sigma} &= F_{0i} F^{0i} + F_{i0} F^{i0} + F_{ij} F^{ij} \\
&= (E_i)(-E_i) + (-E_i)(E_i) + (-\epsilon_{ijk} B_k)(-\epsilon_{ijl} B_l) \\
&= -2\mathbf{E}^2 + 2\mathbf{B}^2,
\end{aligned} \tag{3.21}
$$

---

[5] This is a stronger statement than **gauge invariance** alone. The latter says that a gauge transformation carries one solution to the equations of motion into another solution, but regarding the two solutions as physically equivalent implies that all the physical information is contained in $F^{\mu\nu}$.

[6] Eds. Coleman calls it the "Lorentz" gauge, but it is now more commonly attributed to Ludvig Lorenz, rather than Hendrik Lorentz. See J. D. Jackson and L. B. Okun, *Rev. Mod. Phys.* **73**, 663 (2001).

[7] The Lorenz condition does not uniquely specify the gauge, since one can still perform a transformation such that

$$\Box^2 \chi = 0.$$

so the energy density is[8]

$$T^{00} = F^{0i} F_i{}^0 + \tfrac{1}{4} g^{00} F_{\lambda\sigma} F^{\lambda\sigma} = \mathbf{E}^2 + \tfrac{1}{2}(-\mathbf{E}^2 + \mathbf{B}^2) = \tfrac{1}{2}(\mathbf{E}^2 + \mathbf{B}^2), \qquad (3.22)$$

and the momentum density is

$$T_0{}^i = F_{0j} F^{ji} = (E_j)(-\epsilon_{jik} B_k) = \epsilon_{ijk} E_j B_k = (\mathbf{E} \times \mathbf{B})_i \qquad (3.23)$$

(the **Poynting vector**).

Problem 3.1

In case the passage to Eq. 3.2 was not clear to you (what happened to the 4?), try isolating a particular entry in 3.1—say, $\partial_0 A_1$. Show that

$$\mathcal{L} = \tfrac{1}{2} [(\partial_0 A_1)(\partial_0 A_1) - 2(\partial_0 A_1)(\partial_1 A_0)] + \cdots,$$

where the dots indicate terms *not* involving $\partial_0 A_1$. Now do the variation of $\partial_0 A_1$.

Problem 3.2

(a) Find the canonical stress tensor (2.133) for the Lagrangian (3.1).
(b) Symmetrize it (2.152) to confirm Eq. 3.20. [Use the full machinery of Section 2.4.3 if you like, but you may be able to guess the perfect divergence you need to add.]
(c) Check that $T^{\mu\nu}$ is divergenceless.

3.1.2 Maxwell Field with Source

Suppose the electromagnetic field is coupled to an external source,

$$J^\mu = (\rho, \mathbf{J}), \qquad (3.24)$$

where ρ is the charge density, $\mathbf{J}$ is the current density, and they satisfy the continuity equation (local conservation of charge),

$$\partial_\mu J^\mu = 0. \qquad (3.25)$$

The Lagrangian is now

$$\mathcal{L} = -\tfrac{1}{4} \left(\partial_\mu A_\nu - \partial_\nu A_\mu\right)(\partial^\mu A^\nu - \partial^\nu A^\mu) - A_\mu J^\mu. \qquad (3.26)$$

Varying the field,

$$\delta I = \int d^4 x \left(\partial_\mu F^{\mu\nu} \delta A_\nu - J^\nu \delta A_\nu\right) = 0 \quad \Rightarrow \quad \partial_\mu F^{\mu\nu} - J^\nu = 0. \qquad (3.27)$$

---

[8] Eds. Coleman uses **Heaviside–Lorentz units**, with $\epsilon_0 \to 1$ and $\mu_0 \to 1$ (and, of course, $c = 1$).

For $\nu = 0$ we get

$$\partial_i F^{i0} - J^0 = 0, \quad \text{or} \quad \boxed{\nabla \cdot \mathbf{E} = \rho;} \tag{3.28}$$

for $\nu = i$,

$$\partial_0 F^{0i} + \partial_j F^{ji} - J^i = 0 \quad \Rightarrow \quad -\frac{\partial E_i}{\partial t} + \partial_j(-\epsilon_{jik}B_k) - J^i = 0, \tag{3.29}$$

or

$$\boxed{\frac{\partial \mathbf{E}}{\partial t} - \nabla \times \mathbf{B} = -\mathbf{J}.} \tag{3.30}$$

The homogeneous Maxwell equations (3.13 and 3.15) are unchanged, since they follow from Eq. 3.4:

$$\boxed{\nabla \cdot \mathbf{B} = 0} \quad \text{and} \quad \boxed{\frac{\partial \mathbf{B}}{\partial t} + \nabla \times \mathbf{E} = \mathbf{0}.} \tag{3.31}$$

The boxed results are Maxwell's equations with sources. In the Lorenz gauge (3.17) the equation of motion (3.27) simplifies:[9]

$$\Box^2 A^\mu = J^\mu. \tag{3.32}$$

In the absence of sources, the stress tensor (3.20) is divergenceless (Problem 3.2(c)), but this is no longer true in the presence of J^μ (the external source acts on the fields, and *vice versa*, so it is not surprising that field energy and field momentum are no longer conserved):

$$\begin{aligned}
\partial_\mu T^{\mu\nu} &= \partial_\mu \left[F^{\mu\lambda} F_\lambda{}^\nu + \tfrac{1}{4} g^{\mu\nu} (F_{\lambda\sigma} F^{\lambda\sigma}) \right] \\
&= J^\lambda F_\lambda{}^\nu + F^{\mu\lambda} \partial_\mu F_\lambda{}^\nu + \tfrac{1}{4} \partial^\nu (F_{\lambda\sigma} F^{\lambda\sigma}) \\
&= J^\lambda F_\lambda{}^\nu + F^{\mu\lambda} \left[\partial_\mu F_\lambda{}^\nu + \tfrac{1}{2} \partial^\nu F_{\mu\lambda} \right].
\end{aligned} \tag{3.33}$$

Raising and lowering the dummy indices,

$$\begin{aligned}
F^{\mu\lambda} \left[\partial_\mu F_\lambda{}^\nu + \tfrac{1}{2} \partial^\nu F_{\mu\lambda} \right] &= F_{\mu\lambda} \left[\partial^\mu F^{\lambda\nu} + \tfrac{1}{2} \partial^\nu F^{\mu\lambda} \right] \\
&= F_{\mu\lambda} \left[\partial^\mu \partial^\lambda A^\nu - \partial^\mu \partial^\nu A^\lambda + \tfrac{1}{2} \partial^\nu \partial^\mu A^\lambda - \tfrac{1}{2} \partial^\nu \partial^\lambda A^\mu \right].
\end{aligned} \tag{3.34}$$

Because $F_{\mu\lambda}$ is antisymmetric,

$$F_{\mu\lambda} \left[\partial^\mu \partial^\lambda A^\nu \right] = 0 \quad \text{and} \quad F_{\mu\lambda} \left[-\tfrac{1}{2} \partial^\nu \partial^\lambda A^\mu - \tfrac{1}{2} \partial^\nu \partial^\mu A^\lambda \right] = 0, \tag{3.35}$$

---

[9] Equation 3.32 looks like four equations in four unknowns, which sounds OK; on the other hand, we are ultimately hoping to solve for *six* unknowns (three components each of $\mathbf{E}$ and $\mathbf{B}$). The "extra" two equations are implicit in 3.4. To put it another way, Maxwell's equations are ostensibly *eight* equations for six unknowns—too many. The point is that the components of $\mathbf{E}$ and $\mathbf{B}$ are not all independent. There are really just *four* independent variables in electrodynamics, and A^μ encodes these most efficiently.

so the last part of Eq. 3.33 is in fact zero, and we are left with

$$\partial_\mu T^{\mu\nu} = J_\lambda F^{\lambda\nu}. \tag{3.36}$$

For example, with $\nu = 0$,

$$\partial_\mu T^{\mu 0} = J_i F^{i0} = -J^i F^{i0} = -\mathbf{J} \cdot \mathbf{E} = \partial_0 T^{00} + \partial_i T^{i0}. \tag{3.37}$$

Integrating over all space, at time t,

$$\frac{d}{dt} \int d^3\mathbf{x}\, T^{00} = -\int d^3\mathbf{x}\, \mathbf{J} \cdot \mathbf{E}. \tag{3.38}$$

In words, the rate of change of the energy stored in the fields is minus the work done by the fields on the charges.

3.2 Potentials and Fields of a Point Charge

3.2.1 The Action for a Point Charge

Next we consider an electromagnetic field interacting with a point charge—combining the results of Section 2.3 (mechanics of a particle) and Section 3.1.2 (electromagnetic fields with a source).[10] The action consists of two parts, the free term and the interaction term:

$$I = I_0 + I'. \tag{3.39}$$

The free term is composed of a matter part (the charged particle) and a field part (the electromagnetic field):

$$I_0 = I_0^m + I_0^e, \tag{3.40}$$

where (Eq. 2.85 and Problem 2.6)

$$I_0^m = -\frac{m_0}{2} \int d\tau\, \dot{y}_\mu \dot{y}^\mu. \tag{3.41}$$

Here $y^\mu(\tau)$ is the world line of the particle, parameterized by its proper time, and the dot denotes the τ derivative; m_0 is the **bare mass**—the mass the particle would have in the absence of any interaction (as we shall see, interaction with the

---

[10] Eds. This is known as **classical electron theory**. It doesn't have to be an *electron*, of course—any structureless point charge would do. Coleman wrote a famous treatise on the subject in the summer of 1960, while still a graduate student. It was finally published many years later in *Electromagnetism: Paths to Research*, D. Teplitz, ed., Plenum, New York (1982), Chapter 6. It is also available at www.rand.org/pubs/ research_memoranda/RM2820.html. In the remainder of this chapter Coleman follows that treatment closely.

electromagnetic field modifies the effective mass of the particle). The field part is (Eq. 3.1)

$$I_0^e = -\frac{1}{4} \int d^4x \ (\partial_\mu A_\nu - \partial_\nu A_\mu)(\partial^\mu A^\nu - \partial^\nu A^\mu). \tag{3.42}$$

Finally, the interaction term is

$$I' = -e \int d\tau \int d^4x \ A_\mu(x) \, \delta^4(x - y(\tau)) \, \dot{y}^\mu, \tag{3.43}$$

where e is the electric charge of the particle. Or, if we define the current

$$J^\mu(x) \equiv e \int d\tau \, \delta^4(x - y(\tau)) \, \dot{y}^\mu, \tag{3.44}$$

then

$$I' = -\int d^4x \, J^\mu A_\mu. \tag{3.45}$$

This is from the perspective of the *field* (Eq. 3.26); alternatively (from the point of view of the *particle*), we could do the x integral first:

$$I' = -e \int d\tau \, \dot{y}^\mu \, A_\mu(y) = -e \int dy^\mu \, A_\mu(y). \tag{3.46}$$

The independent dynamical variables here are $y^\mu(\tau)$ and $A_\mu(x)$. Varying A_μ we get (as before, Eq. 3.27)

$$\partial_\mu F^{\mu\nu} - J^\nu = 0. \tag{3.47}$$

For the variation of y^μ, the relevant part of the action is

$$I_p = -\frac{m_0}{2} \int d\tau \, \dot{y}_\mu \, \dot{y}^\mu - e \int d\tau \, \dot{y}^\mu \, A_\mu(y). \tag{3.48}$$

Then

$$\delta I_p = -m_0 \int d\tau \, \dot{y}_\mu \, \delta \dot{y}^\mu - e \int d\tau (\delta \dot{y}_\mu) A^\mu(y) - e \int d\tau \, \dot{y}_\mu (\partial_\nu A^\mu(y)) \delta y^\nu = 0. \tag{3.49}$$

The resulting equation of motion is

$$m_0 \, \ddot{y}_\mu + e \frac{d}{d\tau} A_\mu - e \dot{y}_\nu \, \partial_\mu A^\nu = 0, \quad m_0 \, \ddot{y}^\mu = -e(\partial^\nu A^\mu)\dot{y}_\nu + e(\partial^\mu A^\nu)\dot{y}_\nu, \tag{3.50}$$

or

$$m_0 \, \ddot{y}^\mu = -e F^{\nu\mu} \, \dot{y}_\nu. \tag{3.51}$$

Notice that neither 3.47 nor 3.51 involves the potential A_μ directly—only the field tensor $F^{\mu\nu}$—so the theory remains fully gauge invariant. In the customary 3-dimensional notation, Eq. 3.51 reads

$$m_0 \ddot{y}_0 = e\mathbf{E} \cdot \dot{\mathbf{y}}, \quad m_0 \ddot{\mathbf{y}} = e\left[\dot{y}_0 \mathbf{E} + (\dot{\mathbf{y}} \times \mathbf{B})\right]. \tag{3.52}$$

The former (the $\mu = 0$ component) tells you the power delivered to the charge by the fields; the latter ($\mu = i$) is the **Lorentz force law**[11] telling you the force on the charge.

I won't work out the stress tensor in this theory. It turns out that there is no interaction term[12]—$T_{\mu\nu}$ is the sum of the free particle tensor (2.165) and the free field tensor (3.20),

$$T_{\mu\nu} = T^m_{\mu\nu} + T^e_{\mu\nu}. \tag{3.53}$$

You can check this by showing that $T_{\mu\nu}$ is conserved (Problem 3.3).

Problem 3.3

Check that the stress tensor 3.53 is divergenceless.

3.2.2 Green's Function for the Wave Equation

In the Lorenz gauge the field equation reduces to the **inhomogeneous wave equation** (or rather, *four* of them, one for each component):

$$\Box^2 A^\mu = J^\mu \tag{3.54}$$

(Eq. 3.32). I propose to solve it by the Green's function method. Consider first the *scalar* wave equation,

$$\Box^2 \varphi = \rho. \tag{3.55}$$

Note that the solution is not going to be unique; you can add any solution to the *homogeneous* equation (in practice boundary conditions will select the right one

---

[11] This is the **Minkowski force**, $dp^\mu/d\tau$, not the "ordinary" force $\mathbf{F} = d\mathbf{p}/dt$; recall that the proper velocity $\dot{y}^\mu = \gamma(1, \boldsymbol{\beta})$ and $d\tau = (1/\gamma)dt$, so

$$\mathbf{F} = \frac{d\tau}{dt}e\left[\gamma\mathbf{E} + \gamma(\boldsymbol{\beta} \times \mathbf{B})\right] = e\left[\mathbf{E} + (\boldsymbol{\beta} \times \mathbf{B})\right].$$

[12] The absence of a field–particle cross term in $T^{\mu\nu}$ is a peculiarity of electromagnetism, not shared by other theories of field–particle interactions.

for the context you have in mind). Suppose you could solve Eq. 3.55 for a delta-function source:

$$\Box^2 D(x) = \delta^4(x);$$ (3.56)

then

$$\varphi(x) = \int d^4y \, D(x - y) \, \rho(y).$$ (3.57)

So our problem is to find the **Green's function**, D. This we will do by **Fourier transform**:

$$\delta^4(x) = \int \frac{d^4k}{(2\pi)^4} e^{ik \cdot x}, \quad D(x) = \int \frac{d^4k}{(2\pi)^4} e^{ik \cdot x} \, \tilde{D}(k).$$ (3.58)

Putting these into Eq. 3.56,

$$-k^2 \, \tilde{D}(k) = 1 \quad \Rightarrow \quad \tilde{D}(k) = -\frac{1}{k^2},$$ (3.59)

so

$$D(x) = -\int \frac{d^4k}{(2\pi)^4} e^{ik \cdot x} \frac{1}{k^2} = -\frac{1}{(2\pi)^4} \int d^3k \, e^{-i\mathbf{k} \cdot \mathbf{x}} \left\{ \int dk_0 \frac{e^{ik_0 t}}{k_0^2 - k^2} \right\}.$$ (3.60)

Let's do the k_0 integral first. We need to decide how to skirt the poles (at $k_0 = \pm|\mathbf{k}|$) in the complex k_0 plane. Any prescription will do, because the Green's function is not unique anyhow (different choices will differ by solutions to the homogeneous equation). I'll use the contour that runs *below* both poles:

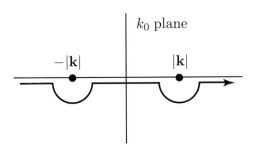

If $t < 0$ we close the contour below (so as to kill the contribution from the semi-circle at negative imaginary k_0), and $D = 0$. Thus only sources at *earlier* times can contribute to φ; we call this the **retarded Green's function**, D_R.[13]

---

[13] To get the **advanced Green's function**, D_A, let the contour go *over* both poles. But D_A is acausal: sources at *later* times affect $\varphi(x)$. For this reason it is ordinarily rejected.

For $t > 0$ we close the contour *above*; letting $\kappa \equiv |\mathbf{k}|$, and using Cauchy's integral formula,

$$\int dk_0 \frac{e^{ik_0 t}}{(k_0 - \kappa)(k_0 + \kappa)} = 2\pi i \frac{1}{2\kappa} \left[e^{i\kappa t} - e^{-i\kappa t} \right]. \tag{3.61}$$

There remains the $\mathbf{k}$ integral:

$$D_R(x) = -\frac{i}{16\pi^3} \int d^3 k \frac{e^{-i\mathbf{k}\cdot\mathbf{x}}}{\kappa} \left[e^{i\kappa t} - e^{-i\kappa t} \right]. \tag{3.62}$$

Adopting spherical coordinates, with the polar axis along $\mathbf{x}$, $|\mathbf{x}| = r$, $\mathbf{k}\cdot\mathbf{x} = \kappa r \cos\theta$, $d^3 k = \kappa^2 \sin\theta \, d\kappa \, d\theta \, d\phi$,

$$D_R(x) = -\frac{i}{16\pi^3} \int_0^{2\pi} d\phi \int_0^\pi d\theta \, \sin\theta \left\{ \int_0^\infty d\kappa \, \kappa \, e^{-i\kappa r \cos\theta} \left[e^{i\kappa t} - e^{-i\kappa t} \right] \right\}. \tag{3.63}$$

Changing variables from θ to $\mu \equiv \cos\theta$,

$$D_R(x) = -\frac{i}{8\pi^2} \int_{-1}^1 d\mu \left\{ \int_0^\infty d\kappa \, \kappa \, e^{-i\kappa r \mu} \left[e^{i\kappa t} - e^{-i\kappa t} \right] \right\}. \tag{3.64}$$

The μ integral is easy:

$$\int_{-1}^1 d\mu \, e^{-i\kappa r \mu} = \frac{1}{i\kappa r} \left(e^{i\kappa r} - e^{-i\kappa r} \right), \tag{3.65}$$

so

$$\begin{aligned}
D_R(x) &= -\frac{1}{8\pi^2 r} \int_0^\infty d\kappa \left(e^{i\kappa r} - e^{-i\kappa r} \right) \left(e^{i\kappa t} - e^{-i\kappa t} \right) \\
&= -\frac{1}{8\pi^2 r} \int_{-\infty}^\infty d\kappa \left[e^{i\kappa(r+t)} - e^{i\kappa(r-t)} \right] = -\frac{2\pi}{8\pi^2 r} \left[\delta(r+t) - \delta(r-t) \right] \\
&= \frac{1}{4\pi r} \delta(r - t)
\end{aligned} \tag{3.66}$$

(the other delta function dies because r and t are both positive). Remember, this was for $t > 0$; if $t < 0$ then $D_R(x) = 0$. In nicer notation,

$$D_R(x) = \frac{1}{2\pi} \delta(r^2 - t^2) \, \theta(t) = \frac{1}{2\pi} \delta(x_\mu x^\mu) \, \theta(x^0). \tag{3.67}$$

Let's explore this Green's function in the vicinity of $t = 0^+$:

$$D_R(\mathbf{x}, 0) = \frac{1}{4\pi r} \delta(r). \tag{3.68}$$

It's zero for $r > 0$, but could there be a delta function lurking at the origin? For any test function $f(\mathbf{x})$ that is not *itself* singular at $r = 0$,

$$\int d^3\mathbf{x}\, D_R(\mathbf{x},0)\, f(\mathbf{x}) = \int_0^{2\pi} d\phi \int_0^\pi d\theta \sin\theta \int_0^\infty dr\, r^2 \frac{1}{4\pi r} \delta(r)\, f(r,\theta,\phi) = 0$$

(3.69)

(the r integral is zero). So, in the language of distributions,

$$D_R(\mathbf{x},0) = 0.$$

(3.70)

How about its time derivative? From Eq. 3.66,

$$\frac{\partial}{\partial t} D_R(\mathbf{x},t)\bigg|_{t=0} = -\frac{1}{4\pi r}\delta'(r).$$

(3.71)

Again, multiplying by a test function and integrating,

$$\int d^3\mathbf{x}\, \frac{\partial}{\partial t} D_R(\mathbf{x},t)\bigg|_{t=0} f(\mathbf{x})$$

$$= -\frac{1}{4\pi}\int_0^{2\pi} d\phi \int_0^\pi d\theta \sin\theta \int_0^\infty dr\, r\, \delta'(r)\, f(r,\theta,\phi).$$

(3.72)

Integrating by parts,

$$\int_0^\infty dr\, r\, \delta'(r)\, f(r,\theta,\phi) = r\,\delta(r)\, f(r,\theta,\phi)\bigg|_0^\infty - \int_0^\infty dr\, \delta(r)\, \frac{\partial}{\partial r}(r\, f)$$

$$= 0 - \int_0^\infty dr\, \delta(r)\left(f + r\frac{\partial f}{\partial r}\right) = -f(0).$$

(3.73)

Thus

$$\frac{\partial}{\partial t} D_R(\mathbf{x},t)\bigg|_{t=0} = \delta^3(\mathbf{x}).$$

(3.74)

We can use $D_R(x)$ to solve the *inhomogeneous* wave equation (Eq. 3.57); it can *also* be used to solve the *homogeneous* wave equation with specified initial conditions.

Example 3.1

Find $f(\mathbf{x},t)$ if

$$\Box^2 f(\mathbf{x},t) = 0, \quad f(\mathbf{x},0) = h(\mathbf{x}), \quad \frac{\partial}{\partial t} f(\mathbf{x},t)\bigg|_{t=0} = g(\mathbf{x})$$

(3.75)

for given functions $h(\mathbf{x})$ and $g(\mathbf{x})$.

Solution:

$$f(\mathbf{x},t) = \int d^3\mathbf{x}'\left\{ D_R(\mathbf{x}-\mathbf{x}',t)\, g(\mathbf{x}') + \left[\frac{\partial}{\partial t} D_R(\mathbf{x}-\mathbf{x}',t)\right] h(\mathbf{x}')\right\}.$$

(3.76)

Proof: In view of Eq. 3.56,

$$\Box^2 D_R(\mathbf{x} - \mathbf{x}', t) = \delta(t)\,\delta^3(\mathbf{x} - \mathbf{x}'), \tag{3.77}$$

so

$$\Box^2 f(\mathbf{x}, t) = \int d^3x' \left\{ \delta(t)\delta^3(\mathbf{x} - \mathbf{x}')\,g(\mathbf{x}') + \frac{\partial}{\partial t}\left[\delta(t)\delta^3(\mathbf{x} - \mathbf{x}')\right]h(\mathbf{x}') \right\}$$

$$= \delta(t)g(\mathbf{x}) + \delta'(t)h(\mathbf{x}), \tag{3.78}$$

so it satisfies the differential equation (3.75) for $t > 0$. In view of Eqs. 3.70 and 3.74,

$$f(\mathbf{x}, 0) = \int d^3x' \left\{ (0)\,g(\mathbf{x}') + \delta^3(\mathbf{x} - \mathbf{x}')h(\mathbf{x}') \right\} = h(\mathbf{x}), \tag{3.79}$$

so it satisfies the first boundary condition. And

$$\left.\frac{\partial}{\partial t}f(\mathbf{x}, t)\right|_{t=0} = \int d^3x' \left\{ \delta^3(\mathbf{x} - \mathbf{x}')g(\mathbf{x}') + \left[\frac{\partial^2}{\partial t^2}D_R(\mathbf{x} - \mathbf{x}', t)\Big|_{t=0}\right]h(\mathbf{x}') \right\}. \tag{3.80}$$

But from Eq. 3.70,

$$\left.\frac{\partial^2}{\partial t^2}D_R(\mathbf{x}, t)\right|_{t=0} = \left.\nabla^2 D_R(\mathbf{x}, t)\right|_{t=0} = \nabla^2 D_R(\mathbf{x}, 0) = \nabla^2\,[0] = 0, \tag{3.81}$$

so

$$\left.\frac{\partial}{\partial t}f(\mathbf{x}, t)\right|_{t=0} = g(\mathbf{x}), \tag{3.82}$$

and therefore Eq. 3.76 also satisfies the second boundary condition.　　QED

Equation 3.76 can be written in a more concise way by introducing the **left–right derivative**,

$$f \overset{\leftrightarrow}{\partial} g \equiv f\,(\partial g) - (\partial f)\,g. \tag{3.83}$$

Thus

$$f(\mathbf{x}, t) = \int d^3x' \left\{ D_R(\mathbf{x} - \mathbf{x}', t - t')\partial_0' f(\mathbf{x}', t') - [\partial_0' D_R(\mathbf{x} - \mathbf{x}', t - t')]\,f(\mathbf{x}', t') \right\}\Big|_{t'=0}$$

$$= \int d^3x' \left\{ D_R(x - x') \overset{\leftrightarrow}{\partial}_0' f(x', t') \right\}\Big|_{t'=0}. \tag{3.84}$$

More generally, if the boundary conditions are not specified on a plane,

$$f(x) = \int_\sigma d\sigma^{\mu'} \left\{ D_R(x - x') \overset{\leftrightarrow}{\partial}_{\mu}' f(x') \right\}\Big|_{(x'\text{ on }\sigma)} \tag{3.85}$$

(this is in essence **Huygens' principle**: you build up the new wave front from the old wave front using each point as a source).

Problem 3.4

Show that $D_A(x) = D_R(-x)$ (footnote 13).

3.2.3 "In" and "Out" Fields

In Eq. 3.32 we reduced Maxwell's electrodynamics to the inhomogeneous wave equation with a specified source:

$$\Box^2 A_\mu = J_\mu. \tag{3.86}$$

In Eq. 3.57 we constructed the general solution to this differential equation, using the retarded Green's function:

$$A_\mu(x) = A_\mu^{in}(x) + \int d^4x'\, D_R(x - x')\, J_\mu(x'), \tag{3.87}$$

where $A_\mu^{in}(x)$ is a solution to the *homogeneous* wave equation:

$$\Box^2 A_\mu^{in}(x) = 0. \tag{3.88}$$

We call A_μ^{in} the "in" (or "incoming") field. The second term (in 3.87) tells us the field generated by the current J_μ; the in field is something that would have been there *even if there had been no source at all*. Informally, we think of it as a kind of "primordial" field, coming in from the distant past, before the current was turned on.[14] In classical electrodynamics we normally[15] assume that

> **all electromagnetic fields are due to charges and currents.** (3.89)

If there's no charge or current anywhere, ever, then the field would be zero. Accordingly, we will stipulate that

$$A_\mu^{in}(x) = 0. \tag{3.90}$$

You can also solve Eq. 3.86 using the *advanced* Green's function (Problem 3.4)—or for that matter with any linear combination of the two, such as the average:

$$A_\mu(x) = A_\mu^{out}(x) + \int d^4x'\, D_A(x - x')\, J_\mu(x') \tag{3.91}$$

$$= \left(\frac{A_\mu^{in} + A_\mu^{out}}{2}\right) + \int d^4x' \left(\frac{D_R + D_A}{2}\right) J_\mu. \tag{3.92}$$

---

[14] Because of charge conservation, it may not be *possible* to turn the current on and off. For instance, it is nonsense to ask what the field would be for a point charge that suddenly materializes from nothing at time $t = 0$; such a source would be incompatible with Maxwell's equations. For more on "in" and "out" fields, see Coleman's treatise, Section 3 (details in footnote 10).

[15] Eds. An exception is **stochastic electrodynamics**, which proposes that the universe is permeated by random sourceless electromagnetic radiation.

The "out" (or "outgoing") field again satisfies the homogeneous wave equation. Informally it is the field that remains after all currents have been turned off[14] (and the integral in 3.91 does not contribute); unlike A_μ^{in}, it is certainly *not* zero. This introduces a time asymmetry into classical electrodynamics: Maxwell's equations are time-reversal invariant, but the standard boundary conditions (in the form of the in and out fields) are *not*: time reversal would take $A^{in} \leftrightarrow A^{out}$, but the first is zero and the second is not.

If you pose the question "What is the source of the observed time irreversibility of the universe?" two possible answers come to mind: (a) statistical mechanics (the second law of thermodynamics) and (b) electrodynamics (an accelerating electron radiates into the future, not into the past). **Feynman–Wheeler electrodynamics**[16] challenges option (b). In their formulation, Maxwellian electrodynamics is correct, but the universe is surrounded by a perfect absorber, so that A^{in} and A^{out} are both zero. The cosmic absorber turns all radiation into heat. Electrons radiate both forward and backward in time (Eq. 3.92); the forward radiation gets to the absorber, which reradiates both forward and backward in time—the backward radiation from the absorber returns to the electron with just the right phase so as to cancel the backward-going radiation from the electron. That's why we don't see the electron radiating backward; according to Feynman–Wheeler, you *can't tell* whether an accelerating electron radiates backward, because—with the absorber out there—it gets canceled anyhow. In Feynman–Wheeler electrodynamics, time asymmetry is due entirely to statistical mechanics.[17]

3.3 Radiation from a Point Charge

3.3.1 The Liénard–Wiechert Potential

Imagine a charged particle moving along its world line, $y(\tau)$:

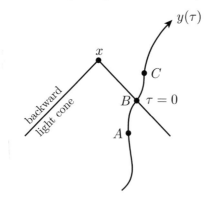

[16] J. A. Wheeler and R. P. Feynman, *Rev. Mod. Phys.* **17**, 157 (1945).
[17] Note that the absorber is not just a mathematical boundary condition, but an extra dynamical system; this is, for most people, the implausible part of the story.

The field it produces at x is determined by the particle's position/velocity/acceleration at the point B, where the backward light cone (of x) intersects the world line of the particle (note that there can be at most *one* such intersection, because the particle cannot travel at or above the speed of light). For simplicity I'll set $\tau = 0$ at point B. In Maxwellian electrodynamics ($A_\mu^{in} = 0$), Eq. 3.87 says

$$A_\mu(x) = \int d^4x'\, D_R(x - x')\, J_\mu(x'), \tag{3.93}$$

where (3.67)

$$D_R(x) = \frac{1}{2\pi}\, \delta(x_\mu x^\mu)\, \theta(x^0) \tag{3.94}$$

(the theta function prevents any influence on *earlier* times—the "message" must arrive at x *after* it left B; the delta function confines any influence to the light cone). Finally, the particle current is given by 3.44:

$$J_\mu(x) \equiv e \int d\tau\, \delta^4(x - y(\tau))\, \dot{y}_\mu(\tau). \tag{3.95}$$

Putting all of this together,

$$A_\mu(x) = \frac{e}{2\pi} \int\int d^4x'\, d\tau\, \delta\left((x - x')^2\right) \delta^4(x' - y(\tau))\theta(x^0 - x^{0'})\, \dot{y}_\mu(\tau). \tag{3.96}$$

First do the x' integral; this simply replaces x' by $y(\tau)$:

$$A_\mu(x) = \frac{e}{2\pi} \int d\tau\, \delta(z^2)\theta(z^0)\, \dot{y}_\mu(\tau), \tag{3.97}$$

where

$$z^\mu(\tau) \equiv x^\mu - y^\mu(\tau), \quad \text{so} \quad \delta(z^2) = \frac{1}{|\partial z^2/\partial \tau|}\delta(\tau). \tag{3.98}$$

Now,

$$\frac{\partial z^2}{\partial \tau} = \frac{\partial}{\partial \tau}\, (x_\nu - y_\nu(\tau))\, (x^\nu - y^\nu(\tau))$$

$$= (x_\nu - y_\nu(\tau))\, (-\dot{y}^\nu(\tau)) + (-\dot{y}_\nu(\tau))\, (x^\nu - y^\nu(\tau)) = -2 z_\nu\, \dot{y}^\nu(\tau). \tag{3.99}$$

What is its sign? At B (light-like separated from x), $z^2 = 0$; at C the separation is *space-like*, so $z^2 < 0$; at A the separation is *time-like*, so $z^2 > 0$. Evidently z^2 is a *decreasing* function of τ, so

$$\left|\frac{\partial z^2}{\partial \tau}\right| = -\frac{\partial z^2}{\partial \tau} = 2 z_\nu\, \dot{y}^\nu \quad \text{and} \quad \delta(z^2) = \frac{1}{2 z_\nu\, \dot{y}^\nu}\,\delta(\tau). \tag{3.100}$$

Therefore

$$A_\mu(x) = \frac{e}{4\pi} \frac{1}{z_\nu \dot{y}^\nu} \dot{y}_\mu \bigg|_{(z^2=0,\, z^0>0)}. \tag{3.101}$$

This is the **Liénard–Wiechert potential** for a point charge in arbitrary motion.

Example 3.2

Coulomb's law. For a particle at rest,

$$y^\nu(\tau) = (\tau, \mathbf{y}), \quad \dot{y}^\nu = (1, \mathbf{0}); \quad A_\mu(x) = \frac{e}{4\pi} \frac{1}{z_0} (1, \mathbf{0}) \bigg|_{(z^2=0,\, z^0>0)}. \tag{3.102}$$

At $z^2 = 0$, $z_0 = \pm|\mathbf{z}| = \pm|\mathbf{x} - \mathbf{y}| = \pm \imath$ (where $\imath \equiv \mathbf{x} - \mathbf{y}$), and because of the theta function we want the *positive* root, $\imath$, the (spatial) distance from the charge to the point $\mathbf{x}$, so

$$A_\mu(x) = \frac{e}{4\pi \imath} (1, \mathbf{0}) \tag{3.103}$$

(the ordinary Coulomb potential, in Heaviside–Lorentz units).

3.3.2 The Fields of a Point Charge

Differentiating Eq. 3.97 (and suppressing the θ-function, which we can take care of at the end),

$$\partial_\nu A_\mu(x) = \frac{e}{2\pi} \int d\tau \, \delta'(z^2) \, 2z_\nu \, \dot{y}_\mu(\tau)$$

$$= \frac{e}{\pi} \int d\tau \, \delta'(z^2) \left(\frac{\partial z^2}{\partial \tau}\right) \left(\frac{\partial z^2}{\partial \tau}\right)^{-1} z_\nu(\tau) \, \dot{y}_\mu(\tau)$$

$$= \frac{e}{\pi} \int d\tau \, \frac{\partial}{\partial \tau} \left[\delta\left(z^2\right)\right] \left(\frac{\partial z^2}{\partial \tau}\right)^{-1} z_\nu(\tau) \, \dot{y}_\mu(\tau)$$

$$= -\frac{e}{\pi} \int d\tau \, \delta\left(z^2\right) \frac{\partial}{\partial \tau} \left[\left(\frac{\partial z^2}{\partial \tau}\right)^{-1} z_\nu(\tau) \, \dot{y}_\mu(\tau)\right]. \tag{3.104}$$

Now

$$\frac{\partial}{\partial \tau} \left[\left(\frac{\partial z^2}{\partial \tau}\right)^{-1} z_\nu \dot{y}_\mu\right] = \left(\frac{\partial z^2}{\partial \tau}\right)^{-1} \left(-\dot{y}_\nu \dot{y}_\mu + z_\nu \ddot{y}_\mu\right) - \left(\frac{\partial z^2}{\partial \tau}\right)^{-2} \frac{\partial^2 z^2}{\partial \tau^2} z_\nu \dot{y}_\mu,$$

$$\tag{3.105}$$

but we are not interested in anything symmetric in $\nu \leftrightarrow \mu$, because these drop out when we calculate $F_{\mu\nu} = \partial_\mu A_\nu - \partial_\nu A_\mu$. So ignore the $\dot{y}_\nu \dot{y}_\mu$ term. From Eq. 3.99,

$$\frac{\partial z^2}{\partial \tau} = -2 z_\nu \dot{y}^\nu \quad \Rightarrow \quad \frac{\partial^2 z^2}{\partial \tau^2} = 2 \dot{y}_\nu \dot{y}^\nu - 2 z_\nu \ddot{y}^\nu = 2 (1 - z_\nu \ddot{y}^\nu). \quad (3.106)$$

Except for the explicitly symmetric term, then,

$$\partial_\nu A_\mu = -\frac{e}{\pi} \int d\tau \, \delta(z^2) \left[\frac{1}{(-2 z_\lambda \dot{y}^\lambda)} z_\nu \ddot{y}_\mu - \frac{1}{(-2 z_\lambda \dot{y}^\lambda)^2} 2(1 - z_\lambda \ddot{y}^\lambda) z_\nu \dot{y}_\mu \right]. \quad (3.107)$$

But (3.100)

$$\delta(z^2) = \frac{1}{2 z_\lambda \dot{y}^\lambda} \delta(\tau), \quad (3.108)$$

so

$$\partial_\nu A_\mu = \frac{e}{4\pi} \int d\tau \, \delta(\tau) \left[\frac{1}{(z_\lambda \dot{y}^\lambda)^2} z_\nu \ddot{y}_\mu + \frac{1}{(z_\lambda \dot{y}^\lambda)^3} (1 - z_\lambda \ddot{y}^\lambda) z_\nu \dot{y}_\mu \right]$$

$$= \frac{e}{4\pi} \left[\frac{1}{(z_\lambda \dot{y}^\lambda)^2} z_\nu \ddot{y}_\mu + \frac{1}{(z_\lambda \dot{y}^\lambda)^3} (1 - z_\lambda \ddot{y}^\lambda) z_\nu \dot{y}_\mu \right], \quad (3.109)$$

with everything evaluated at $\tau = 0$ (which is to say, at the **retarded time** when the "message" left: $z^2 = 0$, $z^0 > 0$). Finally,

$$\boxed{F_{\mu\nu} = \frac{e}{4\pi} \left[\frac{1}{(z_\lambda \dot{y}^\lambda)^2} z_\mu \ddot{y}_\nu + \frac{1}{(z_\lambda \dot{y}^\lambda)^3} (1 - z_\lambda \ddot{y}^\lambda) z_\mu \dot{y}_\nu \right] - (\mu \leftrightarrow \nu).} \quad (3.110)$$

This is the electromagnetic field produced by a point charge in arbitrary motion; it includes a part that goes like $1/\imath^2$ (the second term), known as the **induction field**,[18] and others that go like $1/\imath$, (the first and third), the **radiation field**.[19] Note that if $\ddot{y}_\mu = 0$ (no acceleration), the radiation field vanishes.

Problem 3.5

Use Eq. 3.110 to find **E** and **B** for a point charge at rest.

Incidentally, we could use Eq. 3.91 to determine the A^{out} that corresponds to $A^{\text{in}} = 0$:

$$A_\mu = \int D_R J_\mu = A_\mu^{\text{out}} + \int D_A J_\mu \quad \Rightarrow \quad A_\mu^{\text{out}} = \int [D_R - D_A] J_\mu, \quad (3.111)$$

---

[18] Eds. Coleman's word is unusual; more common is the **generalized Coulomb field** or **velocity field**.
[19] Eds. The radiation field is also called the **acceleration field**.

and interpret this as the "radiation" emitted. Provided the particle acceleration is limited (so it doesn't radiate an infinite amount), the total energy and momentum carried off by the asymptotic field can be calculated[20] from A^{out}. The result is

$$P_\mu^{\text{radiated}} = -\frac{e^2}{6\pi} \int_{-\infty}^{\infty} d\tau \, (\dddot{y}_\lambda \, \dddot{y}^\lambda) \, \dot{y}_\mu. \tag{3.112}$$

It is a miracle that this can be expressed as a single integral; if the photon had mass, it would be a *double* integral.

3.4 Regularization and Renormalization

In principle, our job is done: just plug in the appropriate initial conditions, and solve for the motion of the electron and the fields it generates. But there's a hitch: How does a point charge move under the influence of its *own* field? If you insert $F^{\mu\nu}$ (3.110) into the equation of motion (3.51),

$$m_0 \ddot{y}_\mu = e F_{\mu\nu} \, \dot{y}^\nu, \tag{3.113}$$

you get infinity, because $F_{\mu\nu}$ blows up at the location of the charge.[21] To get around this difficulty we invoke a procedure known as **mass renormalization**.

First we **regularize** the theory: modify it so as to eliminate the infinities. The source of the problem is the interaction Lagrangian (Eq. 3.43):

$$I' = -e \int d\tau \int d^4x \, A_\mu(x) \, \delta^4 \, (x - y(\tau)) \, \dot{y}^\mu(\tau). \tag{3.114}$$

The delta function (representing the point charge) is the culprit. We want to *smear out* the source, but in a Lorentz-invariant way, so the theory remains consistent with relativity (it wouldn't do to make it a 3-dimensional spherical shell, for example). In place of $\delta^4(x - y)$ we'll use $f(x - y)$, where $f \to \delta$ in some suitable limit. Specifically, we start with any old function $F(x)$ such that

$$\int d^4x \, F(x) = 1, \tag{3.115}$$

and let

$$f(x) \equiv \lambda^4 \, F(\lambda x). \tag{3.116}$$

Then as $\lambda \to \infty$, $f \to \delta$.

Eds. This is done in Coleman's treatise, Section 4 (details in footnote 10).
21 The same thing occurs in quantum electrodynamics, but there the divergence is softer: logarithmic, instead of $1/\imath$ (or $1/\imath^2$).

What's the effect of replacing δ by f? It is still true (of course) that $\Box^2 A_\mu = J_\mu$ (that's just Maxwell's equations in the Lorenz gauge, and we're not touching them—only smearing out the source). But J^μ is no longer given by Eq. 3.44,

$$J^\mu = e \int d\tau\, \delta^4(x - y)\, \dot{y}^\mu, \tag{3.117}$$

but rather by

$$\bar{J}^\mu = e \int d\tau\, f(x - y)\, \dot{y}^\mu \tag{3.118}$$

(I'll use an overbar to denote regularized quantities). Thus

$$\begin{aligned}
\bar{J}^\mu &= e \int \int d\tau\, d^4x'\, f(x - x')\, \delta^4(x' - y)\, \dot{y}^\mu(\tau) \\
&= \int d^4x'\, f(x - x')\, J^\mu(x') \equiv f * J^\mu \tag{3.119}
\end{aligned}$$

(the **convolution integral**). This modified current generates a field $\bar{F}_{\mu\nu}$ given by

$$\partial^\mu \bar{F}_{\mu\nu} = \bar{J}_\nu. \tag{3.120}$$

How is it related to the old (*unregularized*) field? We need to solve Eq. 3.120, given that $\bar{J}^\mu = f * J^\mu$ and $\partial^\mu \bar{F}_{\mu\nu} = J_\nu$: What is $\bar{F}_{\mu\nu}$, in terms of $F_{\mu\nu}$? I claim that

$$\bar{F}_{\mu\nu} = f * F_{\mu\nu}. \tag{3.121}$$

Proof: If this is right then

$$\bar{F}_{\mu\nu} = \int d^4x'\, f(x - x')\, F_{\mu\nu}(x'), \tag{3.122}$$

so

$$\begin{aligned}
\partial^\mu \bar{F}_{\mu\nu}(x) &= \int d^4x'\, \left[\partial^\mu_{(x)} f(x - x') \right] F_{\mu\nu}(x') \\
&= \int d^4x'\, \left[-\partial^\mu_{(x')} f(x - x') \right] F_{\mu\nu}(x') \\
&= \int d^4x'\, f(x - x') \left[\partial^\mu F_{\mu\nu}(x') \right] = \int d^4x'\, f(x - x')\, J_\nu(x') \\
&= \bar{J}_\nu(x). \quad \text{QED} \tag{3.123}
\end{aligned}$$

The idea now is to plug 3.110 into 3.122, and expand in inverse powers of λ (the regularization parameter—remember, we want the limit as $\lambda \to \infty$), for x in the immediate vicinity of the particle. I shall not go through the details,[22] but merely quote the answer:

$$\bar{F}_{\mu\nu} = A\lambda + B + C\frac{1}{\lambda} + D\frac{1}{\lambda^2} + \cdots$$

$$= \frac{e}{8\pi}(\dot{y}_\mu\ddot{y}_\nu - \dot{y}_\nu\ddot{y}_\mu)\lambda - \frac{2}{3}\frac{e}{4\pi}(\dot{y}_\mu\dddot{y}_\nu - \dot{y}_\nu\dddot{y}_\mu) + \mathcal{O}\left(\frac{1}{\lambda^2}\right). \tag{3.124}$$

The electron's equation of motion (3.113) is

$$m_0\ddot{y}_\mu = e\bar{F}_{\mu\nu}\dot{y}^\nu + F_\mu^{\text{ext}}. \tag{3.125}$$

The first term (on the right) represents the **self-force** on the charge, due to its *own* electromagnetic field, and F_μ^{ext} stands for any external forces that may be acting on it. Putting in Eq. 3.124, and recalling that $\dot{y}_\nu\dot{y}^\nu = 1$ (1.70) and $\dot{y}_\nu\ddot{y}^\nu = 0$ (1.80),

$$m_0\ddot{y}_\mu = -\frac{e^2\lambda}{8\pi}\ddot{y}_\mu + \frac{2}{3}\frac{e^2}{4\pi}(\dddot{y}_\mu - \dot{y}_\mu\ddot{y}_\nu\dot{y}^\nu) + F_\mu^{\text{ext}}. \tag{3.126}$$

(Because $\dot{y}_\nu\ddot{y}^\nu = 0$, $\ddot{y}_\nu\ddot{y}^\nu = -\dot{y}_\nu\dddot{y}^\nu$, so the second term in parentheses can also be written as $+\dot{y}_\mu\ddot{y}_\nu\ddot{y}^\nu$.)

Of course, the self-force still blows up (as $\lambda \to \infty$), but the infinity has been isolated in a term that multiplies $\ddot{y}_\mu$ and can be combined with the (bare) mass m_0 to define the **renormalized mass**:

$$m \equiv m_0 + \frac{e^2\lambda}{8\pi}. \tag{3.127}$$

Thus

$$\boxed{m\ddot{y}_\mu = \frac{2}{3}\frac{e^2}{4\pi}(\dddot{y}_\mu + \dot{y}_\mu\ddot{y}_\nu\ddot{y}^\nu) + F_\mu^{\text{ext}},} \tag{3.128}$$

and we identify m as the true **physical mass** of the particle. It is awkward that the electromagnetic contribution is infinite, but—who knows?—maybe the bare mass is *minus* infinity (it is not, after all, a measurable quantity, since we cannot strip the electron of its charge). In any case, the remainder of the self-force (the so-called **radiation reaction**) is perfectly finite:

$$F_\mu^{\text{rad}} = \frac{2}{3}\frac{e^2}{4\pi}(\dddot{y}_\mu + \dot{y}_\mu\ddot{y}_\nu\ddot{y}^\nu). \tag{3.129}$$

This formula was first obtained by Dirac, though the nonrelativistic version goes back to Abraham and Lorentz, at the turn of the 20th century.[23]

---

[22] Eds. In Coleman (footnote 10) he calls the calculation "tedious and uninstructive," and relegates it to an appendix.

[23] P. A. M. Dirac, *Proc. Roy. Soc. (London)* **A167**, 148 (1938); H. A. Lorentz, *The Theory of Electrons*, Dover, New York (1952—based on his 1906 lectures at Columbia University).

Feynman–Wheeler electrodynamics revisited: Maxwell's electrodynamics is not time-reversal invariant, because we stipulate that the "in" fields are zero. Restoring them for a moment (and setting aside the external forces), the equation of motion (3.128) becomes

$$m\ddot{y}_\mu = e F^{\text{in}}_{\mu\nu}\,\dot{y}^\nu + \frac{2}{3}\frac{e^2}{4\pi}(\dddot{y}_\mu + \dot{y}_\mu \ddot{y}_\nu \ddot{y}^\nu). \tag{3.130}$$

Because the advanced formulation is the time-reversed version of the retarded formulation, and time reversal switches the sign of the term in parentheses (it's odd in τ), we could as well write

$$m\ddot{y}_\mu = e F^{\text{out}}_{\mu\nu}\,\dot{y}^\nu - \frac{2}{3}\frac{e^2}{4\pi}(\dddot{y}_\mu + \dot{y}_\mu \ddot{y}_\nu \ddot{y}^\nu). \tag{3.131}$$

Averaging these two expressions,

$$m\ddot{y}_\mu = \frac{e}{2}\left(F^{\text{out}}_{\mu\nu} + F^{\text{in}}_{\mu\nu} \right)\dot{y}^\nu. \tag{3.132}$$

But the sum of F^{in} and F^{out} is zero, in the Feynman–Wheeler formulation, and the self-force term cancels out entirely! (A radiating charge does slow down, in the Feynman–Wheeler approach, but that is due to the backward radiation from the absorber, not the influence of the particle's own field.)

Indeed, you can construct an action in which the fields never appear:

$$I = \sum_n I_0^{(n)} + \sum_{n>m} \int d\tau^{(n)}\, d\tau^{(m)}\, \dot{y}^{(n)}_\mu \left(\dot{y}^{(n)}\right)^\mu \bar{D}(y^{(n)} - y^{(m)}), \tag{3.133}$$

where the parenthetical superscripts denote different interacting particles,

$$\bar{D} \equiv \frac{D_A + D_R}{2}, \tag{3.134}$$

and the free particle action is

$$I_0^{(n)} = \frac{m^{(n)}}{2} \int d\tau^{(n)}\, \dot{y}^{(n)}_\mu \left(\dot{y}^{(n)}\right)^\mu. \tag{3.135}$$

This is a retarded action-at-a-distance theory; the charges interact directly, without mediating fields, and with no need for mass renormalization (the only mass that ever appears is the physical mass). Feynman and Wheeler hoped to quantize this system, avoiding the infinities that plague quantum electrodynamics by removing them already at the classical level. Unfortunately, no one has succeeded in quantizing Feynman–Wheeler electrodynamics.

3.4.1 Particle Motion with Radiation Reaction

In the nonrelativistic régime ($|d\mathbf{y}/dt| \ll 1$), the Abraham–Lorentz–Dirac equation (3.128) reduces to

$$m\frac{d^2\mathbf{y}}{dt^2} = \frac{2}{3}\frac{e^2}{4\pi}\frac{d^3\mathbf{y}}{dt^3} + \mathbf{F}^{\text{ext}}. \tag{3.136}$$

Suppose there is a harmonic binding force, $\mathbf{F}^{\text{ext}} = -k\mathbf{y}$:

$$m\frac{d^2\mathbf{y}}{dt^2} = \frac{2}{3}\frac{e^2}{4\pi}\frac{d^3\mathbf{y}}{dt^3} - k\mathbf{y}. \tag{3.137}$$

We'll look for oscillatory solutions:

$$\mathbf{y}(t) = e^{i\omega t}\mathbf{a}. \tag{3.138}$$

Putting this in,

$$m\omega^2 = \frac{2}{3}\frac{e^2}{4\pi}i\omega^3 + k. \tag{3.139}$$

Define

$$\omega_0 \equiv \sqrt{\frac{k}{m}}, \qquad \lambda \equiv \frac{2}{3}\frac{e^2}{4\pi m} \tag{3.140}$$

(not to be confused with the λ in (3.116)!); then

$$\omega^2 = i\lambda\omega^3 + \omega_0^2. \tag{3.141}$$

In the *absence* of any radiation reaction ($\lambda = 0$), $\omega = \pm\omega_0$. If λ is *small*, we can treat it as a perturbation:

$$\omega = \pm\omega_0 + \delta\omega, \tag{3.142}$$

and (dropping terms quadratic and higher in $\delta\omega$)

$$\omega^2 = \omega_0^2 \pm 2\omega_0\,\delta\omega, \qquad \omega^3 = \pm\omega_0^3 + 3\omega_0^2\,\delta\omega. \tag{3.143}$$

So Eq. 3.141 becomes

$$\omega_0^2 \pm 2\omega_0\,\delta\omega = i\lambda(\pm\omega_0^3 + 3\omega_0^2\,\delta\omega) + \omega_0^2, \tag{3.144}$$

or

$$\delta\omega = \frac{i\lambda\omega_0^2}{2 \mp 3i\omega_0\lambda}. \tag{3.145}$$

But we are taking λ to be infinitesimal, so these two solutions coincide (to lowest order):

$$\delta\omega = \frac{i\lambda\omega_0^2}{2}, \tag{3.146}$$

and the motion becomes

$$\mathbf{y}(t) = e^{i(\pm\omega_0+\delta\omega)t}\mathbf{a} = e^{-\lambda\omega_0^2 t/2}e^{\pm i\omega_0 t}\mathbf{a}. \tag{3.147}$$

Not surprisingly, the emission of radiation damps the oscillations (in this context the radiation reaction is known as **radiation damping**).

But 3.141 is a *cubic* equation: What about the *third* root? This one is *not* approximately ω_0 for small λ, so we're not going to get it as a perturbation, but inspection of 3.141 suggests

$$\omega = \frac{1}{i\lambda}. \tag{3.148}$$

It comes in from infinity (as λ increases from zero), so for small λ the ω_0^2 term is negligible. For this root the motion is

$$\mathbf{y}(t) = e^{i\omega t}\mathbf{a} = e^{t/\lambda}\mathbf{a}. \tag{3.149}$$

This is a catastrophe—an *in*creasing exponential, with no oscillations at all. It is known as a **runaway mode**. In spite of appearances, it doesn't actually violate conservation of energy, as we'll see in a moment. But it *is* a huge embarrassment for classical electron theory.

Even the *free* equation of motion admits runaway modes: the general solution to

$$m\frac{d^2\mathbf{y}}{dt^2} = \frac{2}{3}\frac{e^2}{4\pi}\frac{d^3\mathbf{y}}{dt^3} \tag{3.150}$$

is

$$\mathbf{y}(t) = \mathbf{a} + \mathbf{b}t + \mathbf{c}e^{t/\lambda} \tag{3.151}$$

(for constants $\mathbf{a}$, $\mathbf{b}$, and $\mathbf{c}$), as you can check for yourself.

Are runaways perhaps an artifact of the nonrelativistic approximation? Unfortunately, they are *not*. For if we start with the relativistic equation 3.128,

$$m\ddot{y}_\mu = \frac{2}{3}\frac{e^2}{4\pi}\left(\dddot{y}_\mu + \dot{y}_\mu\ddot{y}_\nu\ddot{y}^\nu\right), \tag{3.152}$$

and multiply in $\ddot{y}^\mu$ (remembering that $\dot{y}_\mu\ddot{y}^\mu = 0$), we obtain

$$m\ddot{y}^\mu\ddot{y}_\mu = \frac{2}{3}\frac{e^2}{4\pi}\ddot{y}^\mu\dddot{y}_\mu. \tag{3.153}$$

Define the (real[24]) scalar function

$$f(\tau) \equiv \sqrt{-\ddot{y}_\mu \ddot{y}^\mu}. \tag{3.154}$$

Note that

$$-\frac{d}{d\tau} f^2 = -2f\dot{f} = 2\dddot{y}_\mu \ddot{y}^\mu, \tag{3.155}$$

so 3.153 says

$$-m f^2 = \frac{2}{3} \frac{e^2}{4\pi}(-f\dot{f}) \quad \Rightarrow \quad f = \lambda \frac{df}{d\tau}, \tag{3.156}$$

and therefore

$$f(\tau) = e^{\tau/\lambda} f(0). \tag{3.157}$$

Again, exponential runaway growth (with no external force acting).

How big is λ, for an actual electron? From the definition 3.140 (with three factors of c to give it units of time),

$$\lambda = \frac{e^2}{6\pi m c^3} = \frac{(4.80 \times 10^{-10}\, \text{esu})^2}{(6\pi)(9.11 \times 10^{-28}\, \text{gm})(3.00 \times 10^{10}\, \text{cm/s})^3}$$
$$\approx 5 \times 10^{-25}\, \text{s}. \tag{3.158}$$

Evidently the runaway is extremely fast. This *cannot* be right. Is it perhaps a premonition of quantum mechanics—classical electrodynamics warning us that it cannot be trusted in the realm of the very small?[25]

Why not use a boundary condition to kill the runaway? After all, 3.136 is a *third-order* differential equation, and there will still be two constants left over, to fit the initial position and velocity.[26] This works for a *free* charge (simply stipulate that $\mathbf{c} = \mathbf{0}$ in the general solution 3.151), but if there is an external force acting, the cure can be worse than the disease. Suppose, for example, that the force is a sharp kick at $t = 0$. The equation of motion 3.136 becomes

$$m\frac{d^2\mathbf{y}}{dt^2} = m\lambda\frac{d^3\mathbf{y}}{dt^3} + \mathbf{F}_0\,\delta(t). \tag{3.159}$$

---

[24] In the instantaneous rest frame of the particle, $\ddot{y}^\mu = (0, d^2\mathbf{y}/dt^2)$, so $\ddot{y}^\mu$ is space-like.

[25] Lorentz modeled the electron as a spherical shell of charge, calculating the net self-force due to different parts acting on one another. In the point limit this reproduces the (nonrelativistic) radiation reaction force (3.136). Lorentz could blame the runaways on the model, but we don't have that option. Eds. Interestingly, for sufficiently large spheres the Lorentz model is free of runaways. See, for example, E. Moniz and D. Sharp, *Phys. Rev. D* **15**, 2850 (1977).

[26] This approach was explored by Dirac (see footnote 23).

When $t \neq 0$ the force is zero, so (letting $\mathbf{v} \equiv d\mathbf{y}/dt$)

$$\frac{d\mathbf{v}}{dt} = \lambda \frac{d^2\mathbf{v}}{dt^2} \quad \Rightarrow \quad \frac{d\mathbf{v}}{dt} = \mathbf{a}\, e^{t/\lambda} \quad \Rightarrow \quad \mathbf{v}(t) = \mathbf{a}\,\lambda\, e^{t/\lambda} + \mathbf{b}. \tag{3.160}$$

This holds both *before* the force acts ($t < 0$) and again *afterward* ($t > 0$)—but with different constants $\mathbf{a}$ and $\mathbf{b}$, of course. We might as well assume the particle starts from rest in the distant past, so $\mathbf{b} = \mathbf{0}$ when $t < 0$. And presumably the acceleration, $d\mathbf{v}/dt$, is zero prior to the intervention of $\mathbf{F}$, so $\mathbf{a} = \mathbf{0}$. Then

$$\mathbf{v}(t) = \mathbf{0} \quad (t < 0). \tag{3.161}$$

To determine the constants $\mathbf{a}$ and $\mathbf{b}$ *after* the force acts, integrate 3.159 across the delta function:

$$\int_{-\epsilon}^{\epsilon} dt\, \frac{d\mathbf{v}}{dt} = \lambda \int_{-\epsilon}^{\epsilon} dt\, \frac{d^2\mathbf{v}}{dt^2} + \frac{\mathbf{F}_0}{m} \int_{-\epsilon}^{\epsilon} dt\, \delta(t), \tag{3.162}$$

or

$$\mathbf{v}(\epsilon) - \mathbf{v}(-\epsilon) = \lambda \left(\left.\frac{d\mathbf{v}}{dt}\right|_{\epsilon} - \left.\frac{d\mathbf{v}}{dt}\right|_{-\epsilon} \right) + \frac{\mathbf{F}_0}{m}. \tag{3.163}$$

Thus

$$\mathbf{v}(\epsilon) = \lambda \left.\frac{d\mathbf{v}}{dt}\right|_{\epsilon} + \frac{\mathbf{F}_0}{m}. \tag{3.164}$$

Putting in the general solution (3.160), and taking the limit $\epsilon \to 0$,

$$\mathbf{a}\lambda + \mathbf{b} = \lambda \mathbf{a} + \frac{\mathbf{F}_0}{m}, \quad \text{so} \quad \mathbf{b} = \frac{\mathbf{F}_0}{m}, \tag{3.165}$$

and hence

$$\mathbf{v}(t) = \mathbf{a}\lambda e^{t/\lambda} + \frac{\mathbf{F}_0}{m}. \tag{3.166}$$

Assuming $\mathbf{v}(t)$ is continuous[27] at $t = 0$, it follows that

$$\mathbf{v}(t) = \left(1 - e^{t/\lambda}\right) \frac{\mathbf{F}_0}{m} \quad (t > 0). \tag{3.167}$$

This is the solution we get by assuming that the velocity and the acceleration are zero prior to the intervention of the force; such boundary conditions inevitably excite the runaway.

---

[27] Eds. You can confirm this by treating the delta function as a limit of triangles or rectangles, say. If $v(t)$ had a discontinuity ($\theta(t)$), then dv/dt would carry a delta function, and d^2v/dt^2 a δ', but there is no compensating δ' in 3.159.

To eliminate the runaway, we need $\mathbf{a} = \mathbf{0}$ for $t > 0$. To get the resulting solution we could either run the same argument backward, or (more simply) add a solution $e^{t/\lambda}\,(\mathbf{F}_0/m)$ of the homogeneous equation. Then

$$\mathbf{v}(t) = \begin{cases} e^{t/\lambda}\,\mathbf{F}_0/m & (t < 0), \\ \mathbf{F}_0/m & (t > 0). \end{cases} \tag{3.168}$$

That kills the runaway, but now the particle starts to move *before the force acts*! This **acausal preacceleration** only jumps the gun very briefly, but still, it is clearly unacceptable.

Question: Is it *always* possible to eliminate the runaways in this way (or does it work only for the delta-function force)? The answer is yes, as long as the equations of motion are linear. It is even true in the *non*linear case if the potential V is sufficiently bounded. The general solution far from the scattering center takes the form (3.151)

$$\mathbf{y}(t) = \begin{cases} \mathbf{a} + \mathbf{b}t + \mathbf{c}\,e^{t/\lambda}, & t \to -\infty, \\ \mathbf{a}' + \mathbf{b}'t + \mathbf{c}'\,e^{t/\lambda}, & t \to \infty. \end{cases} \tag{3.169}$$

You can think of the scattering problem as a mapping from $\mathbf{a}, \mathbf{b}, \mathbf{c}$ to $\mathbf{a}', \mathbf{b}', \mathbf{c}'$.

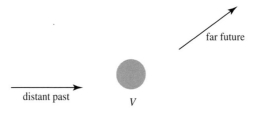

It can be proved that if $|\mathbf{c}| \gg 1$, then $\mathbf{c} \approx \mathbf{c}'$. (The effect of the potential is roughly proportional to the length of time the particle spends in its vicinity, so for high velocities the effect is negligible.) Thus the mapping takes large $\mathbf{c}$ into large $\mathbf{c}'$; it maps large spheres into themselves. According to the **fixed point theorem**, it must map *something* into *zero*, which is to say that there is always a solution with no runaway—you can always choose the preacceleration $\mathbf{c}$ in the past such that there is no runaway in the future. (However, in the case of electron/positron scattering this argument does not hold, since the force is singular at the origin.) Incidentally, one would never know whether there exist runaway modes in *quantum* electrodynamics, because the only method of analysis available is perturbation theory, where they would never appear anyway.

3.4.2 Conservation of Energy

Runaway modes occur when the bare mass m_0 is negative (see Problem 3.7). The total (nonrelativistic) energy of a charged particle (kinetic energy plus energy in its electromagnetic fields, Eq. 3.22) is

$$\frac{1}{2}m_0 v^2 + \frac{1}{2}\int d^3\mathbf{x}\,(\mathbf{E}^2 + \mathbf{B}^2). \tag{3.170}$$

Acceleration pours (positive definite) energy into the fields, but it also increases the magnitude of the kinetic energy. The total energy is not measurable, but it *is* conserved, even in the runaway modes; we are saved by the fact that $m_0 < 0$: the kinetic term gets more and more negative, while the field term gets more and more positive. But this doesn't tell us much, because it makes explicit reference to the unmeasurable bare mass.

Is there a *useful* energy conservation law for this case—one expressed in terms of the *physical* mass, m? Yes, there is. Going back to the relativistic equation of motion (3.128),

$$\frac{d}{d\tau}(m\,\dot{y}_\mu) = F_\mu^{\text{ext}} + \frac{2}{3}\frac{e^2}{4\pi}\left[\dddot{y}_\mu + \dot{y}_\mu\,\ddot{y}^\nu\,\ddot{y}_\nu\right]. \tag{3.171}$$

The $\dddot{y}$ term tells us the rate at which energy is pumped into the induction field; the $\dot{y}\,\ddot{y}\,\ddot{y}$ term is (minus) the rate at which energy/momentum is radiated (Eq. 3.112). In the nonrelativistic régime, the zeroth component reads

$$\mathbf{v}\cdot\mathbf{F}^{\text{ext}} = \frac{d}{dt}\left(\frac{1}{2}mv^2\right) + \frac{2}{3}\frac{e^2}{4\pi}a^2 - \frac{d}{dt}\left[\frac{2}{3}\frac{e^2}{4\pi}\mathbf{a}\cdot\mathbf{v}\right], \tag{3.172}$$

where $\mathbf{v}$ is the velocity and $\mathbf{a}$ is the acceleration. On the left we have the power delivered to the electron by the external force; it is equal to the rate of increase of the kinetic energy, plus the power radiated (the so-called **Larmor formula**), plus the rate of change of the energy stored in the induction field. This, then, is the conservation of energy expressed in terms of measurable quantities. For periodic motion, or motions that begin and end with $\mathbf{a} = \mathbf{0}$, the last term integrates to zero, and in this average sense the work done (by the external force) is equal to the increase in kinetic energy plus the energy radiated. But, in general, energy is also exchanged between the particle and the local fields it drags around with it.

Problem 3.6

Derive 3.172 from 3.171.

Problem 3.7

The physical mass of a charged object at rest is the sum of its bare mass and the energy stored in its electrostatic field (over c^2):

$$m = m_0 + U/c^2. \tag{3.173}$$

Consider the case of a uniformly charged spherical shell, of radius R and charge Q.

(a) What is U in this case?

(b) Find m_0, as a function of the measurable quantities m, Q, and R.

(c) Lorentz contemplated a "purely electromagnetic" particle, whose (physical) mass is *entirely* attributable to energy in the fields (i.e. $m_0 = 0$). The so-called **classical radius** of a particle is the radius at which this occurs. Find the classical radius of a spherical shell (mass m and charge Q). What would this be for an electron (in meters)?

Comment: For particles *smaller* than their classical radius (and *a fortiori* for *point* particles) the bare mass runs negative. Runaways and preacceleration do not occur for particles *larger* than their classical radius (Moniz and Sharp; details in footnote 25). Unfortunately, the electron is known to be much smaller than its classical radius.

3.4.3 Hyperbolic Motion

Recall the twin paradox problem (Section 1.4.2): motion under uniform proper acceleration. The world line is hyperbolic:

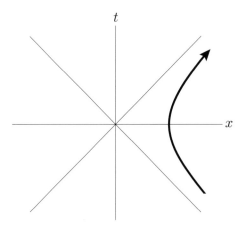

In hyperbolic motion the radiation reaction vanishes.

Proof: The radiation reaction force (Eq. 3.129, with 3.40) is

$$F_\mu^{\text{rad}} = m\lambda \left[\dddot{y}_\mu + \dot{y}_\mu (\ddot{y}_\nu \, \ddot{y}^\nu) \right]. \tag{3.174}$$

First of all, note that F_μ^{rad} is always orthogonal to the 4-velocity:

$$\frac{1}{m\lambda} \dot{y}^\mu F_\mu^{\text{rad}} = \dot{y}^\mu \dddot{y}_\mu + \dot{y}^\mu \dot{y}_\mu (\ddot{y}_\nu \ddot{y}^\nu) = \dot{y}^\mu \dddot{y}_\mu + \ddot{y}_\mu \ddot{y}^\mu$$

$$= \frac{d}{d\tau}(\dot{y}^\mu \ddot{y}_\mu) = 0 \qquad (3.175)$$

(because $\dot{y}^\mu \ddot{y}_\mu = 0$). In the case of hyperbolic motion it is *also* orthogonal to the acceleration:

$$\frac{1}{m\lambda} \ddot{y}^\mu F_\mu^{\text{rad}} = \ddot{y}^\mu \dddot{y}_\mu + \ddot{y}^\mu \dot{y}_\mu (\ddot{y}_\nu \ddot{y}^\nu) = \ddot{y}^\mu \dddot{y}_\mu = \frac{1}{2}\frac{d}{d\tau}(\ddot{y}^\mu \ddot{y}_\mu). \qquad (3.176)$$

But under *uniform* acceleration $\ddot{y}^\mu \ddot{y}_\mu$ is *constant* (1.82). So (in two dimensions) F_μ^{rad} is orthogonal to two mutually orthogonal vectors, and it must be zero.[28]   QED

Does this mean there is no *radiation* from a particle in hyperbolic motion? It does *not* (though Wolfgang Pauli famously drew this erroneous conclusion).[29] According to Eq. 3.111,

$$A_\mu^{\text{out}} = \int [D_R - D_A] J_\mu. \qquad (3.177)$$

The radiated field at X is given by the intersection of the particle trajectory with the backward light cone (via D_R) and the intersection with the *forward* light cone (via D_A). But for hyperbolic motion there *is* no intersection with the forward light cone, only with the backward light cone (at P), so

---

[28] You can check this by the direct "bulldozer" method, of course, since we know the solution to the equations of motion (1.86 and 1.87).

[29] Eds. Part of the problem is simply bad language: it should not be called the "radiation" reaction, but rather the *field* reaction. It is the force exerted by the particle's own fields, acting on the particle itself—not just the fields that ultimately manifest themselves as radiation.

$$A_\mu^{\text{out}} = \int D_R \, J_\mu \neq 0. \tag{3.178}$$

The charge radiates, but it experiences no radiation reaction. Doesn't this violate conservation of energy? No, for the reason discussed in Section 3.4.2.

In fact, if there *were* a radiation reaction force on a particle in hyperbolic motion, then the **principle of equivalence** would not hold. The principle of equivalence says you cannot distinguish uniform acceleration from a uniform gravitational field (with the particle at rest). But a particle at rest does not radiate. If a particle in hyperbolic motion experienced a radiation reaction you would be able to tell a gravitational field from uniform acceleration. *Objection:* Since a particle in hyperbolic motion *does radiate* (even though there is no accompanying radiation reaction), why not measure that *radiation* to demonstrate violation of the equivalence principle? It turns out that what constitutes "radiation" in one reference frame may *not* be radiation in another: the distinction between radiation fields and induction fields is different for accelerating observers.[30] In particular, according to a *stationary* observer a freely falling charge radiates, but according to an observer who is *also* in free fall it does *not*.

[30] T. Fulton and F. Rohrlich, *Ann. Phys.* **9**, 499 (1960).

Part II

General Relativity

4

The Principle of Equivalence

4.1 Gravitational and Inertial Mass

According to Newton's law of **universal gravitation**, any two particles attract one another; the force of attraction is proportional to the product of their masses, and inversely proportional to the square of the distance between them:

$$\mathbf{F} = -G\frac{m_1 m_2}{r^2}\,\hat{\mathbf{r}}. \tag{4.1}$$

Here the **gravitational masses** (m_1 and m_2) are the *same* as the **inertial mass** that goes into Newton's second law of motion[1]—no *new* property (like charge, in the case of Coulomb's law) is involved. Thus the force on m_1 is

$$\mathbf{F} = m_1\mathbf{a} = -G\frac{m_1 m_2}{r^2}\,\hat{\mathbf{r}}; \tag{4.2}$$

the masses (m_1) cancel, and the acceleration of a falling object near the surface of the earth is

$$g = \frac{GM}{R^2} = 9.8\,\text{m/s}^2, \tag{4.3}$$

where M is the mass of the earth and R is its radius. This accounts for Galileo's observation that all objects fall at the same rate, regardless of their mass or what they are made of (neglecting air resistance, of course).

In Newton's theory the equivalence of inertial and gravitational mass is just a coincidence (in truth, an *astounding* coincidence, with no parallel anywhere in physics). Einstein didn't believe in coincidences, and he took the **principle of equivalence** (gravitational mass is "equivalent" to inertial mass) as the foundation for his theory of gravity. But beyond the (perhaps apocryphal) story of Galileo

---

[1] Actually, what we mean here is that gravitational mass is *proportional* to inertial mass, with a proportionality factor (which we choose to absorb into G) that is independent of mass and composition: the same for electrons, bricks, and galaxies.

dropping things from the Leaning Tower of Pisa, what is the experimental evidence for the equality of gravitational and inertial mass?[2]

4.2 The Eötvös Experiment

In 1885, Loránd v. Eötvös[3] began a series of experiments lasting more than 30 years to check the equivalence of gravitational and inertial mass. For an object near the surface of the earth there is a gravitational force pointing toward the center, and (because the earth is rotating) an effective centrifugal force pointing outward from the axis:

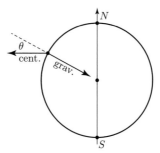

The former is proportional to the gravitational mass, and the latter to the inertial mass. Imagine two objects, made of different materials, suspended from a torsion pendulum:

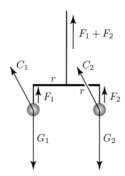

Here G is the gravitational force, C is the centrifugal force (pointing out of the page, at an angle θ to the vertical), and F is the tension in the string. In equilibrium the vertical forces cancel:

---

[2] Eds. In *Principia* (Book III, Proposition VI, Theorem VI) Newton describes an experiment he conducted using a pendulum with a hollow wooden ball that he could fill with various materials—sand, salt, water, lead, etc.—to test whether the ratio of gravitational to inertial mass was the same for all substances. He claimed an accuracy of better than one part in a thousand.

[3] Coleman paused here to explain the proper pronunciation of this Hungarian name, but the result is not transcribable in Roman letters.

$$G_1 - C_1 \cos\theta = F_1, \quad G_2 - C_2 \cos\theta = F_2. \tag{4.4}$$

But if the centrifugal forces are not equal there will be a net torque causing the pendulum to rotate:

$$N = (C_1 - C_2) \, r \sin\theta. \tag{4.5}$$

By using objects made of different materials, and looking for any resulting torque, Eötvös was able to conclude that the ratio of gravitational to inertial mass is the same (for the materials tested) to within one part in 10^8. Later, Dicke[4] improved the precision to one part in 10^{11}.

4.3 Gravitation and Geometry

In Newton's theory, gravity couples to *mass*, but already in special relativity, mass (or rather, mc^2) is just one form of energy. Presumably, therefore, in a relativistic theory the source of gravity would include all forms of energy—gravity should couple to $p^0 = \int d^3\mathbf{x} \, T^{00}$. At least in the static case, then, the source of gravity is T^{00}.[5]

Example 4.1

The gravitational redshift. In a uniform gravitational field the Hamiltonian H picks up a term Vp^0, where V is the gravitational potential gh:

$$H \to H + Vp^0 = H + VH = (1 + V)H. \tag{4.6}$$

In quantum mechanics (or, classically, under Hamilton's equations) the time evolution of a state changes accordingly:

$$e^{-iHt/\hbar} \to e^{-iH(1+V)t/\hbar}. \tag{4.7}$$

Thus we can include gravity explicitly by writing $H \to H(1 + V)$ or implicitly by $t \to (1 + V)t$. In effect, clocks run slow at the bottom of a well (where the gravitational potential is lower), and light emitted from the bottom will be redshifted, as observed from the top.

But if time is modified by gravity, what about space? One way to make a clock is to shine light down a pipe and back (after reflecting off a mirror). If time runs slow at the bottom of a well, then either the pipe must shrink, or else the speed of light is different down there. But whatever happens, the change must be the same

---

[4] P. G. Roll, R. Krotkov, and R. H. Dicke, *Ann. Phys.* **26**, 442 (1964).
[5] In hydrogen the binding energy (13.6 eV) is one part in 10^8 of the total (940 GeV—mostly rest energy of the proton), and for the strong nuclear forces it is much more than that, so the Eötvös and Dicke experiments are already probing the contribution from other forms of energy. We assume this holds as well for the weak interactions.

regardless of what the pipe is made of—*all* clocks, and *all* rods, are affected the same way by gravity. So why not say that the metric ($g_{\mu\nu}$) itself changes—that gravity alters the *geometry* of space and time? We defined the metric such that

$$\int_{s_1}^{s_2} \sqrt{g_{\mu\nu}\dot{x}^\mu\dot{x}^\nu}\,ds \tag{4.8}$$

is the elapsed (proper) time, so in the presence of gravity (at, say, the bottom of a well) the elapsed (proper) time would be

$$\int_{s_1}^{s_2} \sqrt{g'_{\mu\nu}\dot{x}^\mu\dot{x}^\nu}\,ds, \tag{4.9}$$

where g' is the metric as modified by gravity. We could say that 4.8 is "true," but one must then add the effects of gravity, or we could say that 4.9 is right, with no correction necessary. Since the effect of gravity is universal, the latter is a more elegant approach.[6] As Einstein put it,

$$\textbf{Gravitation is geometry}. \tag{4.10}$$

The effects of gravity are encoded in the metric.

4.4 The Equivalence Principle Revisited

Imagine you are in a box (traditionally, an "elevator"), with no windows, and you feel a force downward (your "weight"). By *purely mechanical* measurements you cannot tell whether the box is at rest in a uniform gravitational field, or the box is in empty space and is accelerating upward. Conversely, if you are in an elevator and you feel "weightless," you can't be sure whether the elevator is in free fall, descending with acceleration g, or somebody has turned off gravity and the elevator is at rest.

But what about *light*? You might think that a laser beam would go in a curve if the box is accelerating, but in a straight line if it's at rest in a gravitational field. Einstein extended the principle of equivalence to assert not just that inertial mass is the same as gravitational mass, but that *no* experiment can distinguish uniform acceleration from a uniform gravitational field; *no* experiment performed in a freely falling elevator can distinguish it from one floating in empty space. If this is so, then gravity must influence the propagation of light in just the right way to cancel the

---

[6] If we can treat the effects of gravity as (equivalent to) changes in geometry, why can't you do the same for other forces, such as electromagnetism? The crucial point is that gravity is *universal*; it affects everything in the same way. You can (in principle) make an electrically neutral clock, unaffected by electromagnetic fields, but you cannot make a gravitationally neutral clock, unaffected by gravity.

curvature of the beam (in the freely falling elevator), or to produce the curvature (in the box at rest in a gravitational field).

Of course, this is not true, even in the case of purely mechanical measurements, for a *large* laboratory in a *non*uniform gravitational field. Consider, for example, a huge elevator in free fall above a (spherical) earth:

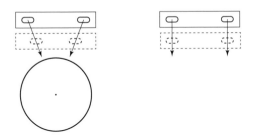

Objects are pulled toward the center, as the elevator falls, so you know there's some kind of gravitational force acting. Evidently we need the box to be *small*—small enough that the gravitational field is effectively uniform throughout.

These considerations lead us to formulate three versions of the equivalence principle:

- **Very weak form:** Gravitational mass equals inertial mass. No purely *mechanical* experiment can distinguish uniform acceleration from a uniform gravitational field.
- **Weak form:** A person in an elevator freely falling through a *uniform* gravitational field cannot (by *any* experiment conducted inside the elevator) tell that the elevator is not at rest in empty space.
- **Strong form:** Even in a *non*uniform gravitational field, it is always possible to make gravity disappear *locally* by going to a freely falling reference frame.

The strong form has an awkward vagueness about it (just *how small* does the box have to be?), but (as we shall see), there are more precise ways to state it: at any point in spacetime, there exists a reference frame[7] in which the metric takes the Minkowski form (1.18) and $\partial_\mu g_{\nu\lambda} = 0$ (at that point).

Einstein based the general theory of relativity on these two insights:

1. **The principle of equivalence** (strong form).
2. **Gravitation is geometry.**

Indeed, general relativity is (essentially) uniquely determined by them.

---

[7] Eds. Coleman calls it the **elevator frame**, but this terminology is not widely used.

5

Differential Geometry

In this chapter we develop the mathematical theory of curved spaces. We will do this in three stages, beginning with the most primitive spaces (manifolds), then adding some structure to allow for parallel transport (affine spaces), and finally introducing a measure of distance (metric spaces).

5.1 Manifolds

Imagine a curved 2-dimensional surface (a soap film, for example). We'll think of it as existing in ordinary 3-space, but for the moment it is endowed with no notion of length, and we are interested in its *intrinsic* geometry, which does not depend on its embedding in a larger space. Suppose we can cover the surface with overlapping patches of stretchable rubber membrane, each carrying (when flat) a grid of Cartesian coordinates:

Suppose, moreover, that these coordinate patches (stretched as necessary) fit smoothly over the whole surface with no holes, tears, or kinks. For example, a sphere can be covered with two such coordinate patches, but not with only one (if you start at the north pole, and wrap it symmetrically, when you get to the south pole the membrane bunches together, and distinct coordinate lines collide in a singular point). Suppose also that in any region where two coordinate patches overlap, one set of coordinates is a differentiable function of the other set. Such a surface, over

which a network of coordinate patches can be stretched, is called a **manifold**.[1]
Informally, a manifold is "like $\mathbb{R}^n$, locally." (A smooth surface is a 2-dimensional
manifold, but the idea generalizes in the obvious way to n dimensions. Note that
the number of dimensions is the same for all points on a manifold—otherwise there
would be a pinch or a tear at the join.)

Manifolds are very primitive spaces, but already they support various constructs.
For instance,

- **scalar fields:** $\phi(x)$ (where x stands for the set of coordinates defining the point
  in question);
- **curves:** $x^\mu(s)$, where s is some suitable parameter;
- **surfaces:** $x^\mu(s_1, s_2)$;
- **tangent vectors:** $\left.\dfrac{dx^\mu}{ds}\right|_{s_0}$ along a curve;
- **tangent spaces:** The collection of tangent vectors for all curves passing through
  a given point (s_0) defines a vector space, the "tangent space" at s_0. Note that the
  tangent space is an ordinary flat vector space, which (as the name suggests) is
  tangent to the surface at the point in question. Thus, for example, the tangent
  space to a sphere at the north pole is a plane, like the mortarboard on the head
  of a graduate. But the tangent space for a point on the equator is a completely
  different plane.

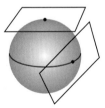

(Note that tangent spaces have the same dimension as the manifold itself.)

In the box below I give a more formal definition of a manifold.

Manifolds

An n-dimensional **manifold** is a mathematical structure consisting of

1. a set M of abstract elements, called **points**, and denoted by P_1, P_2, etc.;
2. a countable family of subsets of M, C_α, called **fundamental coordinate patches**,
   such that the union of all C_α is the whole set M (they "cover" M):

---

[1] Transformations from one system of patches to another are *passive* ("alias") transformations, and cannot be
interpreted in the active ("alibi") sense; they have nothing to do with invariances, and they do not lead to
conservation laws.

$$\bigcup_{\alpha} C_\alpha = M;$$

3. a reversible mapping of each patch onto the interior of the unit "cube" in $\mathbb{R}^n$. An n-tuple of real numbers associated in this way with each point are called the **coordinates** of the point (with respect to the given patch), and denoted by $x(P)$. The inverse mapping is denoted by $P(x)$.

Within a given patch we may use the coordinates to define such concepts as an open subset of a patch (one that is mapped into an open set in $\mathbb{R}^n$), a continuous function on the patch (one that is mapped into a continuous function of the coordinates), etc. In technical language, we can use the coordinates to define a topology for each patch.

The following conditions guarantee that these topologies are consistent when two patches overlap, and that we can speak of open subsets of M, continuous functions on M, etc., without specifying particular patches:

A. The intersection of two patches is open, with respect to either patch.
B. In this intersection, if $x(P)$ and $y(P)$ are the coordinates associated with the two patches, $x(P(y))$ and $y(P(x))$ are continuous functions.
C. Given any two distinct points, there are two open sets, one containing one point and the other containing the other point, such that their intersection is empty.

This completes the definition of a manifold.

An **allowable coordinate patch** is an open subset of M together with an invertible continuous mapping of the subset onto the unit cube in $\mathbb{R}^n$. Clearly, any family of allowable patches that covers M defines the same topology as the original fundamental patches. We will speak loosely and say it "defines the same manifold."

An m-times continuously **differentiable manifold** (a C^m manifold) is defined in the same way as a manifold, except that, in condition B above, the word "continuous" is replaced by "m-times continuously differentiable." A like substitution is made in the definition of an allowable patch.

5.1.1 Vectors

Vectors live in tangent space.[2] If you change coordinates for the manifold $(x^\mu \to x'^\mu)$, the tangent to a curve transforms as

$$\frac{dx'^\mu}{ds} = \frac{\partial x'^\mu}{\partial x^\nu} \frac{dx^\nu}{ds}; \tag{5.1}$$

contravariant vectors transform the same way:

$$\boxed{a'^\mu = \left(\frac{\partial x'^\mu}{\partial x^\nu}\right) a^\nu.} \tag{5.2}$$

---

[2] Notice that x^μ itself (the set of coordinates describing a point in the manifold) is *not* a vector (in spite of appearances!), although the infinitesimal dx^μ is.

Of course, the coefficients $(\partial x'^\mu/\partial x^\nu)$ can vary from point to point in the manifold, so (for example) if you say that something is a "vector field," you mean that it transforms as a vector at every point.

Because vectors associated with different points live in completely different tangent spaces, you cannot add a vector at one point to a vector at some other point. This problem never arose in Minkowski space because it was always possible to *move* a vector from one place to another by **parallel displacement** (keeping the Cartesian components numerically constant). We simply slide one vector over until the two coincide, and add them at the *same* point. But parallel displacement has no intrinsic meaning in a curved manifold. For example, consider a spherical surface; we want to move a vector **A** from the north pole down to the south pole.

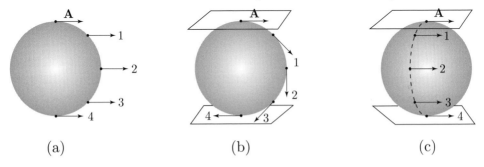

(a)   (b)   (c)

At each point in the journey it must lie in the local tangent space, so (a) is not permitted. We might try moving **A** in such a way as to keep it "as parallel as possible" (but always in the local tangent space); (b) and (c) would be OK, but they yield a *different result*. As we shall see, you have to impose more structure on a manifold before parallel transport is well defined.[3]

Covariant vectors b_μ live in a vector space dual to the tangent space at each point. Like a^μ, b_μ has n components (if it's an n-dimensional manifold). It is defined in such a way that

$$b_\mu a^\mu \tag{5.3}$$

is independent of the coordinates used (which is to say, it's a **scalar**). From this, and 5.2, we can deduce the transformation rule for covariant vectors:

$$b_\mu a^\mu = b'_\mu a'^\mu = b'_\mu \left(\frac{\partial x'^\mu}{\partial x^\nu}\right) a^\nu. \tag{5.4}$$

Evidently

$$\boxed{b'_\mu = \left(\frac{\partial x^\nu}{\partial x'^\mu}\right) b_\nu.} \tag{5.5}$$

---

[3] Actually, a sphere *has* the requisite additional structure, but not if it is considered only as a manifold.

(*Mnemonic:* Always sum over the *old* (unprimed) coordinates; that tells you which way up the derivatives go.)

We can now construct (covariant, contravariant, and mixed) **tensors**, in the usual way; they transform like *products* of vectors. For instance, a (contravariant) second-rank tensor transforms by

$$a'^{\mu\nu} = \left(\frac{\partial x'^{\mu}}{\partial x^{\lambda}} \frac{\partial x'^{\nu}}{\partial x^{\sigma}} \right) a^{\lambda\sigma}; \tag{5.6}$$

for a mixed tensor

$$b'^{\mu}{}_{\nu} = \left(\frac{\partial x'^{\mu}}{\partial x^{\lambda}} \frac{\partial x^{\sigma}}{\partial x'^{\nu}} \right) b^{\lambda}{}_{\sigma}; \tag{5.7}$$

and so on. As before, contraction over an upper and a lower index is an invariant operation (independent of the coordinates used), but we cannot (yet) raise and lower indices, because so far we have no metric tensor—no measure of *distance* between points in the manifold.

5.1.2 Exterior Calculus

Consider a scalar field ϕ on a manifold. Its **gradient** is

$$\frac{\partial \phi}{\partial x^{\mu}}. \tag{5.8}$$

How does it transform, if we change coordinates from x to x'? By the chain rule,

$$\frac{\partial \phi}{\partial x'^{\mu}} = \frac{\partial x^{\nu}}{\partial x'^{\mu}} \frac{\partial \phi}{\partial x^{\nu}}. \tag{5.9}$$

Evidently it is a *covariant* vector; accordingly, we will write

$$\partial_{\mu}\phi \equiv \frac{\partial \phi}{\partial x^{\mu}}. \tag{5.10}$$

(Notice that ϕ itself does not change when we change coordinates; $\phi' = \phi$, since we're evaluating it at the same point. Later, when we consider *vector* fields, the components themselves will mix when we change coordinates.)

For two points s_1 and s_2 in the manifold, and a particular path from one to the other,

$$\int_{s_1}^{s_2} \partial_{\mu}\phi \frac{dx^{\mu}}{ds} ds = \int_{s_1}^{s_2} \partial_{\mu}\phi \, dx^{\mu} \tag{5.11}$$

is an invariant, depending only on the scalar function ϕ and the path taken, but not on the coordinates used, or the parameterization s. More generally, for *any* covariant vector field A_μ, and specified path $\mathcal{S}$,

$$\int_\mathcal{S} A_\mu \, dx^\mu \text{ is invariant.} \tag{5.12}$$

What about *surface* integrals? A surface is a function of *two* variables, s_1 and s_2: $x^\mu(s_1, s_2)$. For any covariant tensor function $T_{\mu\nu}$,

$$T_{\mu\nu} \frac{\partial x^\mu}{\partial s_1} \frac{\partial x^\nu}{\partial s_2} \, ds_1 \, ds_2 \tag{5.13}$$

is invariant under *coordinate* transformations ($x \to x'$), but it *isn't* invariant under a change of *parameterization* ($s_1 \to s_1', s_2 \to s_2'$). Contrast 5.11: if $s \to s'$, then

$$\frac{dx^\mu}{ds} \to \frac{dx^\mu}{ds'} = \frac{dx^\mu}{ds} \frac{ds}{ds'}, \tag{5.14}$$

so

$$\frac{dx^\mu}{ds} ds \to \frac{dx^\mu}{ds'} ds' = \frac{dx^\mu}{ds} \frac{ds}{ds'} ds' = \frac{dx^\mu}{ds} ds \tag{5.15}$$

(it's invariant). But in 5.13,

$$\frac{\partial x^\mu}{\partial s_1} \to \frac{\partial x^\mu}{\partial s_1'} = \frac{\partial x^\mu}{\partial s_1} \frac{\partial s_1}{\partial s_1'} + \frac{\partial x^\mu}{\partial s_2} \frac{\partial s_2}{\partial s_1'}, \tag{5.16}$$

and hence

$$\frac{\partial x^\mu}{\partial s_1} \frac{\partial x^\nu}{\partial s_2} \, ds_1 \, ds_2 \tag{5.17}$$

is *not* invariant—we pick up cross terms.

This suggests that we examine the antisymmetrized form,

$$\frac{1}{2} \left(\frac{\partial x^\mu}{\partial s_1} \frac{\partial x^\nu}{\partial s_2} - \frac{\partial x^\mu}{\partial s_2} \frac{\partial x^\nu}{\partial s_1} \right) ds_1 \, ds_2. \tag{5.18}$$

Under $s_1 \rightarrow s_1', s_2 \rightarrow s_2'$, the term in parentheses becomes

$$\left(\frac{\partial x^\mu}{\partial s_1'} \frac{\partial x^\nu}{\partial s_2'} - \frac{\partial x^\mu}{\partial s_2'} \frac{\partial x^\nu}{\partial s_1'} \right)$$

$$= \left(\frac{\partial x^\mu}{\partial s_1} \frac{\partial s_1}{\partial s_1'} + \frac{\partial x^\mu}{\partial s_2} \frac{\partial s_2}{\partial s_1'} \right) \left(\frac{\partial x^\nu}{\partial s_1} \frac{\partial s_1}{\partial s_2'} + \frac{\partial x^\nu}{\partial s_2} \frac{\partial s_2}{\partial s_2'} \right)$$

$$- \left(\frac{\partial x^\mu}{\partial s_1} \frac{\partial s_1}{\partial s_2'} + \frac{\partial x^\mu}{\partial s_2} \frac{\partial s_2}{\partial s_2'} \right) \left(\frac{\partial x^\nu}{\partial s_1} \frac{\partial s_1}{\partial s_1'} + \frac{\partial x^\nu}{\partial s_2} \frac{\partial s_2}{\partial s_1'} \right)$$

$$= \left(\frac{\partial x^\mu}{\partial s_1} \frac{\partial x^\nu}{\partial s_1} \right) \frac{\partial s_1 \partial s_1}{\partial s_1' \partial s_2'} + \left(\frac{\partial x^\mu}{\partial s_1} \frac{\partial x^\nu}{\partial s_2} \right) \frac{\partial s_1 \partial s_2}{\partial s_1' \partial s_2'} + \left(\frac{\partial x^\mu}{\partial s_2} \frac{\partial x^\nu}{\partial s_1} \right) \frac{\partial s_2 \partial s_1}{\partial s_1' \partial s_2'}$$

$$+ \left(\frac{\partial x^\mu}{\partial s_2} \frac{\partial x^\nu}{\partial s_2} \right) \frac{\partial s_2 \partial s_2}{\partial s_1' \partial s_2'} - \left(\frac{\partial x^\mu}{\partial s_1} \frac{\partial x^\nu}{\partial s_1} \right) \frac{\partial s_1 \partial s_1}{\partial s_2' \partial s_1'} - \left(\frac{\partial x^\mu}{\partial s_1} \frac{\partial x^\nu}{\partial s_2} \right) \frac{\partial s_1 \partial s_2}{\partial s_2' \partial s_1'}$$

$$- \left(\frac{\partial x^\mu}{\partial s_2} \frac{\partial x^\nu}{\partial s_1} \right) \frac{\partial s_2 \partial s_1}{\partial s_2' \partial s_1'} - \left(\frac{\partial x^\mu}{\partial s_2} \frac{\partial x^\nu}{\partial s_2} \right) \frac{\partial s_2 \partial s_2}{\partial s_2' \partial s_1'}$$

$$= \left[\frac{\partial(s)}{\partial(s')} \right] \left(\frac{\partial x^\mu}{\partial s_1} \frac{\partial x^\nu}{\partial s_2} - \frac{\partial x^\mu}{\partial s_2} \frac{\partial x^\nu}{\partial s_1} \right), \tag{5.19}$$

where

$$\frac{\partial(s)}{\partial(s')} \equiv \begin{vmatrix} \dfrac{\partial s_1}{\partial s_1'} & \dfrac{\partial s_1}{\partial s_2'} \\ \dfrac{\partial s_2}{\partial s_1'} & \dfrac{\partial s_2}{\partial s_2'} \end{vmatrix} \tag{5.20}$$

is the Jacobian of the transformation from s_1, s_2 to s_1', s_2'. But

$$ds_1' \, ds_2' = \frac{\partial(s')}{\partial(s)} ds_1 \, ds_2, \tag{5.21}$$

and this Jacobian cancels the one in Eq. 5.19. Accordingly, we define

$$d\sigma^{\mu\nu} \equiv \frac{1}{2} \left(\frac{\partial x^\mu}{\partial s_1} \frac{\partial x^\nu}{\partial s_2} - \frac{\partial x^\mu}{\partial s_2} \frac{\partial x^\nu}{\partial s_1} \right) ds_1 \, ds_2; \tag{5.22}$$

this surface element is invariant under a change of parameterization.

Problem 5.1

Consider a spherical surface, of radius R, centered at the origin in the Cartesian coordinate system $x^1 = x, x^2 = y, x^3 = z$, and parameterized by the ordinary spherical coordinates $s_1 = \theta$ and $s_2 = \phi$. Find all the components of $d\sigma^{\mu\nu}$.

Equation 5.22 generalizes to regions of dimension r:

$$d\sigma^{\mu\nu\lambda\cdots} \equiv \frac{1}{r!} \begin{vmatrix} \dfrac{\partial x^\mu}{\partial s_1} & \dfrac{\partial x^\mu}{\partial s_2} & \cdots \\[2mm] \dfrac{\partial x^\nu}{\partial s_1} & \dfrac{\partial x^\nu}{\partial s_2} & \cdots \\[2mm] \vdots & \vdots & \ddots \end{vmatrix} ds_1\, ds_2 \cdots ds_r, \tag{5.23}$$

giving us the **invariant integrals** (independent of coordinates *and* independent of parameterization) over a fixed r-dimensional region $\mathcal{R}$:

$$\int_{\mathcal{R}} T_{\mu_1\cdots\mu_r}\, d\sigma^{\mu_1\cdots\mu_r}. \tag{5.24}$$

This expression subsumes the invariant line integral of a (covariant) vector, an invariant surface integral of a (covariant) second-rank tensor, and so on. On a 2-dimensional manifold you can construct invariant line and surface integrals; in four dimensions there are four kinds of invariant integrals: line, surface, 3-volume, and 4-volume. Notice that contraction with the fully antisymmetric $d\sigma$ picks out the completely antisymmetric part of T; in particular, the invariant integral of a *symmetric* tensor is always zero (scalars and vectors count as *anti*symmetric). The point, of course, is that $ds_1\, ds_2 \cdots ds_r$ picks up a Jacobian (under a change of parameterization); the determinant in 5.23 is introduced because *it* transforms like the *inverse* of the Jacobian.

We are now in a position to state the generalization of Stokes's theorem. Let $A_{\mu_1\cdots\mu_m}$ be a completely antisymmetric[4] covariant tensor. We define the **exterior derivative**

$$(dA)_{\mu_0\mu_1\cdots\mu_m} \equiv \partial_{\mu_0} A_{\mu_1\mu_2\mu_3\cdots\mu_m} - \partial_{\mu_1} A_{\mu_0\mu_2\mu_3\cdots\mu_m} + \partial_{\mu_2} A_{\mu_0\mu_1\mu_3\cdots\mu_m} - \cdots$$

$$= \sum_P (-1)^P \partial_{\mu_0} A_{\mu_1\mu_2\cdots\mu_m}, \tag{5.25}$$

where the sum is over all distinct permutations P of the indices $\mu_0, \mu_1, \ldots, \mu_m$. Note that there are just $(m+1)$—not $(m+1)!$—distinct permutations: any one of the $m+1$ indices can be the differentiator, but further permutations are accounted for by the antisymmetry of A, and therefore are not to be considered "distinct."

---

[4] That is, its components switch sign under the interchange of *any* two indices; remember that scalars and vectors are to be considered antisymmetric.

Example 5.1

1. Scalar: $(d\phi)_\mu = \partial_\mu \phi$ (think gradient).
2. Vector: $(dA)_{\mu\nu} = \partial_\mu A_\nu - \partial_\nu A_\mu$ (think curl).
3. Tensor: $(dA)_{\mu\nu\lambda} = \partial_\mu A_{\nu\lambda} + \partial_\nu A_{\lambda\mu} + \partial_\lambda A_{\mu\nu}$ (in this case we can choose all the distinct permutations to be even—hence all plus signs).

Problem 5.2

Write out $(dA)_{\mu\nu\lambda\sigma}$. Can it be expressed using only even permutations?

Stokes's theorem: Let $\mathcal{V}$ be a (simply connected[5]) $(r+1)$-dimensional "volume"[6] bounded by the r-dimensional "surface" S. Then

$$\oint_S A_{\mu_1\mu_2\cdots\mu_r}\, d\sigma^{\mu_1\mu_2\cdots\mu_r} = \int_{\mathcal{V}} (dA)_{\mu_0\mu_1\cdots\mu_r}\, d\sigma^{\mu_0\mu_1\cdots\mu_r}, \qquad (5.26)$$

or

$$\oint_S A \cdot d^r\sigma = \int_{\mathcal{V}} (dA) \cdot d^{r+1}\sigma, \qquad (5.27)$$

for short.

Example 5.2

In the case of a vector, the left-hand side becomes a (closed) line integral and the right-hand side a (2-dimensional) surface integral:

$$\int A_\mu \frac{dx^\mu}{ds}\, ds = \int (\partial_{\mu_0} A_\mu - \partial_\mu A_{\mu_0}) \frac{1}{2} \begin{vmatrix} \partial x^{\mu_0}/\partial s_1 & \partial x^{\mu_0}/\partial s_2 \\ \partial x^\mu/\partial s_1 & \partial x^\mu/\partial s_2 \end{vmatrix} ds_1\, ds_2. \qquad (5.28)$$

This is the generalization of Stokes's original theorem to an arbitrary manifold.

---

[5] A **simply connected** space has no "holes" in it. More precisely, if A and B are two points in $\mathcal{V}$, and p and p' are any paths connecting A to B, you can get from p to p' by continuous distortion.

For example, the surface of a cylinder is *not* simply connected, for if the two paths wrap around a different number of times, you can't continuously distort one into the other; likewise, the surface of a torus is not simply connected.

[6] Here $r < n$ (the dimension of the manifold), else there *is* no volume of dimension $r + 1$.

Problem 5.3

Prove Stokes's theorem. *Hint:* Break the volume $\mathcal{V}$ up into many small "cubes," and prove it first for one such cube. Express the right-hand side of 5.26 as

$$\frac{1}{(r+1)!}\int ds_0\, ds_1 \cdots ds_r \left|\frac{\partial x}{\partial s}\right| \sum_P (-1)^P \partial_{\mu_0} A_{\mu_1\cdots\mu_r},$$

and expand the determinant by minors, down the first column. Each minor is the appropriate differential element for a surface of the "cube."

The exterior derivative is antisymmetric by construction, but *is it a tensor?*

Corollary: The exterior derivative is a tensor.

Proof: The left-hand side of 5.26 is coordinate independent (a scalar), so the right-hand side must also be. But $d\sigma$ is a tensor, and we can shrink $\mathcal{V}$ down to a point. If (dA) did *not* transform as a (covariant) tensor at that point, their contraction would not be invariant. Therefore (dA) must in fact be a tensor.   QED

Comment: Without antisymmetrization, the exterior derivative would *not* be a tensor. Antisymmetrization is a device for circumventing the nonadditivity of tensors at different points. Consider the *non*antisymmetrized derivative of a vector, $\partial_\nu A_\mu$. When the coordinates are changed $(x \to x')$,

$$A_\mu \to A'_\mu = \frac{\partial x^\sigma}{\partial x'^\mu} A_\sigma \quad \text{and} \quad \partial_\nu \to \partial'_\nu = \frac{\partial x^\lambda}{\partial x'^\nu} \partial_\lambda, \tag{5.29}$$

so

$$\partial_\nu A_\mu \to (\partial_\nu A_\mu)' = \partial'_\nu A'_\mu = \frac{\partial x^\lambda}{\partial x'^\nu} \partial_\lambda \left(\frac{\partial x^\sigma}{\partial x'^\mu} A_\sigma \right) \tag{5.30}$$

$$= \frac{\partial x^\lambda}{\partial x'^\nu} \frac{\partial x^\sigma}{\partial x'^\mu} \partial_\lambda A_\sigma + \frac{\partial x^\lambda}{\partial x'^\nu} \frac{\partial^2 x^\sigma}{\partial x^\lambda \partial x'^\mu} A_\sigma$$

$$= \left(\frac{\partial x^\lambda}{\partial x'^\nu} \frac{\partial x^\sigma}{\partial x'^\mu} \right) \partial_\lambda A_\sigma + \left(\frac{\partial^2 x^\sigma}{\partial x'^\nu \partial x'^\mu} \right) A_\sigma.$$

The first term is appropriate for tensor transformation, but the second does not belong. If we restrict ourselves to *linear* transformations (as in special relativity) the offending term does not contribute. But in the more general case the best we can do is to exploit the fact that it is *symmetric* to kill it by antisymmetrizing our derivatives.

5.1.3 Tensor Densities

Consider the completely antisymmetric **Levi-Civita symbol** on an n-dimensional manifold: $\varepsilon^{\mu_1\cdots\mu_n}$ switches sign under interchange of any two indices. All elements can be obtained from $\varepsilon^{12\cdots n}$, which we define to be $+1$ in some initial coordinate system. Then (in that system)

$$\varepsilon^{\mu_1\cdots\mu_n} = \begin{cases} +1 & \text{if } \mu_1\cdots\mu_n \text{ is an even permutation of } 1,2,\ldots,n, \\ -1 & \text{if } \mu_1\cdots\mu_n \text{ is an odd permutation of } 1,2,\ldots,n, \\ 0 & \text{if any two indices repeat.} \end{cases} \quad (5.31)$$

How does ε transform? What are its elements in some *other* coordinate system? Let's stipulate that it transforms as a (contravariant) tensor (Eq. 5.6):

$$\varepsilon'^{\mu_1\cdots\mu_n} = \left(\frac{\partial x'^{\mu_1}}{\partial x^{\nu_1}}\frac{\partial x'^{\mu_2}}{\partial x^{\nu_2}}\cdots\frac{\partial x'^{\mu_n}}{\partial x^{\nu_n}}\right)\varepsilon^{\nu_1\cdots\nu_n}. \quad (5.32)$$

This is still totally antisymmetric (since the coefficients in parentheses are totally symmetric). Therefore, to determine it completely we only need to calculate one element:

$$\varepsilon'^{12\cdots n} = \left(\frac{\partial x'^1}{\partial x^{\nu_1}}\frac{\partial x'^2}{\partial x^{\nu_2}}\cdots\frac{\partial x'^n}{\partial x^{\nu_n}}\right)\varepsilon^{\nu_1\cdots\nu_n}$$

$$= \left(\sum_P(-1)^P\frac{\partial x'^1}{\partial x^{\nu_1}}\frac{\partial x'^2}{\partial x^{\nu_2}}\cdots\frac{\partial x'^n}{\partial x^{\nu_n}}\right)\varepsilon^{12\cdots n} = \frac{\partial(x')}{\partial(x)}\varepsilon^{12\cdots n}, \quad (5.33)$$

where $\partial(x')/\partial(x)$ is (again) the Jacobian. Thus

$$\varepsilon'^{\mu_1\cdots\mu_n} = \frac{\partial(x')}{\partial(x)}\varepsilon^{\mu_1\cdots\mu_n}. \quad (5.34)$$

Because each element picks up a "magnification" factor of the Jacobian, ε is *not* numerically invariant (for instance, $\varepsilon^{12\cdots n} = 1 \rightarrow \varepsilon'^{12\cdots n} = \partial(x')/\partial(x)$).

Definition: A (contravariant) **tensor density** of **weight** w is an object (written with a sans-serif letter) that transforms as follows:[7]

$$\mathsf{a}'^{\mu_1\cdots\mu_m} = \left(\frac{\partial x'^{\mu_1}}{\partial x^{\nu_1}}\frac{\partial x'^{\mu_2}}{\partial x^{\nu_2}}\cdots\frac{\partial x'^{\mu_m}}{\partial x^{\nu_m}}\right)\mathsf{a}^{\nu_1\cdots\nu_m}\left|\frac{\partial(x)}{\partial(x')}\right|^w. \quad (5.35)$$

(It need not have the same number of indices as the dimension of the manifold.) An ordinary tensor is a tensor density of weight $w = 0$. We now define $\epsilon^{\mu_1\cdots\mu_n}$ (ϵ, not ε) to be the totally antisymmetric object identical to $\varepsilon^{\mu_1\cdots\mu_n}$ (in the initial coordinate system), but which transforms as a tensor *density* of weight $w = +1$. Thus

$$\epsilon'^{\mu_1\cdots\mu_n} = \epsilon^{\mu_1\cdots\mu_n}. \quad (5.36)$$

---

[7] The determinant of a (proper) Lorentz transformation is 1 (see Eq. 1.49), so in special relativity—leaving aside parity and time reversal—there is no difference between tensors and tensor densities.

The two Jacobians cancel. It's a **numerically invariant** tensor density—each element has the same numerical value ($+1$, -1, or 0, in this instance) no matter what coordinates you use. We also define the covariant counterpart, $\epsilon_{\mu_1\cdots\mu_n}$, which is numerically equal to $\epsilon^{\mu_1\cdots\mu_n}$ in any particular system of coordinates,[8] but transforms as a *covariant* tensor density of weight $w = -1$. It, too, is numerically invariant.

Problem 5.4

In an n-dimensional manifold, show that

(a) $\left(\epsilon^{\kappa_1\kappa_2\kappa_3\cdots\kappa_n}\right)\left(\epsilon_{\kappa_1\kappa_2\kappa_3\cdots\kappa_n}\right) = n!$,

(b) $\left(\epsilon^{\mu\kappa_2\kappa_3\cdots\kappa_n}\right)\left(\epsilon_{\nu\kappa_2\kappa_3\cdots\kappa_n}\right) = (n-1)!\,\delta^\mu_\nu$,

(c) $\left(\epsilon^{\mu\lambda\kappa_3\cdots\kappa_n}\right)\left(\epsilon_{\nu\sigma\kappa_3\cdots\kappa_n}\right) = (n-2)!\,\left(\delta^\mu_\nu\delta^\lambda_\sigma - \delta^\mu_\sigma\delta^\lambda_\nu\right)$.

Consider a totally antisymmetric tensor (weight 0) with n indices (where n is the dimension of the manifold). It has only one independent element (from which all the others can be obtained by antisymmetry), so it can be written

$$A_{\mu_1\cdots\mu_n} = a\epsilon_{\mu_1\cdots\mu_n}, \tag{5.37}$$

where a is a scalar density of weight $+1$. What about a (completely antisymmetric) tensor with $(n-1)$ indices? It has n independent elements, one for each index that can be the "missing" index (but for a *particular* missing index all the elements can be obtained from one of them by invoking antisymmetry). Thus

$$A_{\mu_1\cdots\mu_{n-1}} = a^{\mu_0}\epsilon_{\mu_0\mu_1\cdots\mu_{n-1}}, \tag{5.38}$$

and in fact a^{μ_0} is a *vector density* of weight $+1$.

Proof: Assume a^{μ_0} *is* a vector density of weight 1, so that

$$a^{\mu_0} \rightarrow \left(\frac{\partial x'^{\mu_0}}{\partial x^{\nu_0}}\right) a^{\nu_0} \left[\frac{\partial(x)}{\partial(x')}\right]^{+1}. \tag{5.39}$$

Meanwhile, the covariant ϵ is a tensor density of weight -1,

$$\epsilon_{\mu_0\cdots\mu_{n-1}} \rightarrow \left(\frac{\partial x^{\sigma_0}}{\partial x'^{\mu_0}}\cdots\frac{\partial x^{\sigma_{n-1}}}{\partial x'^{\mu_{n-1}}}\right)\epsilon_{\sigma_0\cdots\sigma_{n-1}}\left[\frac{\partial(x)}{\partial(x')}\right]^{-1}. \tag{5.40}$$

Therefore, since

$$\frac{\partial x'^{\mu_0}}{\partial x^{\nu_0}}\frac{\partial x^{\sigma_0}}{\partial x'^{\mu_0}} = \delta^{\sigma_0}_{\nu_0}, \tag{5.41}$$

---

[8] *Beware:* This *looks* inconsistent with the usual rule for lowering indices using a metric tensor, but (as we shall see) that rule itself is modified for tensor densities.

$A_{\mu_1\cdots\mu_{n-1}}$ would transform as

$$a^{\mu_0}\epsilon_{\mu_0\cdots\mu_{n-1}} \rightarrow \left(\frac{\partial x'^{\mu_0}}{\partial x^{\nu_0}}\frac{\partial x^{\sigma_0}}{\partial x'^{\mu_0}}\frac{\partial x^{\sigma_1}}{\partial x'^{\mu_1}}\cdots\frac{\partial x^{\sigma_{n-1}}}{\partial x'^{\mu_{n-1}}}\right) a^{\nu_0}\epsilon_{\sigma_0\cdots\sigma_{n-1}}$$

$$= \left(\frac{\partial x^{\sigma_1}}{\partial x'^{\mu_1}}\cdots\frac{\partial x^{\sigma_{n-1}}}{\partial x'^{\mu_{n-1}}}\right) a^{\sigma_0}\epsilon_{\sigma_0\cdots\sigma_{n-1}}$$

$$= \left(\frac{\partial x^{\sigma_1}}{\partial x'^{\mu_1}}\cdots\frac{\partial x^{\sigma_{n-1}}}{\partial x'^{\mu_{n-1}}}\right) A_{\mu_1\cdots\mu_{n-1}}$$

$$= A'_{\mu_1\cdots\mu_{n-1}}. \quad \text{QED} \tag{5.42}$$

Thus assumption 5.39 yields the correct transformation law for A, or, running the argument backward, the transformation rule for A confirms that a is a vector density of weight 1.   QED

If *two* indices are "missing," we have a (totally antisymmetric) tensor of rank $n-2$ ($A_{\mu_1\ldots\mu_{n-2}}$). How many independent elements does it carry? There are n ways to choose the first missing index, and that leaves $n-1$ ways to choose the second. But the order doesn't matter (because of antisymmetry), so there are $n(n-1)/2$ distinct elements—just the same number as an antisymmetric second-rank tensor (density). Indeed, we can write

$$A_{\mu_1\cdots\mu_{n-2}} = a^{\mu_{-1}\mu_0}\epsilon_{\mu_{-1}\mu_0\mu_1\cdots\mu_{n-2}}, \tag{5.43}$$

where $a^{\mu\nu}$ is an antisymmetric tensor density of rank 2 and weight 1. And in general there is a duality between totally antisymmetric tensors of rank $(n-r)$ and totally antisymmetric tensor densities of rank r:

$$A_{\mu_1\cdots\mu_{n-r}} = a^{\nu_1\cdots\nu_r}\epsilon_{\nu_1\cdots\nu_r\mu_1\cdots\mu_{n-r}}. \tag{5.44}$$

For instance, in a 4-dimensional manifold (such as the spacetime of general relativity),

$$\begin{aligned}
A_{\mu\nu\lambda\sigma} &\leftrightarrow a, \\
A_{\mu\nu\lambda} &\leftrightarrow a^{\sigma}, \\
A_{\mu\nu} &\leftrightarrow a^{\sigma\lambda}, \\
A_{\mu} &\leftrightarrow a^{\sigma\lambda\nu}, \\
A &\leftrightarrow a^{\sigma\lambda\nu\mu}
\end{aligned} \tag{5.45}$$

(all of them completely antisymmetric).

Problem 5.5

In Eqs. 5.37, 5.38, 5.43, and 5.44 we expressed antisymmetric tensors in terms of complementary tensor densities (and ϵ symbols). You can also go in the other direction, expressing (antisymmetric) tensor densities in terms of (antisymmetric) tensors. Show, for example, that

$$\mathsf{b} \equiv A_{\mu_1 \cdots \mu_n} \epsilon^{\mu_1 \cdots \mu_n} \tag{5.46}$$

is a scalar density of weight $+1$.

We can exploit this duality to extend the theory of integration on a manifold.

Example 5.3

Consider a totally antisymmetric tensor $A_{\mu\nu\lambda\sigma}$ on a 4-dimensional manifold. Using Eq. 5.37 we find

$$\int_{\mathcal{V}} A_{\mu\nu\lambda\sigma} \, d\sigma^{\mu\nu\lambda\sigma} = \int_{\mathcal{V}} \mathsf{a} \epsilon_{\mu\nu\lambda\sigma} \, d\sigma^{\mu\nu\lambda\sigma} = \int_{\mathcal{V}} \mathsf{a} \, d^4\Sigma, \tag{5.47}$$

where

$$d^4\Sigma \equiv \epsilon_{\mu\nu\lambda\sigma} \, d\sigma^{\mu\nu\lambda\sigma}. \tag{5.48}$$

Similarly (using 5.38),

$$\int A_{\mu\nu\lambda} \, d\sigma^{\mu\nu\lambda} = \int \mathsf{a}^\rho \epsilon_{\rho\mu\nu\lambda} \, d\sigma^{\mu\nu\lambda} = \int \mathsf{a}^\rho \, d^3\Sigma_\rho, \tag{5.49}$$

where

$$d^3\Sigma_\rho \equiv \epsilon_{\rho\mu\nu\lambda} \, d\sigma^{\mu\nu\lambda}, \tag{5.50}$$

and so on.

Question: What does the exterior derivative look like in the dual space of tensor densities? *Answer:* The *divergence* of an antisymmetric tensor density is dual to the exterior derivative of the associated tensor. In the dual space $A \to \mathsf{a}$, $d\sigma \to d\Sigma$, and (from 5.25 and 5.44)

$$(dA)_{\mu_0\mu_1\cdots\mu_m} = \sum_P (-1)^P \partial_{\mu_0} A_{\mu_1\dots\mu_m}$$

$$= \sum_P (-1)^P \partial_{\mu_0} \mathsf{a}^{\nu_1\cdots\nu_{n-m}} \epsilon_{\nu_1\cdots\nu_{n-m}\mu_1\cdots\mu_m} \tag{5.51}$$

(everything antisymmetric, of course). Remember that the sum is over the $(m+1)$ distinct permutations of the μ's. Thus

$$(dA) \cdot d^{r+1}\sigma \to (\mathrm{div}\,\mathsf{a}) \cdot d^{r+1}\Sigma. \tag{5.52}$$

This defines the divergence of an antisymmetric tensor density (see Problem 5.6), and we discover that Gauss's theorem is the tensor density statement dual to Stokes's theorem (5.27). Symbolically,

$$\textbf{Stokes} : \oint_{S} A \cdot d^{r}\sigma = \int_{\mathcal{V}} (dA) \cdot d^{r+1}\sigma$$

$$\Updownarrow \qquad\qquad\qquad (5.53)$$

$$\textbf{Gauss} : \oint_{S} \textbf{a} \cdot d^{r}\Sigma = \int_{\mathcal{V}} (\text{div}\,\textbf{a}) \cdot d^{r+1}\Sigma.$$

Example 5.4

Consider integrals over all space in four dimensions. To begin with we need $d\sigma^{\mu\nu\lambda\sigma}$ (Eq. 5.23). A suitable parameterization (s_1, s_2, s_3, s_4) would be (x^1, x^2, x^3, x^4) themselves. In that case,

$$\begin{vmatrix} \dfrac{\partial x^{\mu}}{\partial s_1} & \dfrac{\partial x^{\mu}}{\partial s_2} & \cdots \\[2mm] \dfrac{\partial x^{\nu}}{\partial s_1} & \dfrac{\partial x^{\nu}}{\partial s_2} & \cdots \\[2mm] \vdots & \vdots & \ddots \end{vmatrix} \qquad\qquad (5.54)$$

is a completely antisymmetric object with four indices; the 1234 element is

$$\begin{vmatrix} 1 & 0 & 0 & 0 \\ 0 & 1 & 0 & 0 \\ 0 & 0 & 1 & 0 \\ 0 & 0 & 0 & 1 \end{vmatrix} = 1. \qquad\qquad (5.55)$$

It looks as though the determinant is nothing but $\epsilon^{\mu\nu\lambda\rho}$. But does this transform correctly? Yes: Suppose this is right, and hence

$$d\sigma^{\mu\nu\lambda\rho} = \tfrac{1}{4!}\epsilon^{\mu\nu\lambda\rho}\, dx^1 dx^2 dx^3 dx^4. \qquad\qquad (5.56)$$

The left-hand side is a tensor (5.23), $\epsilon^{\mu\nu\lambda\rho}$ is a tensor density of weight $+1$ (5.36), and of course

$$dx'^1 dx'^2 dx'^3 dx'^4 = \frac{\partial(x')}{\partial(x)}\, dx^1 dx^2 dx^3 dx^4, \qquad\qquad (5.57)$$

so $dx^1\,dx^2\,dx^3\,dx^4$ (the ordinary volume element) is a scalar density of weight -1; it all works out. Moreover (from Eq. 5.48 and Problem 5.4(a)),

$$d^4\Sigma = \epsilon_{\mu\nu\lambda\rho} d\sigma^{\mu\nu\lambda\rho} = \tfrac{1}{4!}\left(\epsilon_{\mu\nu\lambda\rho}\epsilon^{\mu\nu\lambda\rho}\right) dx^1\,dx^2\,dx^3\,dx^4 = dx^1\,dx^2\,dx^3\,dx^4 \quad (5.58)$$

is itself a scalar density of weight -1.

Problem 5.6

On a 3-dimensional Euclidean manifold, four totally antisymmetric tensors are possible: A (scalar), A_μ (vector), $A_{\mu\nu}$ (second-rank tensor), and $A_{\mu\nu\lambda}$ (third-rank tensor). For four or more indices, antisymmetrization yields zero (since at least one index is a repeat). Dual to these there are four antisymmetric tensor densities ($\mathsf{a}^{\mu\nu\lambda}, \mathsf{a}^{\mu\nu}, \mathsf{a}^{\mu}$, and a).

(a) Work out $d\sigma^\mu$, $d\sigma^{\mu\nu}$, and $d\sigma^{\mu\nu\lambda}$ (for the latter, use the parameterization $(s_1, s_2, s_3) = (x^1, x^2, x^3)$). Likewise, work out $d^3\Sigma$, $d^2\Sigma_\mu$, and $d^1\Sigma_{\mu\nu}$.

(b) Construct the exterior derivatives for each of the four antisymmetric tensors. Write out Stokes's theorem for $r = 1$, and compare your result with the familiar vector theorem.[9]

(c) Construct the divergences, for each of the tensor densities. Write out Gauss's theorem for $r = 2$, and compare the familiar vector theorem.

That's as far as we can go, in a general manifold. To proceed, we need to add some geometrical structure to the space.

5.2 Affine Spaces

In this section we endow the manifold with a procedure for **parallel transport**[10] of tensors, from one point to another.

5.2.1 Affine Connections

Definition: An **affinely connected manifold** is a manifold, together with a set of numbers $\Gamma^\mu_{\nu\lambda}$ (they may vary from point to point) that provide a rule for transporting vectors:

$$\delta A^\mu = -\Gamma^\mu_{\nu\lambda} A^\nu \, \delta x^\lambda. \tag{5.59}$$

Here $\delta A^\mu \equiv A^\mu(x + \delta x) - A^\mu(x)$ is the change in the component A^μ when we transport the vector from point x to point $x + \delta x$; Γ is called the **affine connection.**[11] This is the most "natural" rule one could ask for—all it says is that the

---

[9] The exterior derivative of a vector, $\partial_i A_j - \partial_j A_i$, is dual to the curl, $(\nabla \times \mathbf{A})$. We normally think of the latter as a vector, but that's because we usually deal only with proper rotations, for which the Jacobian is 1. If we allow *improper* transformations, for which the Jacobian is -1, then the curl transforms as a vector *density* (a pseudovector).

[10] Eds. Coleman calls it "parallel displacement," but parallel transport is the more conventional term.

[11] It is sometimes written as

$$\left\{ {\mu \atop \nu\lambda} \right\} \quad \text{or} \quad \{\nu\lambda, \mu\}.$$

change in A is proportional to A itself, and to the displacement δx (go twice as far, and you'll get twice the change, at least for infinitesimal displacements). The advantage of introducing Γ is that we can now construct derivatives of vector (and tensor) fields—not just the totally antisymmetric exterior derivatives, which are defined for any old manifold, but *ordinary* derivatives. That's because we will now have a way to add (or rather, subtract) vectors at two adjacent points (which live, therefore, in two different tangent spaces): first "transport" one vector over to the other one's tangent space.

5.2.2 How Γ Transforms

Beware: The Γ's are *not* tensors, in spite of appearances.[12] In fact, we can easily deduce the transformation rule for Γ:

$$A^\mu(x+\delta x) = A^\mu + \delta A^\mu = A^\mu - \Gamma^\mu_{\nu\lambda} A^\nu \, \delta x^\lambda. \tag{5.60}$$

This quantity must transform as a vector at $(x + \delta x)$, while A^μ, A^ν, and δx^λ transform as vectors at x. Now, a vector at x transforms as

$$A^\mu \to A'^\mu = \left(\frac{\partial x'^\mu}{\partial x^\nu}\right) A^\nu, \tag{5.61}$$

but at $(x + \delta x)$ the term in parentheses becomes

$$(\,) \to (\,) + \frac{\partial}{\partial x^\lambda}(\,) \, \delta x^\lambda = \left(\frac{\partial x'^\mu}{\partial x^\nu}\right) + \frac{\partial^2 x'^\mu}{\partial x^\nu \partial x^\lambda} \, \delta x^\lambda. \tag{5.62}$$

So

$$\begin{aligned}
(A^\mu + \delta A^\mu)' &= \left[\left(\frac{\partial x'^\mu}{\partial x^\nu}\right) + \frac{\partial^2 x'^\mu}{\partial x^\nu \partial x^\lambda} \, \delta x^\lambda\right](A^\nu + \delta A^\nu) \\
&= \left(\frac{\partial x'^\mu}{\partial x^\nu}\right) A^\nu + \frac{\partial^2 x'^\mu}{\partial x^\nu \partial x^\lambda} \, \delta x^\lambda \, A^\nu - \left(\frac{\partial x'^\mu}{\partial x^\nu}\right) \Gamma^\nu_{\sigma\lambda} A^\sigma \, \delta x^\lambda
\end{aligned} \tag{5.63}$$

(keeping only terms of first order in δx). But in terms of Γ',

$$\begin{aligned}
(A^\mu + \delta A^\mu)' &= A'^\mu + \delta A'^\mu = A'^\mu - (\Gamma')^\mu_{\nu\tau} A'^\nu \delta x'^\tau \\
&= \left(\frac{\partial x'^\mu}{\partial x^\nu}\right) A^\nu - (\Gamma')^\mu_{\nu\tau} \left(\frac{\partial x'^\nu}{\partial x^\sigma} A^\sigma\right) \left(\frac{\partial x'^\tau}{\partial x^\lambda} \delta x^\lambda\right).
\end{aligned} \tag{5.64}$$

---

[12] This is a reason, perhaps, to prefer the notation in footnote 11.

Comparing 5.63 and 5.64, and dropping the arbitrary factor of δx^λ,

$$- (\Gamma')^\mu_{\nu\tau} \left(\frac{\partial x'^\nu}{\partial x^\sigma} \frac{\partial x'^\tau}{\partial x^\lambda} \right) A^\sigma = \frac{\partial^2 x'^\mu}{\partial x^\sigma \partial x^\lambda} A^\sigma - \left(\frac{\partial x'^\mu}{\partial x^\nu} \right) \Gamma^\nu_{\sigma\lambda} A^\sigma. \tag{5.65}$$

This must hold for *any* vector, so we can cancel A^σ:

$$(\Gamma')^\mu_{\nu\tau} \left(\frac{\partial x'^\nu}{\partial x^\sigma} \frac{\partial x'^\tau}{\partial x^\lambda} \right) = \Gamma^\nu_{\sigma\lambda} \left(\frac{\partial x'^\mu}{\partial x^\nu} \right) - \frac{\partial^2 x'^\mu}{\partial x^\sigma \partial x^\lambda}. \tag{5.66}$$

Now multiply by $(\partial x^\sigma / \partial x'^\alpha \, \partial x^\lambda / \partial x'^\beta)$ and sum over σ and λ:

$$(\Gamma')^\mu_{\alpha\beta} = \left(\frac{\partial x'^\mu}{\partial x^\nu} \frac{\partial x^\sigma}{\partial x'^\alpha} \frac{\partial x^\lambda}{\partial x'^\beta} \right) \Gamma^\nu_{\sigma\lambda} - \frac{\partial^2 x'^\mu}{\partial x^\sigma \partial x^\lambda} \left(\frac{\partial x^\sigma}{\partial x'^\alpha} \frac{\partial x^\lambda}{\partial x'^\beta} \right). \tag{5.67}$$

The first term is just the transformation rule for a mixed tensor, but the second term indicates that Γ does *not* transform as a tensor. (In fact, it is possible for Γ to *vanish* in one coordinate system, but not in another—something a tensor could never do.)

The affine connection $\Gamma^\mu_{\nu\lambda}$ is not (in general) symmetric in its lower indices. But we will restrict our attention to two kinds of spaces: (a) curved spaces embedded in flat (Euclidean) spaces of higher dimension and (b) the 4-dimensional spacetime of general relativity. For such spaces $\Gamma^\mu_{\nu\lambda}$ *is* symmetric in $\nu \leftrightarrow \lambda$, and we shall always assume this from now on.[13] Why *should* Γ be symmetric in general relativity? It follows, really, from the principle of equivalence. In completely empty space, with no gravity, special relativity prevails, and we can use Minkowski coordinates. In that case Γ is zero, and no change of coordinates can give it an antisymmetric part. The principle of equivalence says that free fall in a gravitational field is (locally) Minkowskian (indistinguishable from empty space). So an antisymmetric component would violate the equivalence principle.[14]

Problem 5.7

What are the affine connections for the Euclidean plane, if you use Cartesian coordinates (x, y)? That's trivial, of course: the components don't change at all. But what if you use polar coordinates (r, θ)? Find all (eight) Γ's, using Eq. 5.67. Check your results using 5.59.

---

[13] It would make no (coordinate-independent) sense to say that Γ was *anti*symmetric in $\nu \leftrightarrow \lambda$, since a change of coordinates (5.67) would give it a symmetric part. (The second term in 5.67 is symmetric, while the first term inherits the symmetry or antisymmetry of Γ.) The antisymmetric part $(\Gamma^\mu_{\nu\lambda} - \Gamma^\mu_{\lambda\nu})$ is tensorial, and therefore it cannot be eliminated (or created) by a change of coordinates. So an assertion that Γ is *symmetric* is coordinate independent.

[14] For a more detailed argument, see E. Schrödinger, *Space-Time Structure*, Cambridge University Press, London (1963), page 37.

Problem 5.8

Show that the last term on the right-hand side of Eq. 5.67 can be written as

$$+ \frac{\partial x'^\mu}{\partial x^\lambda} \frac{\partial^2 x^\lambda}{\partial x'^\alpha \partial x'^\beta}.$$

Hint: Use the identity

$$\frac{\partial}{\partial x^\sigma} \left(\frac{\partial x'^\mu}{\partial x^\lambda} \frac{\partial x^\lambda}{\partial x'^\beta} \right) = \frac{\partial}{\partial x^\sigma} (\delta^\mu_\beta) = 0.$$

5.2.3 Parallel Transport of Tensors and Tensor Densities

The parallel-transport rule (5.59) generalizes easily to other (contravariant) tensors. For example, a prototype second-rank tensor is $A^{\mu\nu} \equiv A^\mu A^\nu$. In this case

$$\delta(A^{\mu\nu}) = \delta(A^\mu A^\nu) = A^\mu(\delta A^\nu) + (\delta A^\mu)A^\nu$$
$$= A^\mu \left(-\Gamma^\nu_{\lambda\sigma} A^\lambda \, \delta x^\sigma \right) + \left(-\Gamma^\mu_{\lambda\sigma} A^\lambda \, \delta x^\sigma \right) A^\nu, \qquad (5.68)$$

so, in general,

$$\delta(A^{\mu\nu}) = -\Gamma^\nu_{\lambda\sigma} A^{\mu\lambda} \, \delta x^\sigma - \Gamma^\mu_{\lambda\sigma} A^{\lambda\nu} \, \delta x^\sigma. \qquad (5.69)$$

And the same goes for tensors of higher rank: one gamma hangs on each index.

How about *covariant* vectors? Since $B_\mu A^\mu$ is a scalar, we want $\delta(B_\mu A^\mu) = 0$. (Remember, the issue here is parallel transport of a *given* tensor; don't confuse this with the variation from point to point of a tensor *field*.) Therefore

$$(\delta B_\mu)A^\mu + B_\mu(\delta A^\mu) = 0, \quad (\delta B_\mu)A^\mu = B_\mu \Gamma^\mu_{\lambda\sigma} A^\lambda \, \delta x^\sigma. \qquad (5.70)$$

But A is arbitrary, so

$$\delta B_\mu = \Gamma^\nu_{\mu\sigma} B_\nu \, \delta x^\sigma. \qquad (5.71)$$

The generalization to covariant tensors of higher rank runs the same as before.

We define the parallel transport of tensor *densities* in such a way as to preserve the duality between them and tensors (Eq. 5.44). In particular, the elements of ϵ must not change:

$$0 = \delta(\epsilon^{\mu_1 \cdots \mu_n})$$
$$= \left[-\Gamma^{\mu_1}_{\sigma\lambda} \epsilon^{\sigma\mu_2 \cdots \mu_n} \, \delta x^\lambda - \Gamma^{\mu_2}_{\sigma\lambda} \epsilon^{\mu_1 \sigma\mu_3 \cdots \mu_n} \, \delta x^\lambda - \cdots - \Gamma^{\mu_n}_{\sigma\lambda} \epsilon^{\mu_1 \cdots \mu_{n-1}\sigma} \, \delta x^\lambda \right]$$
$$+ \text{(density terms)}. \qquad (5.72)$$

(The expression in square brackets represents the *tensorial* behavior of ϵ; those in parentheses are whatever else is required for parallel transport of a tensor *density*.)

Antisymmetry is preserved by parallel transport (this is true already of the tensorial terms—check $\mu_1 \leftrightarrow \mu_2$); therefore (once again) we need only consider one element:

$$\delta\epsilon^{12\cdots n} = \left[-\Gamma^1_{1\lambda} \, \delta x^\lambda - \Gamma^2_{2\lambda} \, \delta x^\lambda - \cdots - \Gamma^n_{n\lambda} \, \delta x^\lambda \right] + \text{(density terms)}. \qquad (5.73)$$

Thus, in general,

$$0 = \delta\epsilon^{\mu_1 \cdots \mu_n} = -\Gamma^\sigma_{\sigma\lambda} \epsilon^{\mu_1 \cdots \mu_n} \, \delta x^\lambda + \text{(density terms)}. \qquad (5.74)$$

Evidently

$$\text{(density terms)} = \Gamma^\sigma_{\sigma\lambda} \epsilon^{\mu_1 \cdots \mu_n} \, \delta x^\lambda. \qquad (5.75)$$

By the same argument, the density terms for $\delta\epsilon_{\mu_1 \cdots \mu_n}$ would be

$$-\Gamma^\sigma_{\sigma\lambda} \epsilon_{\mu_1 \cdots \mu_n} \, \delta x^\lambda \qquad (5.76)$$

(opposite sign because covariant—compare Eqs. 5.59 and 5.71). Skipping right to the general case, for any tensor density, the parallel-transport rule is

$$\delta\mathfrak{a}^{\mu_1 \cdots}_{\nu_1 \cdots} = \text{(tensor terms)} + w\Gamma^\sigma_{\lambda\sigma} \, \mathfrak{a}^{\mu_1 \cdots}_{\nu_1 \cdots} \, \delta x^\lambda. \qquad (5.77)$$

5.2.4 Covariant Derivatives

We now have the structure needed to compare tensors at different points, and hence to give a coordinate-independent meaning to "the derivative of a tensor." Recall that $A^\mu(x + \delta x) - A^\mu(x)$ has no invariant meaning (it's not a tensor), so we can't define a derivative in the obvious way. But if we first parallel transport the vector $A^\mu(x)$ up to the point $x + \delta x$, and *then* do the subtraction, the result *is* tensorial:

$$A^\mu(x + \delta x) - \left[A^\mu(x) - \Gamma^\mu_{\nu\lambda} A^\nu(x) \, \delta x^\lambda \right] = \partial_\nu A^\mu(x) \, \delta x^\nu + \Gamma^\mu_{\nu\lambda} A^\nu(x) \, \delta x^\lambda$$
$$= \partial_\nu A^\mu \, \delta x^\nu + \Gamma^\mu_{\nu\lambda} A^\lambda \, \delta x^\nu = \nabla_\nu A^\mu \, \delta x^\nu, \qquad (5.78)$$

where we define the **covariant derivative**[15] as

$$\boxed{\nabla_\nu A^\mu \equiv \partial_\nu A^\mu + \Gamma^\mu_{\nu\lambda} A^\lambda.} \qquad (5.79)$$

The expression on the left *is* tensorial, even though neither term on the right is. Note that the partial derivative ($\partial_\nu = \partial/\partial x^\nu$) is defined in the usual way:[16]

---

[15] This terminology is dangerous, but inescapable: $\partial_\mu \phi$ is a covariant derivative (as opposed to "contravariant" $\partial^\mu \phi$); $\nabla_\mu A^\nu$ is a covariant derivative (as opposed to "partial," $\partial_\mu A^\nu$). Usually, the context will tell you which meaning is intended.

[16] Some authors introduce a shorthand notation for ordinary and covariant derivatives, using a comma (or a |) and a semicolon (or a ||), respectively:

$$\partial_\nu A_\mu \equiv A_{\mu,\nu} = A_{\mu|\nu}, \quad \nabla_\nu A_\mu \equiv A_{\mu;\nu} = A_{\mu||\nu}.$$

Differential Geometry

$$\partial_\nu A^\mu(x)\, \delta x^\nu = A^\mu(x + \delta x) - A^\mu(x). \tag{5.80}$$

Similarly, for covariant vectors and tensors of higher rank,

$$\nabla_\nu A_\mu = \partial_\nu A_\mu - \Gamma^\lambda_{\nu\mu} A_\lambda, \tag{5.81}$$

$$\nabla_\nu A^{\mu\sigma} = \partial_\nu A^{\mu\sigma} + \Gamma^\mu_{\nu\lambda} A^{\lambda\sigma} + \Gamma^\sigma_{\nu\lambda} A^{\mu\lambda}. \tag{5.82}$$

Incidentally, the covariant derivative of a tensor *density* is[17]

$$\nabla_\nu \mathsf{T}^{\cdots}_{\cdots} = (\text{same as for the tensor}) - w\Gamma^\lambda_{\nu\lambda} \mathsf{T}^{\cdots}_{\cdots}. \tag{5.83}$$

Problem 5.9

(a) Check explicitly that $\nabla_\nu A^\mu$ transforms as a tensor.
(b) Show that $\nabla_\mu A_\nu - \nabla_\nu A_\mu = (dA)_{\mu\nu}$, confirming that the exterior derivative (Example 5.1, part 2) is tensorial.
(c) Show that $\nabla_\mu(A^\nu B^\lambda) = (\nabla_\mu A^\nu)B^\lambda + A^\nu(\nabla_\mu B^\lambda)$. (More generally, the product rule holds for all covariant derivatives.)
(d) Work out the expression (analogous to 5.81 and 5.82) for $\nabla_\nu(A^\lambda{}_{\mu\sigma})$.

Notice that *covariant derivatives do not commute*—there is no equality of covariant cross derivatives. In fact, what *is*

$$\nabla_\lambda \nabla_\sigma A^\mu - \nabla_\sigma \nabla_\lambda A^\mu ? \tag{5.84}$$

Let's first calculate $\nabla_\lambda \nabla_\sigma A^\mu$, and then antisymmetrize (I'll cancel symmetric terms as we go along):

$$\nabla_\lambda (\nabla_\sigma A^\mu) = \partial_\lambda (\nabla_\sigma A^\mu) + \Gamma^\mu_{\lambda\rho} (\nabla_\sigma A^\rho) - \cancel{\Gamma^\rho_{\lambda\sigma} (\nabla_\rho A^\mu)} \tag{5.85}$$

$$= \partial_\lambda \left(\partial_\sigma A^\mu + \Gamma^\mu_{\sigma\tau} A^\tau \right) + \Gamma^\mu_{\lambda\rho} \left(\partial_\sigma A^\rho + \Gamma^\rho_{\sigma\tau} A^\tau \right)$$

$$= \cancel{\partial_\lambda \partial_\sigma A^\mu} + \left(\partial_\lambda \Gamma^\mu_{\sigma\tau} \right) A^\tau + \underline{\Gamma^\mu_{\sigma\tau} (\partial_\lambda A^\tau)} + \underline{\Gamma^\mu_{\lambda\rho} (\partial_\sigma A^\rho)} + \Gamma^\mu_{\lambda\rho} \Gamma^\rho_{\sigma\tau} A^\tau.$$

(Of course, the "ordinary" partial derivatives still commute: $\partial_\lambda \partial_\sigma = \partial_\sigma \partial_\lambda$.) Antisymmetrizing in $\sigma \leftrightarrow \lambda$ (note that the result involves derivatives of Γ, but not of A),

$$\nabla_\lambda \nabla_\sigma A^\mu - \nabla_\sigma \nabla_\lambda A^\mu = -R^\mu{}_{\tau\lambda\sigma} A^\tau, \tag{5.86}$$

where

$$\boxed{R^\mu{}_{\tau\lambda\sigma} \equiv -\partial_\lambda \Gamma^\mu_{\sigma\tau} - \Gamma^\mu_{\lambda\rho} \Gamma^\rho_{\sigma\tau} + \partial_\sigma \Gamma^\mu_{\lambda\tau} + \Gamma^\mu_{\sigma\rho} \Gamma^\rho_{\lambda\tau}} \tag{5.87}$$

---

[17] See, for example, Schrödinger, page 33 (details in footnote 14), and Problem 6.8 below.

is the famous **Riemann tensor**.[18] In spite of the partials and the gammas, it is a (fourth-rank) tensor, as you can tell from Eq. 5.86.

Since Γ is zero in Euclidean space with Cartesian coordinates, so too is R, and hence (since it is tensorial), $R^{\mu}{}_{\tau\lambda\sigma}$ is zero in any flat space, regardless of the coordinates used. As we shall see, the Riemann tensor is a coordinate-independent characterization of curvature—it measures the departure from flatness. If Γ is *symmetric* it is always possible to choose **locally geodesic** coordinates, in which $\Gamma = 0$, but evidently it is *not* possible (in general) to choose coordinates in which *both Γ and* its first derivatives vanish, for then the Riemann tensor would also vanish. In locally geodesic coordinates,

$$R^{\mu}{}_{\tau\lambda\sigma} = -\partial_\lambda \Gamma^{\mu}_{\sigma\tau} + \partial_\sigma \Gamma^{\mu}_{\lambda\tau}. \tag{5.88}$$

In the context of general relativity the (strong) equivalence principle guarantees the existence of a reference frame (the "elevator," or "free-fall" frame) that is locally indistinguishable from empty space with no gravity (which is to say, it is locally Minkowskian). This is implemented by assuming the affine connection is symmetric; in relativity, *elevator coordinates are locally geodesic coordinates.*

The Riemann tensor is *antisymmetric* in its last two indices (again, this is clear from 5.86):

$$R^{\mu}{}_{\tau\lambda\sigma} = -R^{\mu}{}_{\tau\sigma\lambda}. \tag{5.89}$$

Less obviously, the cyclic sum over the three lower indices is zero:

$$R^{\mu}{}_{\tau\lambda\sigma} + R^{\mu}{}_{\lambda\sigma\tau} + R^{\mu}{}_{\sigma\tau\lambda} = 0. \tag{5.90}$$

Proof: Since this is a tensorial statement, we need only verify it for one coordinate system; choose locally geodesic coordinates, for which Γ vanishes. Then

$$R^{\mu}{}_{\tau\lambda\sigma} + R^{\mu}{}_{\lambda\sigma\tau} + R^{\mu}{}_{\sigma\tau\lambda} \tag{5.91}$$
$$= -\partial_\lambda \Gamma^{\mu}_{\sigma\tau} + \partial_\sigma \Gamma^{\mu}_{\lambda\tau} - \partial_\sigma \Gamma^{\mu}_{\tau\lambda} + \partial_\tau \Gamma^{\mu}_{\sigma\lambda} - \partial_\tau \Gamma^{\mu}_{\lambda\sigma} + \partial_\lambda \Gamma^{\mu}_{\tau\sigma} = 0. \quad \text{QED}$$

5.3 Riemannian Manifolds

We now introduce a measure of *distance* on a manifold. Again, we are interested in two kinds of spaces: (a) curved spaces embedded in flat (Euclidean) spaces of higher dimension, and (b) the 4-dimensional spacetime of general relativity.[19]

---

[18] *Warning:* There are different sign conventions, and index orderings, in the literature.
[19] In case (a) the measure of length is inherited from the larger embedding space, as applied to paths in the curved subspace; in case (b) we use clocks and meter sticks, as discussed in Section 4.3.

Definition: A **Riemannian manifold** is an affine space equipped as well with a symmetric **metric tensor** $g_{\mu\nu}$.[20] The "length" of a displacement dx^μ is given by

$$\boxed{\text{length}^2 \equiv g_{\mu\nu}\, dx^\mu\, dx^\nu.} \tag{5.92}$$

By extension, the "length" of any (contravariant) vector (A^μ) is given by

$$\text{length}^2 \equiv g_{\mu\nu}\, A^\mu A^\nu. \tag{5.93}$$

Note that length squared need not be positive.[21] We also introduce the (matrix) inverse of the metric:

$$\boxed{g^{\lambda\rho} g_{\rho\mu} = \delta^\lambda_\mu.} \tag{5.94}$$

This defines the *contravariant* metric, in terms of the covariant one (unlike in Minkowski space, they are not—in general—numerically equal). This is the first time we have associated a particular contravariant tensor with a particular covariant one (with the exception of the tensor densities $\epsilon^{\mu_1\cdots\mu_n}$ and $\epsilon_{\mu_1\cdots\mu_n}$). But now, armed with the metric tensors, we can raise and lower indices more generally:

$$A_\rho = g_{\rho\lambda} A^\lambda, \quad A_{\rho\tau} = g_{\rho\lambda} g_{\tau\sigma} A^{\lambda\sigma}, \quad A^\rho = g^{\rho\lambda} A_\lambda, \quad A^\mu{}_\nu = g^{\mu\lambda} A_{\lambda\nu}, \tag{5.95}$$

and so on. This defines the covariant quantity we associate with each contravariant quantity, and *vice versa*.

What happens when we try to raise and lower the indices on a tensor *density*? What, for example, is

$$\epsilon^{\mu_1\mu_2\cdots\mu_n} g_{\mu_1\nu_1} g_{\mu_2\nu_2} \cdots g_{\mu_n\nu_n}? \tag{5.96}$$

It is totally antisymmetric, and covariant, so it must be some multiple of $\epsilon_{\nu_1\cdots\nu_n}$, but what is the coefficient? As always, we need only examine one component, so let $\nu_1 = 1, \nu_2 = 2, \ldots, \nu_n = n$:

$$\epsilon^{\mu_1\cdots\mu_n} g_{\mu_1 1} g_{\mu_2 2} \cdots g_{\mu_n n} = \det(g_{\mu\nu}) \equiv \mathsf{g}; \tag{5.97}$$

it's the *determinant* of the matrix $g_{\mu\nu}$, which we denote by g.[22] Thus

$$\epsilon^{\mu_1\mu_2\cdots\mu_n} g_{\mu_1\nu_1} g_{\mu_2\nu_2} \cdots g_{\mu_n\nu_n} = \mathsf{g}\, \epsilon_{\nu_1\nu_2\cdots\nu_n}. \tag{5.98}$$

Now, $\epsilon^{\mu_1\cdots\mu_n}$ is a tensor density of weight 1, the metric tensors both carry weight 0, and $\epsilon_{\nu_1\cdots\nu_n}$ has weight -1, so

---

[20] Because it is symmetric, $g_{\mu\nu}$ has $n(n+1)/2$ independent components, if n is the dimension of the manifold. Because length should be a function of path only, not the coordinates used, $g_{\mu\nu}$ must be a (covariant) tensor.

[21] In case (a) it is positive definite—all the eigenvalues of the *matrix* $g_{\mu\nu}$ are positive; in case (b) three of the eigenvalues are negative, one is positive, and the sign can be positive, negative, or zero.

[22] Note that g is positive for case (a) and negative for case (b).

$$\mathsf{g} \text{ is a scalar density of weight 2.} \qquad (5.99)$$

For future reference, in general relativity $\sqrt{-\mathsf{g}}$ is a (real) scalar density of weight 1, and hence, if $\mathcal{L}$ is a scalar,

$$\int \mathcal{L} \sqrt{-\mathsf{g}} \, d^4\Sigma \qquad (5.100)$$

(where $d^4\Sigma = dx^0 \, dx^1 \, dx^2 \, dx^3$ has weight -1—see Example 5.4) is a scalar, and therefore has the appropriate form for an **action**.

5.3.1 Relation between Affine Connection and Metric

In either of the cases of interest, (a) and (b), we want parallel transport via $\Gamma^\mu_{\nu\lambda}$ to preserve lengths. Suppose we define the "local vector field" generated by a vector A^μ (residing in the tangent space at point x), when we transport it around in the neighborhood of x:

$$A^\mu(x + \delta x) = A^\mu(x) - \Gamma^\mu_{\nu\lambda}(x)A^\nu(x)\delta x^\lambda \quad \Rightarrow \quad \partial_\lambda A^\mu = -\Gamma^\mu_{\nu\lambda} A^\nu, \qquad (5.101)$$

so

$$\partial_\lambda A^\mu + \Gamma^\mu_{\nu\lambda} A^\nu = 0, \quad \text{or} \quad \nabla_\lambda A^\mu = 0. \qquad (5.102)$$

We demand that $\nabla_\lambda(g_{\mu\nu} A^\mu A^\nu) = 0$:

$$(\nabla_\lambda g_{\mu\nu})A^\mu A^\nu + g_{\mu\nu}(\nabla_\lambda A^\mu)A^\nu + g_{\mu\nu}A^\mu(\nabla_\lambda A^\nu) = 0, \qquad (5.103)$$

and hence[23]

$$\nabla_\lambda g_{\mu\nu} = 0. \qquad (5.104)$$

Writing it out (see Eqs. 5.81 and 5.82),

$$\nabla_\lambda g_{\mu\nu} = \partial_\lambda g_{\mu\nu} - \Gamma^\sigma_{\lambda\mu} g_{\sigma\nu} - \Gamma^\sigma_{\nu\lambda} g_{\mu\sigma} = 0. \qquad (5.105)$$

We now introduce the **Christoffel symbol of the first kind**:[24]

$$\Gamma_{\nu,\lambda\mu} \equiv \Gamma^\sigma_{\lambda\mu} g_{\sigma\nu}. \qquad (5.106)$$

(When a distinction must be made, we call $\Gamma^\mu_{\nu\lambda}$ the **Christoffel symbol of the second kind**.) Thus

$$\partial_\lambda g_{\mu\nu} = \Gamma_{\nu,\lambda\mu} + \Gamma_{\mu,\nu\lambda}. \qquad (5.107)$$

---

[23] This amounts to a fourth axiom about the structure of the space: it's (i) a manifold, with (ii) an affine connection, and (iii) a metric, (iv) such that (ii) and (iii) are related by $\nabla_\lambda(g_{\mu\nu}) = 0$.

[24] In the spirit of footnote 11 some authors write it as $[\lambda\mu, \nu]$.

We want to solve this for Γ. Cyclically permuting the indices,

$$\partial_\mu g_{\nu\lambda} = \Gamma_{\lambda,\mu\nu} + \Gamma_{\nu,\lambda\mu}, \quad \partial_\nu g_{\lambda\mu} = \Gamma_{\mu,\nu\lambda} + \Gamma_{\lambda,\mu\nu}. \tag{5.108}$$

Add these three equations, with the signs indicated:

$$\partial_\lambda g_{\mu\nu} - \partial_\mu g_{\nu\lambda} - \partial_\nu g_{\lambda\mu} = -2\Gamma_{\lambda,\mu\nu}, \tag{5.109}$$

or

$$\Gamma_{\lambda,\mu\nu} = \tfrac{1}{2}\left[\partial_\mu g_{\nu\lambda} + \partial_\nu g_{\lambda\mu} - \partial_\lambda g_{\mu\nu}\right]. \tag{5.110}$$

(*Mnemonic:* All indices are cyclic permutations of those on the left; the minus sign goes with the term whose indices match those of Γ.) To recover the Christoffel symbol of the second kind, note that

$$\Gamma^\lambda_{\mu\nu} = g^{\lambda\rho}\Gamma_{\rho,\mu\nu}, \tag{5.111}$$

so

$$\boxed{\Gamma^\lambda_{\mu\nu} = \tfrac{1}{2}g^{\lambda\rho}\left[\partial_\mu g_{\nu\rho} + \partial_\nu g_{\rho\mu} - \partial_\rho g_{\mu\nu}\right].} \tag{5.112}$$

5.3.2 Symmetries of the Riemann Tensor

Now that we are able to lower indices, we can construct a fully covariant tensor $R_{\mu\nu\lambda\sigma}$, associated with $R^\mu{}_{\nu\lambda\sigma}$.[25] Recall (Eq. 5.89) that $R^\mu{}_{\nu\lambda\sigma}$ is antisymmetric in $\lambda \leftrightarrow \sigma$; obviously, $R_{\mu\nu\lambda\sigma}$ inherits that property. It also inherits the cyclic symmetry (5.90) in the last three indices:

$$R_{\mu\nu\lambda\sigma} + R_{\mu\lambda\sigma\nu} + R_{\mu\sigma\nu\lambda} = 0. \tag{5.113}$$

But the fully covariant Riemann tensor has an additional symmetry *not* directly apparent in $R^\mu{}_{\nu\lambda\sigma}$: it is symmetric under interchange of the first *two* indices and the last two, $\mu\nu \leftrightarrow \lambda\sigma$:

$$R_{\mu\nu\lambda\sigma} = R_{\lambda\sigma\mu\nu} \tag{5.114}$$

(and hence, since it is antisymmetric in the last two, it must also be antisymmetric in the first two). This is in fact the *complete* list of symmetries possessed by $R_{\mu\nu\lambda\sigma}$ (in general).

---

[25] As soon as raising and lowering becomes possible, the position of upper *vis à vis* lower becomes significant. We must adopt a convention for each tensor, specifying which lower position each upper index moves into. I have been writing the Riemann tensor with a gap in the first lower slot, anticipating that μ drops into that position. But this is just a convention—of the four covariant tensors we *could* associate with the Riemann tensor, we have to select one, and stick with it.

Problem 5.10

Confirm explicitly that $R_{\mu\nu\lambda\sigma}$ has each of the symmetries listed. [*Hint:* Use locally geodesic coordinates, in which Γ (but not its derivatives!) vanishes, and (Eq. 5.107) $\partial_\lambda g_{\mu\nu} = 0$ (but not its higher derivatives!). Write out $R_{\mu\nu\lambda\sigma}$ in these coordinates.]

Problem 5.11

Using the tensor transformation law for $g_{\mu\nu}$,

$$g'_{\alpha\beta} = \frac{\partial x^\mu}{\partial x'^\alpha} \frac{\partial x^\nu}{\partial x'^\beta} g_{\mu\nu},$$

confirm that the Christoffel symbol 5.112 satisfies Eq. 5.67.

How many independent components does the Riemann tensor carry? Let's count them.

1. Suppose all four indices are the same (e.g. $\mu\nu\lambda\sigma = 1111$): antisymmetry (in, say, the first two) means there are *no* (nonzero) elements of this form.
2. Three indices are the same (e.g. 1112): antisymmetry $\Rightarrow$ no such terms.
3. Two indices are the same:

   (a) other two also the same:

      (i) (e.g. 1122, i.e. $\mu\mu\nu\nu$ with $\mu \neq \nu$): antisymmetry $\Rightarrow$ 0 terms;
      (ii) (e.g. 1212, i.e. $\mu\nu\mu\nu$ with $\mu \neq \nu$; this also determines $\nu\mu\nu\mu$, $\mu\nu\nu\mu$, and $\nu\mu\mu\nu$): there are n ways to pick μ, and then $n - 1$ ways to pick ν, but which is which doesn't matter, so

$$\left[\frac{n(n - 1)}{2} \right] \quad \text{terms;} \tag{5.115}$$

   (b) other two different (e.g. 1213; this determines all the other orderings): there are n ways to choose the pair, $n - 1$ ways to choose the second, and $n - 2$ ways to choose the third, but the order of the latter two doesn't matter, so

$$\left[\frac{n(n - 1)(n - 2)}{2} \right] \quad \text{terms.} \tag{5.116}$$

4. All four are different: (e.g. 1234; we can always put the 1 first, but 1234, 1324, 1423 are distinct)

$$\left[3\frac{n(n - 1)(n - 2)(n - 3)}{4!} \right] \quad \text{terms.} \tag{5.117}$$

These are all the distinct components consistent with the *symmetries* of R. What about the cyclic sums (5.113)?

- Type 3(a)ii: $R_{1212} + R_{1122} + R_{1221} = 0 \Rightarrow R_{1212} + 0 - R_{1212} = 0$, so no new restriction.
- Type 3(b): $R_{1213} + R_{1132} + R_{1321} = 0 \Rightarrow R_{1213} + 0 - R_{1213} = 0$, so no new restriction.
- Type 4: $R_{1234} + R_{1342} + R_{1423} = 0 \Rightarrow R_{1423} = -R_{1234} + R_{1324}$, so the three combinations are *not* independent, and instead of the factor of 3 in 5.117 we should have a factor of 2.

The total number of independent components in the Riemann tensor is therefore

$$\frac{n(n-1)}{2} + \frac{n(n-1)(n-2)}{2} + 2\frac{n(n-1)(n-2)(n-3)}{(4)(3)(2)} = \frac{n^2(n^2-1)}{12}.$$

(5.118)

The following table lists the number of independent components in a space of n dimensions:

n	components
1	0
2	1
3	6
4	20
5	50
$\vdots$	$\vdots$

Contraction of $R^{\lambda}{}_{\mu\kappa\nu}$ over the first and third[26] indices yields a symmetric second-rank tensor

$$R_{\mu\nu} \equiv R^{\lambda}{}_{\mu\lambda\nu} = g^{\lambda\kappa} R_{\kappa\mu\lambda\nu}$$

(5.119)

known as the **Ricci tensor**. A further contraction produces the **scalar curvature**,

$$R \equiv g^{\mu\nu} R_{\mu\nu},$$

(5.120)

also known as the **Ricci scalar**. In a 2-dimensional Riemann space, R tells you all there is to know (since the full Riemann tensor has only one independent

---

[26] Eds. Some authors prefer to contract the first and *last* indices, which switches the sign of the Ricci tensor (contraction of the first and *second* would give zero, of course). In the first iteration of the course Coleman contracted the first and last indices, but he had defined the Riemann tensor with the opposite sign (in 5.86 and 5.87), so the Ricci tensor came out the same. In the later version he adopted the conventions we use here.

component). Similarly, in a 3-dimensional space $R_{\mu\nu}$ tells all (it has six components, and so does the full tensor). So 4-dimensional spaces are the simplest ones that require the entire machinery of the Riemann tensor to represent their geometry. In this sense there are no nontrivial analogs to general relativity in fewer than four dimensions.

Problem 5.12

Find $g_{\mu\nu}$, $g^{\mu\nu}$, $\Gamma^{\mu}_{\nu\lambda}$, $R^{\mu}_{\ \nu\lambda\sigma}$, $R_{\mu\nu}$, and R on the surface of a sphere (radius r, centered at the origin, in Euclidean 3-space), using spherical coordinates ($x^1 = r\theta$, $x^2 = r\phi$; the factor of r is just to give the coordinates the dimensions of length).

Problem 5.13

Show that the Ricci tensor is symmetric.

Problem 5.14

(a) Show that in a 2-dimensional space the (fully covariant) Riemann tensor can be expressed in terms of the scalar curvature:

$$R_{\mu\nu\lambda\sigma} = \frac{R}{2}(g_{\mu\lambda}g_{\nu\sigma} - g_{\nu\lambda}g_{\mu\sigma}). \tag{5.121}$$

(b) In the same spirit, express the Riemann tensor in a 3-dimensional space in terms of the Ricci tensor and the scalar curvature.

Problem 5.15

(a) Prove the **Bianchi identities**:

$$\nabla_{\rho}R^{\sigma}_{\ \lambda\mu\nu} + \nabla_{\mu}R^{\sigma}_{\ \lambda\nu\rho} + \nabla_{\nu}R^{\sigma}_{\ \lambda\rho\mu} = 0. \tag{5.122}$$

(b) Use the Bianchi identities to find a relation between $\nabla_{\sigma}R^{\sigma}_{\ \rho}$ and $\nabla_{\rho}R$.

5.3.3 Flatness and Curvature

Suppose we transport a contravariant vector A around a closed loop (in a simply connected region). The resulting vector A' will *not* (in general) be the same as A:

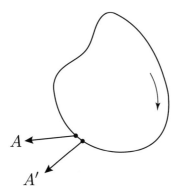

The total change in A is

$$\Delta A^{\mu} \equiv A'^{\mu} - A^{\mu} = \oint dA^{\mu} = -\oint \Gamma^{\mu}_{\nu\lambda} A^{\nu} dx^{\lambda}. \tag{5.123}$$

To calculate ΔA^{μ}, chop the area enclosed by the loop into infinitesimal "rectangles"; the net change in A is the sum of the changes around all of these tiny rectangles (shared sides of adjacent rectangles cancel), so it is enough to treat one of them:

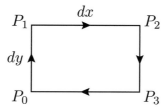

Start at P_0 (using elevator coordinates there, so that $\Gamma(P_0) = 0$):

$$P_0 \rightarrow P_1 : \quad \delta A^{\mu} = 0, \tag{5.124}$$
$$P_1 \rightarrow P_2 : \quad \delta A^{\mu} = -\Gamma^{\mu}_{\nu\lambda}(P_1) A^{\nu}(P_1) dx^{\lambda} = -\left[\partial_{\sigma}(\Gamma^{\mu}_{\nu\lambda}) dy^{\sigma}\right] A^{\nu} dx^{\lambda}, \tag{5.125}$$
$$P_0 \rightarrow P_3 : \quad \delta A^{\mu} = 0, \tag{5.126}$$
$$P_3 \rightarrow P_2 : \quad \delta A^{\mu} = -\Gamma^{\mu}_{\nu\sigma}(P_3) A^{\nu}(P_3) dy^{\sigma} = -\left[\partial_{\lambda}(\Gamma^{\mu}_{\nu\sigma}) dx^{\lambda}\right] A^{\nu} dy^{\sigma}. \tag{5.127}$$

For the whole trip ($P_0 \rightarrow P_1 \rightarrow P_2 \rightarrow P_3 \rightarrow P_0$), then,

$$\delta A^{\mu} = -\left(\partial_{\sigma}\Gamma^{\mu}_{\nu\lambda} - \partial_{\lambda}\Gamma^{\mu}_{\nu\sigma}\right) A^{\nu} dx^{\lambda} dy^{\sigma} = -R^{\mu}_{\nu\lambda\sigma} A^{\nu} dx^{\lambda} dy^{\sigma}. \tag{5.128}$$

(I used 5.88 in the last step.) Because the Riemann tensor is antisymmetric in its last two indices, we can replace $dx^{\lambda} dy^{\sigma}$ by $d\sigma^{\lambda\sigma}$ (Eq. 5.22); moreover, since this is now a tensorial statement, it is valid in any coordinates. Adding up all the infinitesimal rectangles, I conclude that[27]

$$\boxed{\Delta A^{\mu} = - \int R^{\mu}{}_{\nu\lambda\sigma} A^{\nu} d\sigma^{\lambda\sigma}.} \qquad (5.129)$$

The Riemann tensor tells us how different the transported vector is from the original. In a flat space the two would be identical; in this sense the Riemann tensor measures the departure from flatness.

Flatness (and its opposite, **curvature**) are potentially misleading terms. In the first place, we are talking about the *intrinsic* geometry of the space, not its structure as viewed from some larger space in which it may be embedded. Intrinsic geometry is not affected by bending (as opposed to stretching or pinching). For example, a patch of surface on a cylinder or a cone is intrinsically flat: if the cylinder is made out of paper you could cut out the patch and it would lie flat. People living in the surface would have no way of distinguishing it (locally) from a patch of plane (of course, if they went *all the way around* and found themselves back where they started, they would know something was fishy, but that's another matter). By contrast, a sphere is *intrinsically* curved: cut out a patch and you will *not* be able to make it lie flat (as cartographers discovered long ago). And every 1-dimensional manifold is (intrinsically) flat (you can always unbend an arc and make it straight).

A Riemannian manifold (or a region within it) is said to be **flat** if there exist coordinates such that the metric is constant:

$$\partial_{\lambda} g_{\mu\nu} = 0 \text{ (throughout a neighborhood } \mathcal{R}) \iff \text{ the space is flat in } \mathcal{R}. \qquad (5.130)$$

Of course, it is *always* the case that $\nabla_{\lambda} g_{\mu\nu} = 0$ (Eq. 5.104). But if $\partial_{\lambda} g_{\mu\nu} = 0$ then $\Gamma = 0$ (5.112). *Note:* You aren't allowed to change the coordinates used from point to point, in the definition 5.130. For instance, a sphere is *not* flat (obviously), yet for points on the equator $\partial_{\lambda} g_{\mu\nu}$ is zero, and Γ *does* vanish. If we were permitted to redefine our coordinates for each point, we could set *every* point on an "equator" and conclude that the sphere is flat. But there exists *no* coordinate system on the sphere such that $\partial_{\lambda} g_{\mu\nu} = 0$ at *every* point in some finite neighborhood, and therefore the sphere is *not* flat. To be sure, the definition allows flat *regions* within curved spaces, but the equator would not count—it's a *line*, not an entire neighborhood, and while $\Gamma = 0$ there, its *derivatives* do *not* vanish, and $R^{\mu}{}_{\nu\lambda\sigma} \neq 0$. Flatness *in a region* requires that $\partial_{\lambda} g_{\mu\nu} = 0$ throughout the region, so $\Gamma = 0$ there, but also $\partial\Gamma = 0$, and hence

$$R^{\mu}{}_{\lambda\sigma\nu} = 0 \quad \text{in a flat region.} \qquad (5.131)$$

---

[27] This whole argument amounts to an application of Stokes's theorem (though the integrand in 5.123 is nontensorial).

Thus a *necessary* condition for flatness is that the Riemann tensor vanishes. I will now show that this is also *sufficient*.

> **Theorem:** *Let $\mathcal{R}$ be a simply connected region in a Riemannian manifold, and let $R^{\mu}{}_{\nu\lambda\sigma} = 0$ throughout $\mathcal{R}$. Then $\mathcal{R}$ is flat.*

Proof: If $R^{\mu}{}_{\nu\lambda\sigma} = 0$ we can define *distant* parallel transport unambiguously (that is, independent of the path taken from point A to point B). Equation 5.129 says that the change in A^{μ} around a *closed loop* is zero; it follows that the change from A to B is the same for any two paths connecting them:

$$\Delta A^{\mu}_{\text{loop}} = \Delta A^{\mu}_p - \Delta A^{\mu}_{p'} = 0 \quad \Rightarrow \quad \Delta A^{\mu}_p = \Delta A^{\mu}_{p'}. \tag{5.132}$$

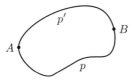

Let the manifold have dimension n, and let e^{α}_{μ} be n linearly independent covariant vectors, at some point in $\mathcal{R}$. (*Note:* α is an index running from 1 to n; it is *not* a contravariant vector index.) By distant parallel transport we continue the e^{α}_{μ} to cover the entire region $\mathcal{R}$ (that is, form the vector fields of the parallel-transported vectors). By construction, the *actual* change in e^{α}_{μ} in going from x to $x + \delta x$ is equal to the change under parallel transport (5.71):

$$\frac{\partial e^{\alpha}_{\mu}}{\partial x^{\sigma}} = \frac{\delta(e^{\alpha}_{\mu})}{\delta x^{\sigma}} = \Gamma^{\nu}_{\mu\sigma} e^{\alpha}_{\nu}, \tag{5.133}$$

and hence

$$\partial_{\lambda} e^{\alpha}_{\mu} - \partial_{\mu} e^{\alpha}_{\lambda} = \Gamma^{\sigma}_{\mu\lambda} e^{\alpha}_{\sigma} - \Gamma^{\sigma}_{\lambda\mu} e^{\alpha}_{\sigma} = 0. \tag{5.134}$$

Since the curl of each vector field is zero, we can write e^{α}_{μ} as a gradient:

$$e^{\alpha}_{\mu} = \partial_{\mu} y^{\alpha}, \tag{5.135}$$

for n scalar functions y^{α}. In terms of y^{α}, Eq. 5.133 becomes

$$\frac{\partial^2 y^{\alpha}}{\partial x^{\sigma}\, \partial x^{\mu}} = \Gamma^{\nu}_{\mu\sigma} \frac{\partial y^{\alpha}}{\partial x^{\nu}}. \tag{5.136}$$

We'll take y^{α} as our new coordinates; i.e. label each point by the n values of y^{α}. *Question:* What are the Christoffel symbols, Γ', in these new coordinates? According to Eq. 5.67,

$$(\Gamma')^{\mu}_{\nu\lambda} = \left(\frac{\partial y^{\mu}}{\partial x^{\tau}} \frac{\partial x^{\sigma}}{\partial y^{\nu}} \frac{\partial x^{\phi}}{\partial y^{\lambda}} \right) \Gamma^{\tau}_{\sigma\phi} - \frac{\partial^2 y^{\mu}}{\partial x^{\sigma} \partial x^{\phi}} \left(\frac{\partial x^{\sigma}}{\partial y^{\nu}} \frac{\partial x^{\phi}}{\partial y^{\lambda}} \right). \tag{5.137}$$

But in view of 5.136 the second term is

$$-\Gamma^\tau_{\phi\sigma} \frac{\partial y^\mu}{\partial x^\tau} \left(\frac{\partial x^\sigma}{\partial y^\nu} \frac{\partial x^\phi}{\partial y^\lambda} \right), \tag{5.138}$$

and the two terms cancel. In this system, then, $\Gamma^\mu_{\nu\lambda} = 0$. So there exists a set of coordinates such that Γ vanishes *throughout* the region $\mathcal{R}$. (There *always* exists a system for which $\Gamma = 0$ at *any given point*—to wit, locally geodesic coordinates—but only when $R^\mu{}_{\nu\lambda\sigma} = 0$ *everywhere* in $\mathcal{R}$ can we be sure that Γ vanishes in the whole *region*.) Since $\Gamma = 0$, we conclude (5.107) that $\partial_\lambda g_{\mu\nu} = 0$ in these coordinates, and hence that the region $\mathcal{R}$ is flat.   QED

A region is flat, then, if and only if $R^\mu{}_{\nu\lambda\sigma} = 0$ throughout; the Riemann tensor is a measure of the departure from flatness—which is to say, it is a measure of **curvature**.

Problem 5.16

Let a vector on the 2-dimensional surface of a unit sphere in Euclidean 3-space be parallel transported around a closed path which does not cross itself. Show that when the vector returns to its starting point it has been rotated through an angle equal to the solid angle subtended by the portion of the sphere enclosed by the path. [If you seem to be getting *zero* rotation, try expanding 5.59 to second order, which you can do by using the fact that parallel transport preserves the length of a vector.]

Problem 5.17

In Euclidean n-space, using the ordinary (Cartesian) coordinates, $g^{\mu\nu} = \delta^{\mu\nu}$. **Conformal coordinates** are coordinates x^μ such that

$$g^{\mu\nu} = \rho^2(x)\, \delta^{\mu\nu}.$$

Find all possible forms of the function ρ, (a) for $n = 2$ and (b) for $n > 2$. [*Hint:* Whatever the coordinates, the space is still flat.][28]

---

[28] A classic theorem of Liouville states that for $n > 2$ all conformal transformations can be generated by taking products of the following transformations:

1. Euclidean transformations (translations and rotations);
2. dilations;
3. inversions in a unit sphere centered at the origin.

By contrast, in two dimensions there is a conformal transformation for every harmonic function (that is, for any function satisfying Laplace's equation)—including the real and imaginary parts of any meromorphic function (analytic or at most pole singularities). That's why conformal mappings are so useful in 2-dimensional potential theory, and practically useless for 3-dimensional potential theory.

6

Gravity

6.1 Motion in Curved Spacetime

6.1.1 Program for a Theory of Gravity

I can now make more precise the slogan "Gravitation is geometry": matter curves spacetime. In a Riemannian manifold, geometry is encoded in the metric; from $g_{\mu\nu}$ we get $\Gamma^{\mu}_{\nu\lambda}$ (Eq. 5.112), and from $\Gamma^{\mu}_{\nu\lambda}$ we get $R^{\mu}_{\nu\lambda\sigma}$ (Eq. 5.87), the measure of curvature. A theory of gravity, then, consists of two parts:

1. how matter affects the metric;
2. how things move in a nonflat metric.

In its logical structure this is just like classical electrodynamics: (1) Maxwell's equations tell us how matter (electric charge, in that case) produces fields and (2) the Lorentz force law tells us how (charged) objects move in the presence of those fields. In this section I'll start with the easy part, item (2). The remainder of the chapter will be devoted to item (1): obtaining Einstein's equation (the gravitational analog to Maxwell's).

6.1.2 Classical Equations in Covariant Form

We begin with the ordinary (flat space) Lorentz-covariant equations of classical physics and recast them in generally covariant form. This involves two steps: (1) we express them in first-order form (that is to say, using only first derivatives) and (2) we replace ordinary partial derivatives with covariant[1] derivatives.

---

[1] Eds. You will have noticed that the word covariant is used in several different senses: (a) a *tensor* is covariant (as opposed to contravariant) if it transforms according to Eq. 5.5; (b) an *equation* is "covariant" if the two sides transform in the same way; (c) a *derivative* is "covariant" (∇_{μ}, as opposed to ∂_{μ}) as defined by Eq. 5.79, and (d) a *theory* is "covariant" if it is consistent with special relativity ("Lorentz covariance") or invariant under *any* change of coordinates ("general covariance").

- **The Klein–Gordon equation:** The Klein–Gordon equation for a free scalar field ϕ with mass μ says (Eq. 2.97)

$$\partial_\mu \partial^\mu \phi = -\mu^2 \phi. \tag{6.1}$$

We can render it as a pair of coupled first-order equations:

$$\partial_\mu \pi^\mu = -\mu^2 \phi, \quad \text{where} \quad \pi^\mu \equiv g^{\mu\nu} \partial_\nu \phi. \tag{6.2}$$

Replacing the partial derivative in the first equation by a covariant derivative,

$$\nabla_\mu \pi^\mu = -\mu^2 \phi, \quad \pi^\mu = g^{\mu\nu} \partial_\nu \phi \tag{6.3}$$

(we don't have to modify the second equation because ϕ is a scalar). These are now coordinate-independent equations that reduce to the Klein–Gordon equation in the flat space (Minkowski) limit.
- **Maxwell's equations:** In free space, Maxwell's equations read

$$\partial^\mu F_{\mu\nu} = 0, \quad \text{where} \quad F_{\mu\nu} \equiv \partial_\mu A_\nu - \partial_\nu A_\mu. \tag{6.4}$$

The second of these is already tensorial (it's an exterior derivative, and can be written either with partials or with ∇'s), but in the first equation we change to a covariant derivative:

$$\nabla_\mu F^{\mu\nu} = 0, \quad F_{\mu\nu} = \partial_\mu A_\nu - \partial_\nu A_\mu. \tag{6.5}$$

- **Action principles:** In (special) relativistic field theory the action principle takes the form 2.98:

$$\delta I = \delta \int \mathcal{L}(\phi^\alpha, \partial_\mu \phi^\alpha) \, d^4 x = 0. \tag{6.6}$$

As we have already found (Eq. 5.100), in generally covariant form this becomes

$$\delta I = \delta \int \mathcal{L}(\phi^\alpha, \nabla_\mu \phi^\alpha) \sqrt{-g} \, d^4 \Sigma = 0. \tag{6.7}$$

Remember, $\sqrt{-g}$ is a (real) scalar density of weight $+1$ (5.99) and $d^4\Sigma = dx^0 \, dx^1 \, dx^2 \, dx^3$ is a scalar density of weight -1 (5.58), so 6.7 is manifestly tensorial, and in Minkowski space $g = -1$, so it reduces correctly to the flat space expression.[2]

---

[2] There can be equivalent ways of writing the flat space equations that do not lead to the same generally covariant form. For instance, you can throw in extra terms involving the Riemann tensor that vanish trivially in flat space (where $R^\mu{}_{\nu\lambda\sigma} = 0$). And if the equation involves $\partial_\mu \partial_\nu$, the tensor form ($\nabla_\mu \nabla_\nu$) will depend on the order of the indices.

Problem 6.1

Suppose you want to integrate some scalar function $\phi(\mathbf{r})$ over a certain volume $\mathcal{V}$ in ordinary Euclidean 3-space. The general expression is

$$\int_{\mathcal{V}} \phi(\mathbf{r}) \sqrt{g}\, dx^1\, dx^2\, dx^3. \tag{6.8}$$

Work out g in Cartesian coordinates $(x^1, x^2, x^3) = (x, y, z)$, cylindrical coordinates (ρ, ϕ, z), and spherical coordinates (r, θ, ϕ), and confirm that Eq. 6.8 reproduces the usual expressions.

- **Free particles:** In the absence of gravity, a free particle (no forces) travels in a straight line at constant speed, which is to say, its acceleration is zero. How would it move in a spacetime curved by gravity, where there *are* no straight lines? Proper velocity,[3]

$$v^\mu(s) \equiv \frac{dx^\mu}{ds}, \tag{6.9}$$

is already tensorial and requires no modification. But proper *acceleration*,

$$\frac{dv^\mu}{ds} = \frac{dx^\nu}{ds}\frac{\partial v^\mu}{\partial x^\nu} = v^\nu \partial_\nu v^\mu, \tag{6.10}$$

is *not* tensorial, because of the ordinary derivative of v^μ, which we need to replace by the *covariant* derivative. For a vector field $A^\mu(x)$ we introduce the **covariant path derivative**, D_s; using Eq. 5.79,

$$D_s A^\mu \equiv \frac{dx^\nu}{ds}\nabla_\nu A^\mu = v^\nu\left(\partial_\nu A^\mu + \Gamma^\mu_{\nu\lambda} A^\lambda\right) = \frac{dA^\mu}{ds} + \Gamma^\mu_{\nu\lambda} v^\nu A^\lambda, \tag{6.11}$$

which *is* tensorial. In particular,

$$D_s v^\mu = \frac{dv^\mu}{ds} + \Gamma^\mu_{\nu\lambda} v^\nu v^\lambda = \frac{d^2 x^\mu}{ds^2} + \Gamma^\mu_{\nu\lambda}\frac{dx^\nu}{ds}\frac{dx^\lambda}{ds}. \tag{6.12}$$

This (with $s \to \tau$) is the tensorial generalization of proper acceleration.

For a free particle, then,

$$D_s v^\mu = 0. \tag{6.13}$$

Equation 6.12 suggests that the Christoffel symbols play the role of gravitational "forces" (more precisely, *pseudo*forces); since they are derivatives of the metric (5.112), we sometimes call $g_{\mu\nu}$ the "gravitational potential."

Curves satisfying 6.13 are known as **geodesics**. As I'll prove in the following theorem, geodesics are the next best thing to straight lines, in a curved space:

---

[3] Actually, proper velocity is the derivative with respect to *proper time*, τ, but for this argument the generic parameter s will do.

paths for which the distance (elapsed proper time, in our case) is extremal.[4] In
the presence of gravity, then, free particles travel along geodesics. This is, if you
like, the generalization of Newton's first law, and the geodesic equation itself is
the apotheosis of Fermat's principle.

Theorem: *A geodesic is the path of extremal distance between two points.*

Proof: The extremal distance between fixed points s_1 and s_2 is given by 5.92:

$$\delta \int_{s_1}^{s_2} \left(\frac{dx^\mu}{ds} \frac{dx^\nu}{ds} g_{\mu\nu} \right)^{1/2} ds = 0, \tag{6.14}$$

or

$$\int \frac{1}{2} \left(\frac{dx^\mu}{ds} \frac{dx^\nu}{ds} g_{\mu\nu} \right)^{-1/2} \left[2 \frac{dx^\sigma}{ds} g_{\sigma\lambda} \delta \left(\frac{dx^\lambda}{ds} \right) + \frac{dx^\sigma}{ds} \frac{dx^\kappa}{ds} \delta(g_{\sigma\kappa}) \right] ds$$

$$= 0. \tag{6.15}$$

Write

$$\delta \left(\frac{dx^\nu}{ds} \right) = \frac{d(\delta x^\nu)}{ds}, \quad \text{and} \quad \delta(g_{\mu\nu}) = (\partial_\lambda g_{\mu\nu}) \delta x^\lambda, \tag{6.16}$$

and integrate the first term by parts:

$$\int \left\{ -\frac{d}{ds} \left[\left(\frac{dx^\mu}{ds} \frac{dx^\nu}{ds} g_{\mu\nu} \right)^{-1/2} \frac{dx^\sigma}{ds} g_{\sigma\lambda} \right] \delta x^\lambda \right.$$

$$\left. + \frac{1}{2} \left[\left(\frac{dx^\mu}{ds} \frac{dx^\nu}{ds} g_{\mu\nu} \right)^{-1/2} \frac{dx^\sigma}{ds} \frac{dx^\kappa}{ds} \partial_\lambda g_{\sigma\kappa} \right] \delta x^\lambda \right\} ds = 0. \tag{6.17}$$

Therefore

$$\frac{d}{ds} \left[\left(\frac{dx^\mu}{ds} \frac{dx^\nu}{ds} g_{\mu\nu} \right)^{-1/2} \frac{dx^\sigma}{ds} g_{\sigma\lambda} \right]$$

$$= \frac{1}{2} \left[\left(\frac{dx^\mu}{ds} \frac{dx^\nu}{ds} g_{\mu\nu} \right)^{-1/2} \frac{dx^\sigma}{ds} \frac{dx^\kappa}{ds} \partial_\lambda g_{\sigma\kappa} \right]. \tag{6.18}$$

Now let s be the parameterization (proper time) for which

$$\left(\frac{dx^\mu}{ds} \frac{dx^\nu}{ds} g_{\mu\nu} \right) = 1. \tag{6.19}$$

---

[4] Typically, in spacetime, geodesics *maximize* the proper time (Problem 1.1), but they can also be inflections
(saddle points, as opposed to mountain tops). In curved spaces embedded in Euclidean space, geodesics
typically *minimize* the distance.

(You can carry the argument through in a parameter-free way if you like, but this is a little simpler.) Then

$$\frac{d^2 x^\sigma}{ds^2} g_{\sigma\lambda} + \frac{dx^\sigma}{ds} \frac{d}{ds} g_{\sigma\lambda} = \frac{1}{2} \frac{dx^\sigma}{ds} \frac{dx^\kappa}{ds} \partial_\lambda g_{\sigma\kappa}. \tag{6.20}$$

But

$$\frac{d}{ds} g_{\sigma\lambda} = \frac{dx^\nu}{ds} \partial_\nu g_{\sigma\lambda}, \tag{6.21}$$

so

$$\begin{aligned}
g_{\sigma\lambda} \frac{d^2 x^\sigma}{ds^2} &= \frac{1}{2} \frac{dx^\sigma}{ds} \frac{dx^\kappa}{ds} \partial_\lambda g_{\sigma\kappa} - \frac{dx^\sigma}{ds} \frac{dx^\nu}{ds} \partial_\nu g_{\sigma\lambda} \\
&= \frac{1}{2} \frac{dx^\mu}{ds} \frac{dx^\nu}{ds} \partial_\lambda g_{\mu\nu} - \frac{dx^\mu}{ds} \frac{dx^\nu}{ds} \left(\frac{1}{2} \partial_\nu g_{\mu\lambda} + \frac{1}{2} \partial_\mu g_{\nu\lambda} \right) \\
&= -\frac{1}{2} \frac{dx^\mu}{ds} \frac{dx^\nu}{ds} \left(\partial_\mu g_{\nu\lambda} + \partial_\nu g_{\lambda\mu} - \partial_\lambda g_{\mu\nu} \right) \\
&= -\frac{dx^\mu}{ds} \frac{dx^\nu}{ds} \Gamma_{\lambda,\mu\nu} \tag{6.22}
\end{aligned}$$

(5.110), and hence (multiplying by $g^{\lambda\rho}$)

$$g_{\sigma\lambda} g^{\lambda\rho} \frac{d^2 x^\sigma}{ds^2} = -v^\mu v^\nu \Gamma_{\lambda,\mu\nu} g^{\lambda\rho}. \tag{6.23}$$

Finally, then,

$$\frac{d^2 x^\rho}{ds^2} = \frac{dv^\rho}{ds} = -v^\mu v^\nu \Gamma^\rho_{\mu\nu}, \quad \text{or} \quad D_s v^\rho = 0, \tag{6.24}$$

which is the equation for a geodesic (6.13).   QED

Example 6.1

Gravitational precession of a gyroscope. A gyroscope carries an associated space-like unit vector e^μ (the direction of its axis) that is constrained to be perpendicular to its velocity. In the absence of gravity, e^μ is a constant:

$$\frac{de^\mu}{ds} = 0. \tag{6.25}$$

In the presence of gravity this generalizes to

$$D_s e^\mu = \frac{de^\mu}{ds} + \Gamma^\mu_{\nu\lambda} v^\nu e^\lambda = 0. \tag{6.26}$$

6.1.3 Tidal Forces

I'd like now to make quantitative the notion of "tidal" forces in a nonuniform gravitational field, which we touched on briefly in Section 4.4. Imagine two nearby particles in free fall. The figure below shows their world lines, which we'll parameterize by τ, their proper times. A space-like geodesic (call it $\tau = 0$) intersects the two world lines, and the arc length between them, along that geodesic, is ds.

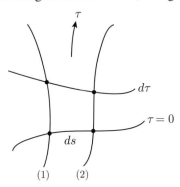

The (infinitesimal) vector

$$\delta^\mu(\tau) \equiv dx^\mu = x^\mu_{(2)} - x^\mu_{(1)} \tag{6.27}$$

is the coordinate difference between the particles, and we'll take $D_\tau^2 \delta^\mu$ (the acceleration of their separation) as a measure of the tidal "force." Applying 6.11 twice,[5]

$$\begin{aligned}
D_\tau^2 \delta^\mu &= D_\tau \left(\frac{d\delta^\mu}{d\tau} + \Gamma^\mu_{\nu\lambda} v^\nu \delta^\lambda \right) \\
&= \frac{d^2\delta^\mu}{d\tau^2} + \frac{d}{d\tau} \left(\Gamma^\mu_{\nu\lambda} v^\nu \delta^\lambda \right) + \Gamma^\mu_{\sigma\phi} v^\sigma \left(\frac{d\delta^\phi}{d\tau} + \Gamma^\phi_{\nu\lambda} v^\nu \delta^\lambda \right).
\end{aligned} \tag{6.28}$$

In elevator coordinates (where Γ vanishes, but not its derivatives), this simplifies considerably:

$$D_\tau^2 \delta^\mu = \frac{d^2\delta^\mu}{d\tau^2} + \partial_\sigma \left(\Gamma^\mu_{\nu\lambda} \right) v^\sigma v^\nu \delta^\lambda. \tag{6.29}$$

Meanwhile, the two particles are in free fall, so their world lines are both geodesics:

$$D_\tau^2 x^\mu = \frac{d^2 x^\mu}{d\tau^2} + \Gamma^\mu_{\sigma\nu} v^\sigma v^\nu = 0. \tag{6.30}$$

---

[5] Of course, there are two different trajectories here, and two different velocities, but that's a second-order correction.

Subtracting two copies of 6.28 (at x_2 and x_1),

$$\frac{d^2\delta^\mu}{d\tau^2} + \left[\Gamma^\mu_{\sigma\nu 2} v^\sigma_2 v^\nu_2 - \Gamma^\mu_{\sigma\nu 1} v^\sigma_1 v^\nu_1\right] = 0. \tag{6.31}$$

Let x_1 be the point where $\Gamma^\mu_{\sigma\nu} = 0$; then

$$\Gamma^\mu_{\sigma\nu 1} = 0, \quad \Gamma^\mu_{\sigma\nu 2} = \partial_\lambda\left(\Gamma^\mu_{\sigma\nu}\right)\delta^\lambda + \cdots \tag{6.32}$$

(with the derivative evaluated at x_1). So

$$\frac{d^2\delta^\mu}{d\tau^2} = -(\partial_\lambda\Gamma^\mu_{\sigma\nu})\,\delta^\lambda v^\sigma v^\nu. \tag{6.33}$$

(The velocities are actually evaluated at x_2, but again, the correction is second order in δ.) Substituting this into Eq. 6.29, and using 5.88, we conclude that

$$D^2_\tau\,\delta^\mu = \partial_\sigma\left(\Gamma^\mu_{\nu\lambda}\right)v^\sigma v^\nu\delta^\lambda - (\partial_\lambda\Gamma^\mu_{\sigma\nu})\delta^\lambda v^\sigma v^\nu = R^\mu_{\ \nu\lambda\sigma} v^\sigma v^\nu\delta^\lambda. \tag{6.34}$$

Tidal forces, then, are directly related to the Riemann curvature tensor; they are manifestations of gravity that *cannot* be transformed away by adopting free-fall (elevator) coordinates.

6.2 The Gravitational Field

We have seen how to modify the equations of motion to include the effects of gravity-induced curvature: express them in covariant (tensorial) form, and use the appropriate metric (which encodes the curvature). Now we need to describe the gravitational field itself: How does the presence of matter affect the metric? In principle we could use the same method as in Section 6.1.2: express the classical field equations in fully covariant form. The trouble is, we don't *know* the classical field equations for gravity. (Newton gave us the analog to Coulomb's law, but we need the analog to Maxwell's equations.) So the first order of business is to deduce those equations. We'll start out in a region of empty space, and then we'll figure out how things change in the presence of matter.

6.2.1 Einstein's Equation in Empty Space

If the gravitational field is weak, the departure of $g_{\mu\nu}$ from the Minkowski metric (1.18, which we will now denote by $\eta_{\mu\nu}$) is very small; we'll treat it as a perturbation:

$$g_{\mu\nu} = \eta_{\mu\nu} + h_{\mu\nu}. \tag{6.35}$$

To first order in $h_{\mu\nu}$,

$$g^{\mu\lambda}h_{\lambda\nu} = h^{\mu}{}_{\nu} = \eta^{\mu\lambda}h_{\lambda\nu}, \qquad (6.36)$$

and we define

$$h \equiv h^{\mu}{}_{\mu}. \qquad (6.37)$$

I shall assume that the Lagrangian is quadratic in $h_{\mu\nu}$, and involves no higher than first derivatives. The most general scalar form is then

$$\mathcal{L} = a(\partial_{\lambda}h_{\mu\nu})(\partial^{\lambda}h^{\mu\nu}) + b(\partial^{\mu}h_{\mu\nu})(\partial_{\lambda}h^{\lambda\nu}) + c(\partial^{\nu}h)(\partial^{\mu}h_{\mu\nu})$$
$$+ d(\partial_{\mu}h)(\partial^{\mu}h) + e(h_{\mu\nu}h^{\mu\nu}) + fh^{2}, \qquad (6.38)$$

where $a, b, c, d, e,$ and f are some constants, to be determined. Our strategy will be to treat this as a flat (Minkowski) space Lagrangian representing weak gravity in empty space.

The variation of the action is

$$\delta I = \delta \int \mathcal{L} d^{4}x = \int (\delta\mathcal{L}) d^{4}x. \qquad (6.39)$$

Taking the terms one at a time (and integrating by parts where appropriate),

$$\delta\mathcal{L}_a = a\left[\delta(\partial_{\lambda}h_{\mu\nu})\partial^{\lambda}h^{\mu\nu} + \partial_{\lambda}h_{\mu\nu}\delta(\partial^{\lambda}h^{\mu\nu})\right]$$
$$= -a\left[\delta(h_{\mu\nu})\partial_{\lambda}\partial^{\lambda}h^{\mu\nu} + \partial^{\lambda}\partial_{\lambda}h_{\mu\nu}\delta(h^{\mu\nu})\right] = \left[-2a\partial_{\lambda}\partial^{\lambda}h_{\mu\nu}\right]\delta h^{\mu\nu}, \qquad (6.40)$$

$$\delta\mathcal{L}_b = b\left[\delta(\partial^{\mu}h_{\mu\nu})\partial_{\lambda}h^{\lambda\nu} + \partial^{\mu}h_{\mu\nu}\delta(\partial_{\lambda}h^{\lambda\nu})\right]$$
$$= -b\left[\delta(h_{\mu\nu})\partial^{\mu}\partial_{\lambda}h^{\lambda\nu} + \partial_{\lambda}\partial^{\mu}h_{\mu\nu}\delta(h^{\lambda\nu})\right]$$
$$= -b\left[\partial_{\mu}\partial^{\lambda}h_{\lambda\nu}\delta(h^{\mu\nu}) + \partial_{\mu}\partial^{\lambda}h_{\lambda\nu}\delta(h^{\mu\nu})\right] = \left[-2b\partial_{\mu}\partial^{\lambda}h_{\lambda\nu}\right]\delta h^{\mu\nu}, \qquad (6.41)$$

$$\delta\mathcal{L}_c = c\left[\delta(\partial^{\nu}h)\partial^{\mu}h_{\mu\nu} + \partial^{\nu}h\delta(\partial^{\mu}h_{\mu\nu})\right]$$
$$= -c\left[\delta(h)\partial^{\nu}\partial^{\mu}h_{\mu\nu} + \partial^{\mu}\partial^{\nu}h\delta(h_{\mu\nu})\right] = \left[-c\partial_{\mu}\partial_{\nu}h\right]\delta h^{\mu\nu} - \left[c\partial^{\mu}\partial^{\nu}h_{\mu\nu}\right]\delta h$$
$$\text{(but } h = h^{\mu}{}_{\mu} = \eta_{\mu\nu}h^{\mu\nu} \text{ to first order, so } \delta h = \eta_{\mu\nu}\delta(h^{\mu\nu}))$$
$$= \left[-c\partial_{\mu}\partial_{\nu}h\right]\delta h^{\mu\nu} + \left[-c\eta_{\mu\nu}\partial^{\sigma}\partial^{\lambda}h_{\sigma\lambda}\right]\delta h^{\mu\nu}, \qquad (6.42)$$

$$\delta\mathcal{L}_d = d\left[\delta(\partial_{\mu}h)\partial^{\mu}h + \partial_{\mu}h\delta(\partial^{\mu}h)\right] = \left[-2d\partial_{\mu}\partial^{\mu}h\right]\delta h$$
$$= \left[-2d\eta_{\mu\nu}\partial_{\sigma}\partial^{\sigma}h\right]\delta h^{\mu\nu}, \qquad (6.43)$$

$$\delta\mathcal{L}_e = e\left[\delta(h_{\mu\nu})h^{\mu\nu} + h_{\mu\nu}\delta(h^{\mu\nu})\right] = \left[2eh_{\mu\nu}\right]\delta h^{\mu\nu}, \qquad (6.44)$$

$$\delta\mathcal{L}_f = f\left[2h\delta(h)\right] = \left[2f\eta_{\mu\nu}h\right]\delta h^{\mu\nu}. \qquad (6.45)$$

Only the part symmetric in $\mu \leftrightarrow \nu$ will survive contraction with $\delta h^{\mu\nu}$. All terms *are* symmetric except those in b, so in that term we symmetrize. Canceling the arbitrary $\delta h^{\mu\nu}$, this yields the equation of motion:

$$-2a\Box^2 h_{\mu\nu} - b\partial_\mu\partial^\lambda h_{\lambda\nu} - b\partial_\nu\partial^\lambda h_{\lambda\mu} - c\partial_\mu\partial_\nu h - c\eta_{\mu\nu}\partial^\sigma\partial^\lambda h_{\sigma\lambda}$$
$$- 2d\eta_{\mu\nu}\Box^2 h + 2e h_{\mu\nu} + 2f h\eta_{\mu\nu} = 0. \tag{6.46}$$

There are six free parameters here, but we want a unique theory. First we consider a (small) change in coordinates,

$$x^\mu \to x'^\mu = x^\mu + A^\mu(x). \tag{6.47}$$

To first order in A,

$$\frac{\partial x^\mu}{\partial x'^\nu} = \delta^\mu_\nu - \partial_\nu A^\mu, \tag{6.48}$$

so (using the transformation law for a covariant tensor, Eq. 5.5)

$$\eta'_{\mu\nu} = \left(\frac{\partial x^\lambda}{\partial x'^\mu}\frac{\partial x^\sigma}{\partial x'^\nu}\right)\eta_{\lambda\sigma} = \left(\delta^\lambda_\mu - \partial_\mu A^\lambda\right)\left(\delta^\sigma_\nu - \partial_\nu A^\sigma\right)\eta_{\lambda\sigma}$$
$$= \delta^\lambda_\mu\delta^\sigma_\nu\eta_{\lambda\sigma} - \delta^\sigma_\nu\left(\partial_\mu A^\lambda\right)\eta_{\lambda\sigma} - \delta^\lambda_\mu\left(\partial_\nu A^\sigma\right)\eta_{\lambda\sigma}$$
$$= \eta_{\mu\nu} - \partial_\mu A_\nu - \partial_\nu A_\mu. \tag{6.49}$$

Thus anything of the form

$$h_{\mu\nu} = -(\partial_\mu A_\nu + \partial_\nu A_\mu) \tag{6.50}$$

must satisfy Eq. 6.46, to first order in A, since it represents the *same system*—it's just a change of coordinates.[6] Now, terms $a, b, c,$ and d involve *third* derivatives of A, whereas e and f involve just *first* derivatives; since A is arbitrary, these must separately vanish for 6.50. The e and f terms are

$$e h_{\mu\nu} + f h\eta_{\mu\nu} = -e(\partial_\mu A_\nu + \partial_\nu A_\mu) - 2f(\partial_\lambda A^\lambda)\eta_{\mu\nu} = 0, \tag{6.51}$$

and this must hold for *any* (small) A, so

$$e = f = 0. \tag{6.52}$$

The $a, b, c,$ and d terms are

$$2a\Box^2(\partial_\mu A_\nu + \partial_\nu A_\mu) + b\left[\partial_\mu\Box^2 A_\nu + \partial_\nu\Box^2 A_\mu + 2\partial_\mu\partial_\nu(\partial_\lambda A^\lambda)\right]$$
$$+ c\eta_{\mu\nu}(2\Box^2\partial^\lambda A_\lambda) + c\partial_\mu\partial_\nu(2\partial^\lambda A_\lambda) + 2d\eta_{\mu\nu}(2\Box^2\partial_\lambda A^\lambda) = 0. \tag{6.53}$$

---

[6] This is the gravitational analog to a **gauge transformation** (Eq. 3.16), and what we are saying is that the equations must be **gauge invariant**.

Collecting like terms,

$$\Box^2(\partial_\mu A_\nu + \partial_\nu A_\mu) \quad \text{terms} \Rightarrow 2a + b = 0, \tag{6.54}$$

$$\partial_\mu \partial_\nu(\partial^\lambda A_\lambda) \quad \text{terms} \Rightarrow 2b + 2c = 0, \tag{6.55}$$

$$\eta_{\mu\nu}\Box^2(\partial_\lambda A^\lambda) \quad \text{terms} \Rightarrow 2c + 4d = 0. \tag{6.56}$$

It follows that $b = -2a$, $c = 2a$, and $d = -a$, so this requirement alone makes the Lagrangian unique. Since any overall constant in the action is arbitrary, we might as well take $a = -\frac{1}{2}$, $b = 1$, $c = -1$, and $d = \frac{1}{2}$. Then

$$\boxed{\Box^2 h_{\mu\nu} - \partial^\lambda(\partial_\mu h_{\lambda\nu} + \partial_\nu h_{\lambda\mu}) + \eta_{\mu\nu}\partial^\lambda\partial^\sigma h_{\lambda\sigma} + \partial_\mu\partial_\nu h - \eta_{\mu\nu}\Box^2 h = 0.} \tag{6.57}$$

This is the analog to Maxwell's equations (in empty space), for a weak gravitational field.

What are the affine connections and the Riemann tensor, to first order in the perturbation $h_{\mu\nu}$ (6.35)? According to Eq. 5.110,

$$\Gamma_{\lambda,\mu\nu} = \frac{1}{2}\left(\partial_\mu h_{\nu\lambda} + \partial_\nu h_{\lambda\mu} - \partial_\lambda h_{\mu\nu}\right), \tag{6.58}$$

and then 5.112 says

$$\Gamma^\lambda_{\mu\nu} = \frac{1}{2}\left(\partial_\mu h^\lambda{}_\nu + \partial_\nu h^\lambda{}_\mu - \partial^\lambda h_{\mu\nu}\right). \tag{6.59}$$

Since Γ is already first order, the Riemann tensor (5.87) is

$$R^\lambda{}_{\mu\sigma\nu} = \partial_\nu\Gamma^\lambda_{\sigma\mu} - \partial_\sigma\Gamma^\lambda_{\nu\mu} \tag{6.60}$$

$$= \frac{1}{2}\left(\partial_\nu\partial_\sigma h^\lambda{}_\mu + \partial_\nu\partial_\mu h^\lambda{}_\sigma - \partial_\nu\partial^\lambda h_{\sigma\mu} - \partial_\sigma\partial_\nu h^\lambda{}_\mu - \partial_\sigma\partial_\mu h^\lambda{}_\nu + \partial_\sigma\partial^\lambda h_{\nu\mu}\right).$$

The contracted curvature tensors are then (5.119)

$$R_{\mu\nu} = R^\lambda{}_{\mu\lambda\nu} = \frac{1}{2}(\partial_\mu\partial_\nu h + \Box^2 h_{\mu\nu} - \partial_\mu\partial_\lambda h^\lambda{}_\nu - \partial_\nu\partial_\lambda h^\lambda{}_\mu) \tag{6.61}$$

and (5.120)

$$R = R^\mu{}_\mu = \Box^2 h - \partial_\mu\partial_\nu h^{\mu\nu}. \tag{6.62}$$

In particular,

$$R_{\mu\nu} - \frac{1}{2}\eta_{\mu\nu}R \tag{6.63}$$

$$= \frac{1}{2}\left[\partial_\mu\partial_\nu h + \Box^2 h_{\mu\nu} - \partial_\mu\partial_\lambda h^\lambda{}_\nu - \partial_\nu\partial_\lambda h^\lambda{}_\mu - \eta_{\mu\nu}\Box^2 h + \eta_{\mu\nu}\partial_\lambda\partial_\sigma h^{\lambda\sigma}\right],$$

and the term in square brackets is *zero*, according to 6.57. In generally covariant form, then, the gravitational field equation can be written

$$\boxed{R_{\mu\nu} - \frac{1}{2}g_{\mu\nu}R = 0.} \tag{6.64}$$

This is **Einstein's equation** for the gravitational field in empty space.[7] Although it was *motivated* by considering the weak-field limit, once it is expressed in the generally covariant form (6.64) it makes no reference to the magnitude of the field, and holds no matter how strong the gravity may be.

The logical structure of this section is subtle, so let me recapitulate: In *flat* space ($\eta_{\mu\nu}$) both $R_{\mu\nu}$ and R are trivially zero. When we turn on a *weak* gravitational field ($h_{\mu\nu}$), in empty space (enforced by 6.39, in which no interaction Lagrangian is included), the special combination $R_{\mu\nu} - \frac{1}{2}g_{\mu\nu}R$ *remains* zero. We have thus accomplished for gravity (in empty space) what we did for the other phenomena in Section 6.1: recasting the classical equation of motion (6.57) in fully covariant form (6.64), and we take this to be the correct expression regardless of the strength of the gravitational field.

6.2.2 Alternative Theories

The argument in Section 6.2.1 seems to lead unambiguously to 6.64, but there are loopholes.[8] You can, of course, allow higher derivatives in the Lagrangian. A more interesting possibility is to reject the assumption (built into 6.35) that flat (Minkowski) space is a solution. To be sure, global Poincaré invariance would require that flat space be a solution, but we don't really know that the universe is globally invariant—only that it is *locally* invariant. Specifically, local Poincaré invariance does not require that the generators of translations commute. Here's an analogy: Suppose you are sitting at the north pole; rotations about the earth's axis will still look exactly like rotations, but rotations about an axis through the equator will look (locally) like translations. However, they don't quite commute (as true translations must) because in fact they are rotations. If the earth is large enough, the distinction may not be noticeable. Similarly, it might be that what we think of as translations, governed by the Poincaré group, are only approximately so when considered locally. Perhaps we actually inhabit a space that is only *locally* Poincaré invariant.

There exist two such spaces that are locally Poincaré invariant everywhere.[9] They are called **de Sitter spaces**. There is the 5-dimensional space with an extra "spatial" coordinate and the invariant quadratic form

---

[7] Note the distinction between *flat* space, where the gravitational field is *zero*, and *empty* space, in which there may very well be a gravitational field, due to sources external to the region in question.

[8] Eds. For a useful discussion of alternatives to general relativity (and experimental tests of the theory) see C. Will, *Theory and Experiment in Gravitational Physics*, 2nd ed., Cambridge U. P. (2018).

[9] Of course, even *local* Poincaré invariance has only been tested in our small portion of the universe; conceivably it might not hold elsewhere.

$$x_0^2 - \sum_{i=1}^{3} x_i^2 - x_4^2 = \sum_{\mu,\nu=0}^{4} g_{\mu\nu}^5 x^\mu x^\nu \quad \text{with} \quad g_{\mu\nu}^5 = (+1, -1, -1, -1, -1).$$

$$(6.65)$$

This symmetry group is called O(4,1). The other possibility is O(3,2), with an extra "temporal" axis, and the invariant form

$$x_0^2 - \sum_{i=1}^{3} x_i^2 + x_4^2 = \sum_{\mu,\nu=0}^{4} g_{\mu\nu}^5 x^\mu x^\nu \quad \text{with} \quad g_{\mu\nu}^5 = (+1, -1, -1, -1, +1).$$

$$(6.66)$$

In either case, we presumably occupy a (curved) 4-dimensional invariant subspace defined by

$$\sum_{\mu,\nu=0}^{4} g_{\mu\nu}^5 x^\mu x^\nu = \Lambda^2,$$

$$(6.67)$$

instead of the (flat) 4-dimensional Minkowski space. Taking this as the presumptive vacuum solution leads to

$$R_{\mu\nu} - \tfrac{1}{2} g_{\mu\nu} R \pm \Lambda g_{\mu\nu} = 0$$

$$(6.68)$$

in place of 6.64, with one sign for (4,1) and the other for (3,2). Historically, Λ is known as the **cosmological constant**; it is certainly very small, and although there are sound arguments against (4,1), there is nothing wrong in principle with (3,2).

6.2.3 The Source of Gravity

The combination

$$G_{\mu\nu} \equiv R_{\mu\nu} - \tfrac{1}{2} g_{\mu\nu} R$$

$$(6.69)$$

is called the **Einstein tensor**. In a region empty of "matter,"

$$G_{\mu\nu} = 0,$$

$$(6.70)$$

as we found in Section 6.2.1. In the *presence* of a source (including, for instance, an electromagnetic field) $G_{\mu\nu}$ is *not* zero. The equivalence principle suggests that the source of gravity is $T_{\mu\nu}$, the energy/momentum/stress tensor (see Section 4.1), so let's try something of the form

$$G_{\mu\nu} = -\alpha T_{\mu\nu} - \beta g_{\mu\nu} T^\lambda_{\ \lambda}.$$

$$(6.71)$$

Now, for a weak gravitational field $h_{\mu\nu}$, Eqs. 6.61 and 6.62 give

$$\partial^\nu G_{\mu\nu} = \partial^\nu R_{\mu\nu} - \tfrac{1}{2}\partial_\mu R$$
$$= \tfrac{1}{2}\partial^\nu \left[\partial_\mu \partial_\nu h + \Box^2 h_{\mu\nu} - \partial_\mu \partial_\lambda h^\lambda{}_\nu - \partial_\nu \partial_\lambda h^\lambda{}_\mu\right] - \tfrac{1}{2}\partial_\mu \left[\Box^2 h - \partial_\nu \partial_\lambda h^{\nu\lambda}\right]$$
$$= \tfrac{1}{2}\left[\Box^2 \partial_\mu h + \Box^2 \partial^\nu h_{\mu\nu} - \partial_\mu \partial_\lambda \partial_\nu h^{\lambda\nu} - \Box^2 \partial^\nu h_{\nu\mu} - \Box^2 \partial_\mu h + \partial_\mu \partial_\nu \partial_\lambda h^{\nu\lambda}\right]$$
$$= 0. \tag{6.72}$$

So Eq. 6.71 will require

$$-\alpha \partial^\nu T_{\mu\nu} - \beta \partial_\mu T^\lambda{}_\lambda = 0. \tag{6.73}$$

But $T_{\mu\nu}$ is divergenceless (at least, to lowest order in $h_{\mu\nu}$), so either $\beta = 0$, or else $\partial_\mu T^\lambda{}_\lambda = 0$. The latter is impossible, in general, as we can see from a simple counterexample: Suppose there is a point mass m at rest at the center of the universe. Far away, $T^\lambda{}_\lambda$ surely goes to zero, and hence the law would say that $T^\lambda{}_\lambda = 0$ *everywhere*. But $T^\lambda{}_\lambda$ certainly does not vanish at the center (where $T_{00} = m\,\delta^3(\mathbf{x})$ and the other elements are zero). So it must be that $\beta = 0$, leaving

$$\boxed{G_{\mu\nu} = -\alpha T_{\mu\nu}.} \tag{6.74}$$

This is **Einstein's field equation** in the presence of matter. It plays the role (in general relativity) of Maxwell's equations (in electrodynamics). It tells us precisely how matter (in the form of the stress tensor) curves spacetime (in the form of the Einstein tensor). We'll fix the constant α in due course, by comparison with the Newtonian limit.

6.2.4 Action Principle Formulation

In this section we'll obtain Einstein's equation from an action principle. I'll start with the empty space equation 6.64 for regions where $T_{\mu\nu} = 0$. The action takes the universal form (5.100):

$$I = \int \mathcal{L}_g \sqrt{-g}\, d^4x, \tag{6.75}$$

where $\mathcal{L}_g$ is the Lagrangian for the gravitational field. The action has to be a scalar (to guarantee covariance of the field equations), so let's try a multiple of the curvature scalar as our Lagrangian:

$$\mathcal{L}_g = \kappa R \tag{6.76}$$

(where κ is a constant to be determined[10]). To get δR we first calculate $\delta R_{\mu\nu}$. From Eqs. 5.87 and 5.119,

$$R_{\mu\nu} = R^\lambda_{\ \mu\lambda\nu} = -\partial_\lambda \Gamma^\lambda_{\nu\mu} + \partial_\nu \Gamma^\lambda_{\lambda\mu} - \Gamma^\lambda_{\lambda\rho} \Gamma^\rho_{\nu\mu} + \Gamma^\lambda_{\nu\rho} \Gamma^\rho_{\lambda\mu}, \tag{6.77}$$

so

$$\delta R_{\mu\nu} = -\partial_\lambda \delta(\Gamma^\lambda_{\nu\mu}) + \partial_\nu \delta(\Gamma^\lambda_{\lambda\mu})$$
$$- \Gamma^\lambda_{\lambda\rho} \delta(\Gamma^\rho_{\nu\mu}) - \delta(\Gamma^\lambda_{\lambda\rho}) \Gamma^\rho_{\nu\mu} + \Gamma^\lambda_{\nu\rho} \delta(\Gamma^\rho_{\lambda\mu}) + \delta(\Gamma^\lambda_{\nu\rho}) \Gamma^\rho_{\lambda\mu}. \tag{6.78}$$

Now $\delta\Gamma^\lambda_{\mu\nu}$ (unlike $\Gamma^\lambda_{\mu\nu}$ itself) is a tensor,[11] so (Eq. 5.81 and Problem 5.9(d))

$$\nabla_\nu(\delta\Gamma^\lambda_{\mu\lambda}) = \partial_\nu(\delta\Gamma^\lambda_{\mu\lambda}) - \Gamma^\tau_{\nu\mu}(\delta\Gamma^\lambda_{\tau\lambda}), \tag{6.79}$$

$$\nabla_\lambda(\delta\Gamma^\lambda_{\mu\nu}) = \partial_\lambda(\delta\Gamma^\lambda_{\mu\nu}) - \Gamma^\tau_{\lambda\mu}(\delta\Gamma^\lambda_{\tau\nu}) - \Gamma^\tau_{\lambda\nu}(\delta\Gamma^\lambda_{\mu\tau}) + \Gamma^\lambda_{\lambda\tau}(\delta\Gamma^\tau_{\mu\nu}). \tag{6.80}$$

Thus (comparing 6.78 in the final step),

$$\nabla_\nu(\delta\Gamma^\lambda_{\mu\lambda}) - \nabla_\lambda(\delta\Gamma^\lambda_{\mu\nu}) = \partial_\nu(\delta\Gamma^\lambda_{\lambda\mu}) - \Gamma^\tau_{\nu\mu}(\delta\Gamma^\lambda_{\tau\lambda}) - \partial_\lambda(\delta\Gamma^\lambda_{\mu\nu}) \tag{6.81}$$
$$+ \Gamma^\tau_{\lambda\mu}(\delta\Gamma^\lambda_{\tau\nu}) + \Gamma^\tau_{\lambda\nu}(\delta\Gamma^\lambda_{\tau\mu}) - \Gamma^\lambda_{\lambda\tau}(\delta\Gamma^\tau_{\mu\nu}) = \delta R_{\mu\nu}.$$

How about the variation of $\sqrt{-g}$? For any matrix A,[12]

$$\det A = e^{\text{Tr}(\ln A)}. \tag{6.82}$$

Using this,

$$\delta(\det A) = \delta\left(e^{\text{Tr}(\ln A)}\right) = e^{\text{Tr}(\ln A)} \delta\left(\text{Tr}(\ln A)\right). \tag{6.83}$$

But

$$\delta\left(\text{Tr}(\ln A)\right) = \text{Tr}\left(\delta(\ln A)\right) = \text{Tr}\left(\frac{1}{A}\delta A\right). \tag{6.84}$$

Now,

$$\delta(A^{-1}A) = 0 = \frac{1}{A}\delta A + (\delta A^{-1})A, \tag{6.85}$$

---

[10] Of course, an *overall* constant in the Lagrangian is of no significance, but we shall soon be adding the field Lagrangian to the matter Lagrangian, and their *relative* size fixes the strength of the gravitational interaction.

[11] The last term in the transformation rule for $\Gamma^\lambda_{\mu\nu}$ (Eq. 5.67) depends only on the coordinate change, not at all on the affine connection itself. If we vary the connection slightly, therefore, the final term is unchanged, and drops out in the subtraction to get $\delta\Gamma$.

[12] Eds. This is easy to prove for *diagonalizable* matrices, but it is actually true in *general*. See, for example, B. C. Hall, *Lie Groups, Lie Algebras, and Representations: An Elementary Introduction*, 2nd ed., Springer, New York (2015) Theorem 2.12.

so

$$\text{Tr}\left(\frac{1}{A}\delta A\right) = -\text{Tr}\left((\delta A^{-1})A\right) = -\text{Tr}\left(A\delta(A^{-1})\right), \tag{6.86}$$

and hence

$$\delta(\det A) = -(\det A)\,\text{Tr}\left(A\delta(A^{-1})\right). \tag{6.87}$$

In the present case $A = g_{\mu\nu}$ (written as a matrix), and $A^{-1} = g^{\mu\nu}$, so

$$\delta g = -g\,g_{\mu\nu}\delta g^{\mu\nu}, \tag{6.88}$$

and, finally,

$$\delta(\sqrt{-g}) = \frac{1}{2}\frac{1}{\sqrt{-g}}(-\delta g) = \frac{1}{2}\frac{1}{\sqrt{-g}}g\,g_{\mu\nu}\,\delta g^{\mu\nu} = -\frac{1}{2}\sqrt{-g}\,g_{\mu\nu}\,\delta g^{\mu\nu}. \tag{6.89}$$

Define the tensor *density*

$$\mathsf{g}^{\mu\nu} \equiv \sqrt{-g}\,g^{\mu\nu}. \tag{6.90}$$

Then

$$\delta\mathsf{g}^{\mu\nu} = \sqrt{-g}\,(\delta g^{\mu\nu}) + \left(-\tfrac{1}{2}\sqrt{-g}\,g_{\lambda\sigma}\,\delta g^{\lambda\sigma}\right)g^{\mu\nu}. \tag{6.91}$$

The proposed action (6.75) can be written as

$$I = \kappa\int R_{\mu\nu}\,\mathsf{g}^{\mu\nu}\,d^4x, \tag{6.92}$$

and its variation is

$$\delta I = \kappa\int(\delta R_{\mu\nu})\,\mathsf{g}^{\mu\nu}\,d^4x + \kappa\int R_{\mu\nu}(\delta\mathsf{g}^{\mu\nu})\,d^4x. \tag{6.93}$$

Now (6.81)

$$\delta R_{\mu\nu} = \nabla_\nu(\delta\Gamma^\lambda_{\mu\lambda}) - \nabla_\lambda(\delta\Gamma^\lambda_{\mu\nu}), \tag{6.94}$$

and we integrate the first term by parts:

$$\int\nabla_\nu(\delta\Gamma^\lambda_{\mu\lambda})\mathsf{g}^{\mu\nu}\,d^4x = \int\nabla_\nu\left(\delta\Gamma^\lambda_{\mu\lambda}\mathsf{g}^{\mu\nu}\right)d^4x - \int\delta\Gamma^\lambda_{\mu\lambda}\left(\nabla_\nu\mathsf{g}^{\mu\nu}\right)d^4x. \tag{6.95}$$

But $\nabla_\nu\mathsf{g}^{\mu\nu} = 0$ (Problem 6.2), while[13]

$$\nabla_\nu\left(\delta\Gamma^\lambda_{\mu\lambda}\mathsf{g}^{\mu\nu}\right) = \partial_\nu\left(\delta\Gamma^\lambda_{\mu\lambda}\mathsf{g}^{\mu\nu}\right); \tag{6.96}$$

---

[13] This is the (covariant) divergence of a vector *density*; combining 5.79 with 5.83, the Γ's cancel, and it is equivalent to an ordinary divergence (∂ in place of ∇).

this turns the first term on the right in 6.95 into a surface integral, which vanishes. The same thing happens for the second term in 6.94. Thus the first term in 6.93 is zero, and there remains (inserting 6.91)

$$\delta I = \kappa \int R_{\mu\nu}(\delta g^{\mu\nu})\, d^4x = \kappa \int R_{\mu\nu}\left(\delta g^{\mu\nu} - \tfrac{1}{2}g^{\mu\nu}g_{\lambda\sigma}\,\delta g^{\lambda\sigma}\right)\sqrt{-g}\,d^4x$$

$$= \kappa \int \left(R_{\mu\nu} - \tfrac{1}{2}R\,g_{\mu\nu}\right)\delta g^{\mu\nu}\sqrt{-g}\,d^4x = 0. \tag{6.97}$$

So

$$R_{\mu\nu} - \tfrac{1}{2}R\,g_{\mu\nu} = G_{\mu\nu} = 0, \tag{6.98}$$

reproducing the field equation 6.64. Evidently my proposed Lagrangian (6.76) works, for the gravitational field in empty space.

Problem 6.2

Show that $\nabla_\nu g^{\mu\lambda} = 0$. *Hint:* Remember that covariant derivatives obey the product rule. Using 5.104, show that $\nabla_\nu\sqrt{-g} = 0$, using locally geodesic coordinates and 5.83.

Now let's include the *source* of gravity: *matter* (which is to say, everything *except* gravity—particles, fields, etc.),

$$I = I_g + I_m. \tag{6.99}$$

We'll assume $\mathcal{L}_g$ is still κR; the action for matter takes the standard form (5.100)

$$I_m = \int \mathcal{L}_m\sqrt{-g}\,d^4x, \tag{6.100}$$

with $\mathcal{L}_m$ the classical Lagrangian for the system in question. Varying the metric (Eq. 6.97),

$$\delta I = \kappa \int G_{\mu\nu}\sqrt{-g}\,\delta g^{\mu\nu}\,d^4x + \int \frac{\partial(\mathcal{L}_m\sqrt{-g})}{\partial g^{\mu\nu}}\,\delta g^{\mu\nu}\,d^4x = 0. \tag{6.101}$$

Thus

$$G_{\mu\nu} = -\frac{1}{\kappa}\left[\frac{\partial(\mathcal{L}_m\sqrt{-g})}{\partial g^{\mu\nu}}\right]\frac{1}{\sqrt{-g}}. \tag{6.102}$$

In order to recover Einstein's equation $G_{\mu\nu} = -\alpha T_{\mu\nu}$ (6.74), it must be the case that

$$T_{\mu\nu} = \frac{1}{\alpha\kappa}\left[\frac{\partial(\mathcal{L}_m\sqrt{-g})}{\partial g^{\mu\nu}}\right]\frac{1}{\sqrt{-g}}. \tag{6.103}$$

That such an identification is legitimate is by no means obvious; after all, the stress tensor $T_{\mu\nu}$ has already been defined, independently, in special relativity. It is possible (though tedious) to obtain 6.103 from our earlier definition of $T_{\mu\nu}$, but I'll just *check* it for two special cases.

- **The Klein–Gordon field.** The Lagrangian is (2.95)[14]

$$\mathcal{L}_m = \tfrac{1}{2}(g^{\mu\nu}\,\partial_\mu\phi\,\partial_\nu\phi - \mu^2\phi^2). \tag{6.104}$$

According to 6.103, then,

$$
\begin{aligned}
T_{\mu\nu} &= \frac{1}{\alpha\kappa}\frac{\partial}{\partial g^{\mu\nu}}\left[\frac{1}{2}\left(g^{\lambda\sigma}\partial_\lambda\phi\,\partial_\sigma\phi - \mu^2\phi^2\right)\sqrt{-g}\right]\frac{1}{\sqrt{-g}}\\
&= \frac{1}{2\alpha\kappa}\frac{1}{\sqrt{-g}}\left[\sqrt{-g}\left(\partial_\mu\phi\,\partial_\nu\phi\right) + \left(g^{\lambda\sigma}\partial_\lambda\phi\,\partial_\sigma\phi - \mu^2\phi^2\right)\frac{\partial}{\partial g^{\mu\nu}}\sqrt{-g}\right]\\
&= \frac{1}{2\alpha\kappa}\left[\partial_\mu\phi\,\partial_\nu\phi - \frac{1}{2}g_{\mu\nu}\left(g^{\lambda\sigma}\partial_\lambda\phi\,\partial_\sigma\phi - \mu^2\phi^2\right)\right]. \tag{6.105}
\end{aligned}
$$

(In the last step I used

$$\frac{\partial}{\partial g^{\mu\nu}}\sqrt{-g} = -\frac{1}{2}\sqrt{-g}\,g_{\mu\nu}, \tag{6.106}$$

which follows from 6.89.) But the (canonical) stress tensor from special relativity is (2.133)

$$T_{\mu\nu} = -\mathcal{L}g_{\mu\nu} + \pi_\mu(\partial_\nu\phi), \quad \text{where} \quad \pi_\mu \equiv \frac{\partial\mathcal{L}}{\partial(\partial^\mu\phi)} = \partial_\mu\phi, \tag{6.107}$$

so

$$T_{\mu\nu} = \partial_\mu\phi\,\partial_\nu\phi - \tfrac{1}{2}g_{\mu\nu}g^{\lambda\sigma}\partial_\lambda\phi\,\partial_\sigma\phi + \tfrac{1}{2}g_{\mu\nu}\mu^2\phi^2. \tag{6.108}$$

In this case, at least, 6.103 holds, with $2\alpha\kappa = 1$.

- **The free electromagnetic field.** The Lagrangian for electrodynamics in empty space is (3.1)[15]

$$\mathcal{L}_{em} = -\frac{1}{4}F_{\mu\nu}F^{\mu\nu} = -\frac{1}{4}F_{\omega\sigma}F_{\rho\tau}g^{\omega\rho}g^{\sigma\tau}. \tag{6.109}$$

---

[14] In implementing 6.103 it is important to identify any "hidden" $g^{\mu\nu}$ in the Lagrangian—hence the cumbersome form in 6.104.

[15] In the Klein–Gordon case there was no question of general covariance to worry about, since the field is a *scalar* (so $\partial = \nabla$). Here the fields are *not* scalars, but since $F_{\mu\nu} \equiv \partial_\mu A_\nu - \partial_\nu A_\mu$ is an exterior derivative, in this case too the equations are automatically generally covariant.

Using 6.106,

$$\frac{1}{\alpha\kappa}\frac{1}{\sqrt{-g}}\left[\frac{\partial(\mathcal{L}_{em}\sqrt{-g})}{\partial g^{\mu\nu}}\right] = -\frac{1}{4\alpha\kappa}\frac{1}{\sqrt{-g}}\left[\left(F_{\omega\sigma}F_{\rho\tau}g^{\omega\rho}g^{\sigma\tau}\right)\left(-\frac{1}{2}\sqrt{-g}\,g_{\mu\nu}\right)\right.$$
$$\left.+\sqrt{-g}\left(\frac{\partial}{\partial g^{\mu\nu}}\right)\left(F_{\omega\sigma}F_{\rho\tau}g^{\omega\rho}g^{\sigma\tau}\right)\right]. \qquad (6.110)$$

Now[16]

$$\delta\left(F_{\omega\sigma}F_{\rho\tau}g^{\omega\rho}g^{\sigma\tau}\right) = F_{\omega\sigma}F_{\rho\tau}g^{\omega\rho}(\delta g^{\sigma\tau}) + F_{\omega\sigma}F_{\rho\tau}g^{\sigma\tau}(\delta g^{\omega\rho}). \qquad (6.111)$$

So

$$\frac{\partial}{\partial g^{\mu\nu}}\left(F_{\omega\sigma}F_{\rho\tau}g^{\omega\rho}g^{\sigma\tau}\right) = F_{\omega\mu}F_{\rho\nu}g^{\omega\rho} + F_{\mu\sigma}F_{\nu\tau}g^{\sigma\tau} = -2F_{\mu\sigma}F^{\sigma}{}_{\nu}. \qquad (6.112)$$

Thus 6.103 says

$$T_{\mu\nu} = \frac{1}{4\alpha\kappa}\left[2F_{\mu\sigma}F^{\sigma}{}_{\nu} + \frac{1}{2}g_{\mu\nu}F_{\lambda\sigma}F^{\lambda\sigma}\right]. \qquad (6.113)$$

Or, letting $2\alpha\kappa = 1$, as before,

$$T_{\mu\nu} = F_{\mu\sigma}F^{\sigma}{}_{\nu} + \frac{1}{4}g_{\mu\nu}F_{\lambda\sigma}F^{\lambda\sigma}, \qquad (6.114)$$

in agreement with Eq. 3.20.

This confirms 6.103 for two special cases, and fixes $\kappa = 1/(2\alpha)$:

$$\boxed{T_{\mu\nu} = \frac{2}{\sqrt{-g}}\frac{\partial\left(\mathcal{L}_m\sqrt{-g}\right)}{\partial g^{\mu\nu}}.} \qquad (6.115)$$

In practice, Eq. 6.115 is the fastest way to *get* $T_{\mu\nu}$, in most cases, and it automatically delivers a symmetric tensor, avoiding the awkward symmetrization procedure (Eq. 2.152).[17]

Problem 6.3

Prove the identity

$$\Gamma^{\mu}_{\mu\lambda} = \partial_{\lambda}(\ln\sqrt{-g}). \qquad (6.116)$$

Hint: From 6.88 it follows that $\partial_{\lambda}g = -g\,g_{\mu\nu}\,\partial_{\lambda}g^{\mu\nu}$.

---

[16] You might expect two more terms here (and another in 6.105), on the ground that we cannot vary $g^{\mu\nu}$ without simultaneously varying $g^{\nu\mu}$, to preserve symmetry. A small change in $g^{\mu\nu}$ would in this sense produce a double effect. But this is *not* the interpretation to be attached to $\partial/\partial g^{\mu\nu}$ in 6.101; the derivatives are *not* to be taken so as to preserve symmetry. This is a *convention*: we could do it the other way, but then a factor of 2 would be needed in 6.101.

[17] In the case of the Klein–Gordon field and the free electromagnetic field, the canonical stress tensor is already symmetric, so this virtue of 6.115 does not reveal itself.

6.3 Linearized Gravity

6.3.1 Simplifying the Field Equation

We return to the case of weak gravity (small perturbations from flat Minkowski space)—but this time including matter. As before (6.35) we assume

$$g_{\mu\nu} = \eta_{\mu\nu} + h_{\mu\nu}, \tag{6.117}$$

but now the field equation has a source term (6.74); referring to 6.63 and 6.69,

$$\tfrac{1}{2}\left[\Box^2 h_{\mu\nu} - \partial^\lambda(\partial_\mu h_{\lambda\nu} + \partial_\nu h_{\lambda\mu}) + \eta_{\mu\nu}\partial^\lambda\partial^\sigma h_{\lambda\sigma} + \partial_\mu\partial_\nu h - \eta_{\mu\nu}\Box^2 h\right] = -\alpha T_{\mu\nu}. \tag{6.118}$$

This can be simplified by a substitution and a change of coordinates:

- **Substitution:** Let

$$\gamma_{\mu\nu} \equiv h_{\mu\nu} - \tfrac{1}{2}\eta_{\mu\nu}h \tag{6.119}$$

(as before, $h \equiv h^\mu{}_\mu$). The trace of $\gamma_{\mu\nu}$ is

$$\gamma \equiv \gamma^\mu{}_\mu = h^\mu{}_\mu - \tfrac{1}{2}\delta^\mu_\mu h = h - \tfrac{1}{2}(4)h = -h, \tag{6.120}$$

so

$$h_{\mu\nu} = \gamma_{\mu\nu} - \tfrac{1}{2}\eta_{\mu\nu}\gamma. \tag{6.121}$$

With this, the term in square brackets in 6.118 becomes

$$\Box^2 \gamma_{\mu\nu} - \partial^\lambda(\partial_\mu\gamma_{\lambda\nu} + \partial_\nu\gamma_{\lambda\mu}) + \eta_{\mu\nu}\partial^\lambda\partial^\sigma\gamma_{\lambda\sigma} - \cancel{\partial_\mu\partial_\nu\gamma} + \cancel{\eta_{\mu\nu}\Box^2\gamma}$$
$$- \cancel{\tfrac{1}{2}\eta_{\mu\nu}\Box^2\gamma} + \tfrac{1}{2}\partial^\lambda(\cancel{\partial_\mu\eta_{\lambda\nu}\gamma + \partial_\nu\eta_{\lambda\mu}\gamma}) - \cancel{\tfrac{1}{2}\eta_{\mu\nu}\partial^\lambda\partial^\sigma\eta_{\lambda\sigma}\gamma}. \tag{6.122}$$

This removes the trace terms, leaving

$$\tfrac{1}{2}\left[\Box^2\gamma_{\mu\nu} - \partial^\lambda(\partial_\mu\gamma_{\lambda\nu} + \partial_\nu\gamma_{\lambda\mu}) + \eta_{\mu\nu}\partial^\lambda\partial^\sigma\gamma_{\lambda\sigma}\right] = -\alpha T_{\mu\nu}. \tag{6.123}$$

- **Coordinate change:** We are always free to make a coordinate change (as long as it keeps $h_{\mu\nu}$ infinitesimal):

$$x^\mu \to x^\mu + A^\mu(x). \tag{6.124}$$

By the argument leading to Eq. 6.49, this entails

$$h_{\mu\nu} \to h_{\mu\nu} - (\partial_\mu A_\nu + \partial_\nu A_\mu), \quad h \to h - 2\partial^\lambda A_\lambda, \tag{6.125}$$

and

$$\gamma_{\mu\nu} \to h_{\mu\nu} - (\partial_\mu A_\nu + \partial_\nu A_\mu) - \tfrac{1}{2}\eta_{\mu\nu}h + \eta_{\mu\nu}\partial^\lambda A_\lambda$$
$$= \gamma_{\mu\nu} - (\partial_\mu A_\nu + \partial_\nu A_\mu) + \eta_{\mu\nu}\partial^\lambda A_\lambda, \tag{6.126}$$
$$\partial^\nu\gamma_{\mu\nu} \to \partial^\nu\gamma_{\mu\nu} - \partial_\mu\partial^\nu A_\nu - \Box^2 A_\mu + \partial_\mu\partial^\lambda A_\lambda = \partial^\nu\gamma_{\mu\nu} - \Box^2 A_\mu. \tag{6.127}$$

(Equation 6.127 is valid to first order—there are, of course, corrections of second order due to the replacement of x by $x + A$ in the derivative and $g_{\mu\nu}$ by $\eta_{\mu\nu}$ in raising and lowering indices.)

So far A is arbitrary; now we choose

$$\Box^2 A_\mu = \partial^\nu\gamma_{\mu\nu}. \tag{6.128}$$

This defines the coordinate transformation (it fixes the **gauge** we work in and is analogous to choosing the Lorenz gauge in electrodynamics). In these coordinates[18]

$$\partial^\nu\gamma_{\mu\nu} = 0, \tag{6.129}$$

and the linearized field equation (6.123) reduces to

$$\Box^2\gamma_{\mu\nu} = -2\alpha T_{\mu\nu}. \tag{6.130}$$

We recognize here the inhomogeneous wave equation (3.54). Structurally, the linearized form of Einstein's equation closely resembles Maxwell's equations, $\Box^2 A_\mu = J_\mu$ (3.32), but with tensors in place of vectors.

6.3.2 Recovering Newton's Law

Does this theory contain Newton's law for the gravitational force between distant bodies? Suppose there is a "particle" of mass M, at rest at the origin. What is its gravitational field? The stress tensor is (2.169)

$$T_{00} = M\,\delta^3(\mathbf{x}) \quad \text{(all other components zero).} \tag{6.131}$$

There is no time dependence, so 6.130 says

$$\Box^2\gamma_{00} = -\nabla^2\gamma_{00} = -2\alpha M\,\delta^3(\mathbf{x}). \tag{6.132}$$

---

[18] Eds. The gauge in which 6.129 holds has many names in the literature: Einstein, Hilbert, de Donder, Fock, and even Loren(t)z, and the coordinates are called **harmonic**. See Sean Carroll: ned.ipac.caltech.edu/level5/March01/Carroll3/Carroll6.html.

This is Poisson's equation for a point source, and we know from classical electro-dynamics that the solution is[19]

$$\gamma_{00} = -\frac{\alpha M}{2\pi r} \tag{6.133}$$

(while all the other components of $\gamma_{\mu\nu}$ are zero). It follows that

$$\gamma = \gamma^0{}_0 = -\frac{\alpha M}{2\pi r}, \tag{6.134}$$

and hence

$$h_{00} = \gamma_{00} - \frac{1}{2}\eta_{00}\gamma = \gamma_{00} - \frac{1}{2}\gamma = -\frac{\alpha M}{4\pi r}. \tag{6.135}$$

In this space a "free" particle travels along a geodesic (6.12):

$$\frac{dv^\mu}{d\tau} + \Gamma^\mu{}_{\lambda\sigma} v^\lambda v^\sigma = 0, \quad \text{with} \quad v^\mu = \frac{dx^\mu}{d\tau}. \tag{6.136}$$

In particular, the spatial components are[20]

$$\frac{dv^i}{d\tau} = -\Gamma^i{}_{00}. \tag{6.137}$$

Now (5.110)

$$\Gamma_{i,00} = \frac{1}{2}(\partial_0 g_{0i} + \partial_0 g_{i0} - \partial_i g_{00}) = -\frac{1}{2}\partial_i g_{00} = -\frac{1}{2}\partial_i h_{00}$$

$$= \frac{\alpha M}{8\pi}\partial_i\left(\frac{1}{r}\right), \tag{6.138}$$

so (5.111)

$$\Gamma^i{}_{00} = \eta^{ij}\Gamma_{j,00} = -\frac{\alpha M}{8\pi}\partial_i\left(\frac{1}{r}\right) = -\frac{dv^i}{d\tau}. \tag{6.139}$$

The acceleration of the particle is therefore

$$\mathbf{a} = \frac{\alpha M}{8\pi}\nabla\left(\frac{1}{r}\right), \tag{6.140}$$

and if we take

$$\boxed{\alpha = 8\pi G,} \tag{6.141}$$

---

[19] Eds. See, for example, D. J. Griffiths, *Introduction to Electrodynamics*, 4th ed., Cambridge U. P., Cambridge, UK (2017), Eq. 1.102.

[20] Note that Γ is already first order, and for comparison with Newton's law we want nonrelativistic motion, so terms involving $v^i v^j$ and $v^0 v^i$ are negligible.

the gravitational "force" is

$$\mathbf{F} = m\mathbf{a} = -\frac{GmM}{r^2}\,\hat{\mathbf{r}}. \tag{6.142}$$

Thus we recover Newton's law, in the weak gravity, nonrelativistic limit, and 6.141 fixes the value of α in Einstein's field equation.

With the constants κ and α determined (in terms of Newton's G), I can finally write the Lagrangian for the gravitational field (6.76) and Einstein's equation (6.74):

$$\boxed{\mathcal{L}_g = \frac{1}{16\pi G}\,R,} \tag{6.143}$$

$$\boxed{G_{\mu\nu} = -8\pi G\,T_{\mu\nu}.} \tag{6.144}$$

Example 6.2

Consider a static localized source ($T_{\mu\nu} = 0$ outside some finite sphere). In the linearized approximation (6.130 and 6.141), $\nabla^2 \gamma_{\mu\nu} = 16\pi G\,T_{\mu\nu}$. This is again Poisson's equation, and from electrodynamics we know the solution:

$$\gamma_{\mu\nu}(\mathbf{r}) = -4G \int \frac{T_{\mu\nu}(\mathbf{r}')}{|\mathbf{r} - \mathbf{r}'|}\, d^3r'. \tag{6.145}$$

At large distances, $r \gg r'$ for all points in the domain of integration, so

$$\gamma_{\mu\nu}(\mathbf{r}) \approx -\frac{4G}{r} \int T_{\mu\nu}(\mathbf{r}')\, d^3r'. \tag{6.146}$$

In particular,

$$\gamma_{00} \approx -\frac{4GM}{r}, \tag{6.147}$$

where $M \equiv \int T_{00}\, d^3r$ is the total mass of the source. The other components of $\gamma_{\mu\nu}$ are all zero.

Proof: Since the stress tensor is divergenceless,

$$0 = \partial^0 \int r_j T_{0\mu}\, d^3r = \int r_j \partial^0 T_{0\mu}\, d^3r = -\int r_j \partial^i T_{i\mu}\, d^3r = \int (\partial^i r_j) T_{i\mu}\, d^3r$$

$$= \int \delta^i_j T_{i\mu}\, d^3r = \int T_{j\mu}\, d^3r. \quad \text{QED} \tag{6.148}$$

Thus

$$\gamma_{\mu\nu} = -\frac{4GM}{r}\,\delta^0_\mu \delta^0_\nu + \mathcal{O}\!\left(\frac{1}{r^2}\right). \tag{6.149}$$

Problem 6.4

In 6.132 we considered just *one* component ($\mu\nu = 00$) of the field equation. What about all the others?

(a) Show that $\gamma_{\mu\nu} = 0$ unless μ and ν are both zero.

(b) Find all the elements of $h_{\mu\nu}$.

(c) Show that the **Newtonian line element** is

$$ds^2 = \left(1 - \frac{2GM}{r}\right) dt^2 - \left(1 + \frac{2GM}{r}\right)(dx^2 + dy^2 + dz^2). \tag{6.150}$$

(d) Check that $\partial^\nu \gamma_{\mu\nu} = 0$, so the coordinates used in Section 6.3.2 are consistent with the coordinate change in Section 6.3.1.

Problem 6.5

Find the next term in the expansion 6.149. That is, find an expression that is accurate if terms of order $1/r^3$ are neglected. Show that this term can be expressed in terms of the total angular momentum. Choose the center of coordinates such that

$$\int \mathbf{r}\, T_{00}\, d^3r = \mathbf{0}.$$

6.3.3 Gravity Waves

In empty space ($T_{\mu\nu} = 0$) the linearized field equations (6.130) reduce to

$$\Box^2 \gamma_{\mu\nu} = 0, \tag{6.151}$$

provided (6.129) we use harmonic coordinates, such that

$$\partial^\nu \gamma_{\mu\nu} = 0. \tag{6.152}$$

These equations support the propagation of **gravitational waves**, just as Maxwell's equations (in vacuum) support electromagnetic waves. Let's first review that familiar case.

Electromagnetic waves. The field equations are

$$\partial_\mu F^{\mu\nu} = 0 \quad \Rightarrow \quad \Box^2 A_\mu - \partial_\mu \partial^\nu A_\nu = 0. \tag{6.153}$$

The fields ($F^{\mu\nu}$) are unchanged under a **gauge transformation**

$$A_\mu \to A_\mu + \partial_\mu \lambda, \tag{6.154}$$

where λ is any scalar function. Use this to choose the Lorenz gauge:

$$\partial^\nu A_\nu = 0. \tag{6.155}$$

Then the field equations reduce to

$$\Box^2 A_\mu = 0 \tag{6.156}$$

(this decouples the four components). There remains some gauge freedom: we can still perform a gauge transformation with any λ that satisfies

$$\Box^2 \lambda = 0. \tag{6.157}$$

We look for **plane-wave** solutions of the form

$$A_\mu = e_\mu e^{i\kappa \cdot x} \tag{6.158}$$

(the physical solution is the real or imaginary part of this). Equations 6.155 and 6.156 stipulate (respectively)

$$\kappa^\mu e_\mu = 0 \quad \text{and} \quad \kappa^2 = 0. \tag{6.159}$$

There are three possible choices for e_μ (three directions perpendicular to κ^μ). We'll take κ_μ to be

$$\kappa_\mu = (1,0,0,1) \tag{6.160}$$

(representing plane waves propagating in the z direction). In this case we can choose the three perpendicular vectors to be

$$\begin{aligned}
e_\mu^{(1)} &= (0,1,0,0) = (0,\hat{\mathbf{x}}), \\
e_\mu^{(2)} &= (0,0,1,0) = (0,\hat{\mathbf{y}}), \\
e_\mu^{(3)} &= (1,0,0,1) = (1,\hat{\mathbf{z}}) = \kappa_\mu.
\end{aligned} \tag{6.161}$$

But the third of these is spurious, because a gauge transformation (6.154) with

$$\lambda = i e^{i\kappa \cdot x} \tag{6.162}$$

(which satisfies 6.157) kills it:

$$A_\mu \to A_\mu + \partial_\mu \lambda = \kappa_\mu e^{i\kappa \cdot x} + i\left(i\kappa_\mu e^{i\kappa \cdot x}\right) = 0. \tag{6.163}$$

Thus there are just *two* linearly independent polarizations for plane electromagnetic waves.

We can also construct states of circular polarization, by forming appropriate linear combinations. Rotation by an angle θ about the z-axis takes

$$\begin{aligned}
\hat{\mathbf{x}} &\to \hat{\mathbf{x}} \cos\theta - \hat{\mathbf{y}} \sin\theta, \\
\hat{\mathbf{y}} &\to \hat{\mathbf{x}} \sin\theta + \hat{\mathbf{y}} \cos\theta,
\end{aligned} \tag{6.164}$$

so

$$\begin{aligned}
\hat{\mathbf{e}}^{(\pm)} \equiv \hat{\mathbf{x}} \pm i \hat{\mathbf{y}} &\to \hat{\mathbf{x}} (\cos\theta \pm i \sin\theta) + \hat{\mathbf{y}} (-\sin\theta \pm i \cos\theta) \\
&= e^{\pm i\theta} (\hat{\mathbf{x}} \pm i \hat{\mathbf{y}}) = e^{\pm i\theta} \hat{\mathbf{e}}^{(\pm)}.
\end{aligned} \tag{6.165}$$

If we were to quantize this system,[21] the de Broglie formula would say that the momentum is related to the wavelength:

$$p = \hbar\kappa, \quad p^2 = m^2 = \hbar^2\kappa^2 = 0, \tag{6.166}$$

so this represents a particle (the **photon**) of mass zero. It admits two helicity states (6.165), with eigenvalues ± 1, so it carries spin 1.[22]

How do you know there are no gauge transformations (like 6.162) that would eliminate $e_\mu^{(1)}$ or $e_\mu^{(2)}$? Look at the field tensor

$$F_{\mu\nu}^{(n)} = i e^{i\kappa \cdot x}(\kappa_\mu e_\nu^{(n)} - \kappa_\nu e_\mu^{(n)}); \tag{6.167}$$

$F_{\mu\nu}^{(1)}$ and $F_{\mu\nu}^{(2)}$ are independent solutions that cannot be transformed away (since $F_{\mu\nu}$ is gauge invariant), whereas

$$F_{\mu\nu}^{(3)} = i e^{i\kappa \cdot x}(\kappa_\mu\kappa_\nu - \kappa_\nu\kappa_\mu) = 0, \tag{6.168}$$

so this one is indeed spurious.

Gravitational waves. Now, by close analogy, we develop the theory of gravity waves. In this case a gauge transformation is a change of coordinates (6.124, 6.126, and 6.127):

$$\begin{aligned}
x^\mu &\to x^\mu + A^\mu, \\
\gamma_{\mu\nu} &\to \gamma_{\mu\nu} - (\partial_\mu A_\nu + \partial_\nu A_\mu) + \eta_{\mu\nu}\partial^\lambda A_\lambda, \\
\partial^\mu\gamma_{\mu\nu} &\to \partial^\mu\gamma_{\mu\nu} - \Box^2 A_\nu,
\end{aligned} \tag{6.169}$$

and we choose the "gauge function" A (not to be confused with the electromagnetic vector potential A!) to satisfy $\Box^2 A_\nu = \partial^\mu\gamma_{\mu\nu}$, so that $\partial^\mu\gamma_{\mu\nu} = 0$ (harmonic coordinates). But (as in electrodynamics) we can make a *further* gauge transformation, provided

$$\Box^2 A_\nu = 0. \tag{6.170}$$

Again, we look for plane-wave solutions,

$$\gamma_{\mu\nu} = E_{\mu\nu}e^{i\kappa \cdot x}. \tag{6.171}$$

Since $\gamma_{\mu\nu}$ is symmetric, we start with ten independent E's. But the field equation (6.151) and the gauge condition (6.152) require (respectively)

$$\kappa^2 = 0, \quad \kappa^\nu E_{\mu\nu} = 0. \tag{6.172}$$

---

[21] A note about units: potential energy GM^2/r has the same units as rest energy Mc^2, so if we set $G = 1$ and $c = 1$, then mass is measured in meters. But if we set $\hbar = 1$ we get *another* (incompatible) way to measure mass in meters (or rather, m^{-1}): $h\nu = hc/\lambda \sim mc^2$. *Conclusion:* You can't set all three constants to 1.

[22] Eds. Massless particles with spin $s > 0$ allow for only two helicity states, unlike massive particles, which have $2s + 1$. This was proved in a famous paper by V. Bargmann and E. Wigner, *Proc. Nat. Acad. Sci.* **34**, 211 (1948), available at www.ncbi.nlm.nih.gov/pmc/articles/PMC1079095/.

The latter amounts to four linear equations—one for each μ—which reduce the number of independent E's from ten to six (the analogous step in electrodynamics reduced the number of e's from four to three). Again, we'll take κ to be

$$\kappa_\mu = (1,0,0,1), \tag{6.173}$$

representing plane waves propagating in the z direction. From the three vectors in 6.161 we can construct six symmetric tensors, all of them perpendicular to κ:

$$E^{(1)}_{\mu\nu} = e^{(1)}_\mu e^{(1)}_\nu,$$

$$E^{(2)}_{\mu\nu} = e^{(2)}_\mu e^{(2)}_\nu,$$

$$E^{(3)}_{\mu\nu} = e^{(3)}_\mu e^{(3)}_\nu,$$

$$E^{(4)}_{\mu\nu} = \tfrac{1}{2}\left(e^{(1)}_\mu e^{(2)}_\nu + e^{(1)}_\nu e^{(2)}_\mu\right),$$

$$E^{(5)}_{\mu\nu} = \tfrac{1}{2}\left(e^{(2)}_\mu e^{(3)}_\nu + e^{(2)}_\nu e^{(3)}_\mu\right),$$

$$E^{(6)}_{\mu\nu} = \tfrac{1}{2}\left(e^{(3)}_\mu e^{(1)}_\nu + e^{(3)}_\nu e^{(1)}_\mu\right). \tag{6.174}$$

But by exploiting the remaining gauge freedom (6.170), using

$$A^{(n)}_\mu = -\frac{i}{2}e^{(n)}_\mu e^{i\kappa\cdot x}, \text{ where } n = 1,2,3, \text{ or } 4, \text{ with } e^{(4)}_\mu \equiv (1,0,0,-1), \tag{6.175}$$

we can "gauge away" four of the six (6.169). Specifically, $n = 1$ eliminates $E^{(6)}$:

$$\gamma_{\mu\nu} \rightarrow E^{(6)}_{\mu\nu}e^{i\kappa\cdot x} - (\partial_\mu A^{(1)}_\nu + \partial_\nu A^{(1)}_\mu) + \eta_{\mu\nu}\partial^\lambda A^{(1)}_\lambda$$

$$= \frac{1}{2}\left(e^{(3)}_\mu e^{(1)}_\nu + e^{(3)}_\nu e^{(1)}_\mu\right)e^{i\kappa\cdot x} - \left(\frac{-i}{2}\right)\left(e^{(1)}_\nu (i\kappa_\mu) + e^{(1)}_\mu (i\kappa_\nu)\right)e^{i\kappa\cdot x}$$

$$+ \eta_{\mu\nu}\left(\frac{-i}{2}\right)\left(e^{(1)}_\lambda (i\kappa^\lambda)\right)e^{i\kappa\cdot x}$$

$$= \frac{1}{2}\left[\left(\kappa_\mu e^{(1)}_\nu + \kappa_\nu e^{(1)}_\mu\right) - \left(e^{(1)}_\nu \kappa_\mu + e^{(1)}_\mu \kappa_\nu\right) + \eta_{\mu\nu}(0)\right]e^{i\kappa\cdot x} = 0. \tag{6.176}$$

Similarly, $n = 2$ eliminates $E^{(5)}$. And $n = 3$ kills $E^{(3)}$:

$$\gamma_{\mu\nu} \rightarrow E^{(3)}_{\mu\nu}e^{i\kappa\cdot x} - (\partial_\mu A^{(3)}_\nu + \partial_\nu A^{(3)}_\mu) + \eta_{\mu\nu}\partial^\lambda A^{(3)}_\lambda$$

$$= \left[\kappa_\mu \kappa_\nu - \left(\frac{-i}{2}\right)\left(\kappa_\nu (i\kappa_\mu) + \kappa_\mu (i\kappa_\nu)\right) + \eta_{\mu\nu}\left(\frac{-i}{2}\right)\kappa_\lambda (i\kappa^\lambda)\right]e^{i\kappa\cdot x}$$

$$= 0. \tag{6.177}$$

The last one ($n=4$) is more tricky—it eliminates the linear combination $(E^{(1)} + E^{(2)})$:

$$\gamma_{\mu\nu} \rightarrow \left(E^{(1)}_{\mu\nu} + E^{(2)}_{\mu\nu}\right) e^{i\kappa \cdot x} - \left(\partial_\mu A^{(4)}_\nu + \partial_\nu A^{(4)}_\mu\right) + \eta_{\mu\nu} \partial^\lambda A^{(4)}_\lambda$$

$$= \left[\left(e^{(1)}_\mu e^{(1)}_\nu + e^{(2)}_\mu e^{(2)}_\nu\right) - \left(\frac{-i}{2}\right)\left(e^{(4)}_\nu (i\kappa_\mu) + e^{(4)}_\mu (i\kappa_\nu)\right)\right.$$

$$\left. + \eta_{\mu\nu}\left(\frac{-i}{2}\right) e^{(4)}_\lambda (i\kappa^\lambda)\right] e^{i\kappa \cdot x}$$

$$= \frac{1}{2}\left[2\left(e^{(1)}_\mu e^{(1)}_\nu + e^{(2)}_\mu e^{(2)}_\nu\right) - \left(\kappa_\mu e^{(4)}_\nu + \kappa_\nu e^{(4)}_\mu\right) + \eta_{\mu\nu}\left(\kappa \cdot e^{(4)}\right)\right] e^{i\kappa \cdot x}. \tag{6.178}$$

Now $\kappa \cdot e^{(4)} = 2$, so

$$\kappa_\mu e^{(4)}_\nu + \kappa_\nu e^{(4)}_\mu - \eta_{\mu\nu}(\kappa \cdot e^{(4)}) = \begin{pmatrix} 0 & 0 & 0 & 0 \\ 0 & 2 & 0 & 0 \\ 0 & 0 & 2 & 0 \\ 0 & 0 & 0 & 0 \end{pmatrix}$$

$$= 2\left(e^{(1)}_\mu e^{(1)}_\nu + e^{(2)}_\mu e^{(2)}_\nu\right), \tag{6.179}$$

and $\gamma_{\mu\nu} \rightarrow 0$. This leaves the orthogonal combination,

$$E^{(7)}_{\mu\nu} \equiv \tfrac{1}{2}\left(e^{(1)}_\mu e^{(1)}_\nu - e^{(2)}_\mu e^{(2)}_\nu\right), \tag{6.180}$$

together with $E^{(4)}$, as the surviving polarization states. They correspond to plane polarization. The circular modes are linear combinations:

$$E^{(\pm)}_{\mu\nu} \equiv e^{(\pm)}_\mu e^{(\pm)}_\nu = \left(e^{(1)}_\mu \pm i e^{(2)}_\mu\right)\left(e^{(1)}_\nu \pm i e^{(2)}_\nu\right)$$

$$= \left(e^{(1)}_\mu e^{(1)}_\nu - e^{(2)}_\mu e^{(2)}_\nu\right) \pm i\left(e^{(1)}_\mu e^{(2)}_\nu + e^{(1)}_\nu e^{(2)}_\mu\right)$$

$$= 2\left(E^{(7)}_{\mu\nu} \pm i E^{(4)}_{\mu\nu}\right). \tag{6.181}$$

Under a rotation through angle θ, each $e^{(\pm)}$ picks up a factor of $e^{\pm i\theta}$ (6.165), so

$$E^{(\pm)} \rightarrow e^{\pm 2i\theta} E^{(\pm)}. \tag{6.182}$$

If we were to quantize this system,

$$p = \hbar\kappa, \quad p^2 = m^2 = \hbar^2 \kappa^2 = 0. \tag{6.183}$$

So it represents a particle (the **graviton**) of mass zero; it admits two helicity states (with eigenvalues ± 2), so it carries spin 2.

How do we know there are no gauge transformations (beyond 6.175) that would eliminate one or both of the remaining polarizations? We need to construct a *gauge-invariant* quantity (analogous to $F_{\mu\nu}$ in electrodynamics) that we can show admits two independent solutions. The Riemann tensor would be a natural thing to try. I'll let you work out the details (Problem 6.6).

Comments:

- In electrodynamics an oscillating monopole (a charged spherical shell, for example, whose radius changes sinusoidally) does not radiate, because the spin of the photon is 1; the dominant radiation is typically dipole. In general relativity an oscillating dipole cannot radiate, because the spin of the graviton is 2; quadrupole oscillation is the lowest order that can produce gravitational waves.
- The solutions we have studied are not *exact*, since they were based on a linearized approximation to Einstein's equation. (By contrast, the electromagnetic waves we considered earlier *are* exact solutions to Maxwell's equations.) The missing nonlinear terms reflect the fact that energy in the gravitational field generates its *own* gravity. But presumably the exact solutions reduce to these (or wave packets built from them) in the weak-field limit.

Problem 6.6

Show that the Riemann tensor, to lowest order in $h_{\mu\nu}$, is

$$R_{\mu\nu\lambda\sigma} = \tfrac{1}{2}(\partial_\sigma \partial_\nu h_{\mu\lambda} - \partial_\sigma \partial_\mu h_{\lambda\nu} - \partial_\lambda \partial_\nu h_{\mu\sigma} + \partial_\lambda \partial_\mu h_{\sigma\nu}). \tag{6.184}$$

(Locally geodesic coordinates will do—they differ from those stipulated in 6.129 only in second order.) Check that it is gauge invariant. For the circularly polarized gravitational waves

$$\gamma_{\mu\nu} = E^{(\pm)}_{\mu\nu} e^{i\kappa \cdot x}, \tag{6.185}$$

show that

$$R_{\mu\nu\lambda\sigma} = \tfrac{1}{2} e^{i\kappa \cdot x} \left(\kappa_\mu e^{(\pm)}_\nu - \kappa_\nu e^{(\pm)}_\mu \right) \left(\kappa_\sigma e^{(\pm)}_\lambda - \kappa_\lambda e^{(\pm)}_\sigma \right), \tag{6.186}$$

and confirm that these solutions cannot be eliminated by a gauge transformation. Contract twice, to get $R_{\mu\nu}$ and R: Are they what you expected?

Problem 6.7

We found in Section 6.2.3 that the tidal force (the differential acceleration between two particles separated by a small vector δ^μ) is given by

$$D^2_\tau \delta^\mu = R^\mu{}_{\nu\sigma\lambda} v^\nu \delta^\lambda v^\sigma, \tag{6.187}$$

where v^μ is the velocity of the particles. Show that for the plane waves in Section 6.3.3,

$$\gamma_{\mu\nu} = E_{\mu\nu} e^{i\kappa \cdot x}, \tag{6.188}$$

and for particles at rest,

$$D_s^2 \delta^\mu \propto E^{\mu\nu} \delta_\nu. \tag{6.189}$$

(Work only to lowest order in h throughout.) Note that the 3-dimensional transversality of E implies that the tidal force acts only in the plane orthogonal to the direction of propagation of the wave, and the tracelessness of E implies that when the particles are being squeezed together in one direction, they are being pushed apart in the orthogonal direction.[23]

Problem 6.8

Suppose the field $\phi(x)$ is a scalar density of weight 1 (5.35), so that under a change in coordinates it transforms according to

$$\phi(x) \rightarrow \phi'(x') = J\phi(x),$$

where $J \equiv |\partial(x)/\partial(x')|$ is the Jacobian determinant of the transformation matrix $\partial x^\alpha / \partial x'^\beta$. The derivative $\partial_\mu \phi$ does *not* transform as a vector density, because

$$\partial_\mu \phi \rightarrow \partial'_\mu \phi' = \frac{\partial x^\alpha}{\partial x'^\mu} \partial_\alpha (J\phi) \neq J \frac{\partial x^\alpha}{\partial x'^\mu} \partial_\alpha \phi.$$

To get around this, we introduce a "gauge field" B_μ, and an associated "gauge-covariant derivative" D_μ,

$$D_\mu \phi \equiv \partial_\mu \phi + B_\mu \phi,$$

such that

$$D_\mu \phi \rightarrow D'_\mu \phi' = \partial'_\mu \phi' + B'_\mu \phi' = J \frac{\partial x^\alpha}{\partial x'^\mu} (\partial_\alpha \phi + B_\alpha \phi).$$

That is, the gauge-covariant derivative $D_\mu \phi$ transforms as a covariant vector density of weight 1.

(a) Show that the transformation law for the field B_μ is

$$B_\mu \rightarrow B'_\mu = \frac{\partial x^\alpha}{\partial x'^\mu} (B_\alpha - J^{-1} \partial_\alpha J).$$

(b) Using the identity 6.87, show that

$$\partial_\mu J = J \left(\frac{\partial x'^\alpha}{\partial x^\beta} \right) \partial_\mu \left(\frac{\partial x^\beta}{\partial x'^\alpha} \right).$$

---

[23] Eds. It is precisely this characteristic squeeze/stretch motion that was measured by LIGO on September 14, 2015, in the first direct observation of gravity waves: B. P. Abbott *et al.*, *Phys. Rev. Lett.* **116** (6): 061102 (2016). Indirect evidence for their existence was obtained earlier, from the decay of the orbit of binary pulsars: J. H. Taylor and J. M. Weisberg, *ApJ* **253** 908 (1982).

(c) Show, using 5.67 and Problem 5.8, that under a coordinate transformation,

$$\Gamma^\lambda_{\mu\lambda} \rightarrow \Gamma'^\lambda_{\mu\lambda} = \frac{\partial x^\nu}{\partial x'^\mu}\left(\Gamma^\lambda_{\nu\lambda} + \frac{\partial x'^\alpha}{\partial x^\beta}\frac{\partial^2 x^\beta}{\partial x^\nu \partial x'^\alpha}\right),$$

and hence an appropriate choice of B_μ is $-\Gamma^\alpha_{\mu\alpha}$. Thus the gauge-covariant derivative of a tensor density is nothing but the usual covariant derivative:[24]

$$D_\mu\phi = \nabla_\mu\phi = \partial_\mu\phi - \Gamma^\alpha_{\mu\alpha}\phi.$$

The extension to derivatives of densities of weight w is straightforward, given the identity 6.85; if C^μ is a vector density of weight w, then

$$\nabla_\alpha C^\mu = \partial_\alpha C^\mu + \Gamma^\mu_{\alpha\nu}C^\nu - w\Gamma^\beta_{\alpha\beta}C^\mu.$$

---

[24] Eds. The close analogy between covariant derivatives in general relativity and gauge-covariant derivatives in electromagnetism was pointed out by Weyl in 1918, and again in 1929. Following the groundbreaking paper of Yang and Mills in 1954 on a gauge theory based on the non-Abelian group SU(2), Utiyama pointed out in 1956 that general relativity could be regarded as a gauge theory of the Lorentz group. See L. O'Raifeartaigh, *The Dawning of Gauge Theory*, Princeton U. P. (1997) (which includes reprints of the Weyl, Utiyama, and Yang–Mills papers); for gravity considered as a gauge theory, as well as relevant references, see F. W. Hehl (2014), arXiv:1204.3672v2 [gr-qc].

7

The Schwarzschild Solution

The next step is to look for *exact* solutions to Einstein's equation. Not many of these are known, but already in 1916 Schwarzschild obtained the gravitational analog to Coulomb's law: the metric produced by a point mass M at rest at the origin (or external to a stationary spherical mass). Presumably the metric is *static* (independent of time) and *spherically symmetric*; this suggests we try something of the form

$$ds^2 = f(r)\,dt^2 - g(r)\,dr^2 - h(r)\,d\Omega^2, \tag{7.1}$$

where $d\Omega^2 \equiv (d\theta^2 + \sin^2\theta\,d\phi^2)$. But in a theory that allows us to use any old coordinates, what's static to one person may not be static to somebody else (who defines time differently), and it's not clear what r *means* (it's not necessarily the distance from the origin). So I'm going to take a moment to discuss **isometries** (coordinate-independent symmetries), to put the *ansatz* 7.1 on a firmer foundation.[1]

7.1 Isometries

Einstein's equation

$$G_{\mu\nu} = -8\pi G\,T_{\mu\nu} \tag{7.2}$$

is an inhomogeneous differential equation for $G_{\mu\nu}$, which is a symmetric tensor, so there are ten unknown functions of four variables (x^μ) to be determined. (Ultimately we will extract from $G_{\mu\nu}$ the metric, $g_{\mu\nu}$, but that is still ten functions of four variables.)[2] To simplify matters, we look for solutions with a high degree

---

[1] If you are prepared to accept 7.1, feel free to skip the next section.
[2] Eds. Actually, Coleman's count is not quite right. Local conservation of energy–momentum requires that the (covariant) divergence of the stress tensor vanish, which imposes four constraints on the elements of $T_{\mu\nu}$, and the Bianchi identities guarantee the matching zero divergence of $G_{\mu\nu}$ (Problem 7.2(c)). So in fact there are only six independent functions to be determined by Einstein's equation. As for the metric, it is subject to fourfold gauge invariance (Eq. 6.47)—again, just six independent functions.

of symmetry. For instance, we might have reason to believe that the metric (in some particular context) should be constant in time, or spherically symmetrical. The trouble is, what is constant (or a function of r) to one person may *not* be to somebody else (who uses a different set of coordinates). First, then, we need a coordinate-free definition of symmetry.

Definition: An **isometry** is a mapping from a Riemannian manifold to itself,

$$T : \mathcal{M} \to \mathcal{M}, \tag{7.3}$$

that is bijective, continuous, has continuous derivatives, and *preserves lengths*.

For example, rotations are isometries of a Euclidean manifold, and Lorentz transformations are isometries of the Minkowski manifold; the group of isometries of an infinite circular cylinder would consist of longitudinal translations, rotations about the axis, and mirror reflections in a plane perpendicular to the axis. We are particularly interested in solutions exhibiting three isometries: time translation invariance, time-reversal invariance, and rotational invariance.

(1) Time translation invariance

Suppose we have a continuous group of isometries, parameterized by a variable τ, on the spacetime manifold:

$$T(\tau_1)\, T(\tau_2) = T(\tau_1 + \tau_2). \tag{7.4}$$

An **orbit** is a set of points (in the manifold) transformed, one into the next, by a succession of these isometries, as τ advances. The tangent vectors to an orbit form a continuous vector field. Suppose that all these tangent vectors are *time-like*. Spacetime consists of a bundle of such orbits—one through each point. Pick a surface through which each orbit passes just once.[3] This generates another such surface for each τ, and a succession of them fills out a volume in the manifold.

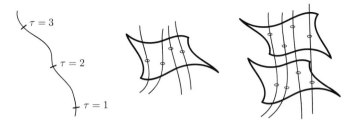

---

[3] In principle an orbit might double back, penetrating the surface again. This is impossible for time-like curves in Minkowski space (and it would raise profound philosophical problems), but it can occur in curved spacetimes. If necessary we'll stick to a local region so as to eliminate the second crossing.

In this case we can define the time coordinate as the value of τ: $t \equiv \tau$. We introduce the spatial coordinates on one (arbitrarily chosen) surface (say, the one at $t = 0$); then operation by T defines the coordinates on the subsequent surfaces. In these coordinates, the element of length takes the form

$$ds^2 = g_{00}\, dt^2 + 2g_{0i}\, dx^i\, dt + g_{ij}\, dx^i\, dx^j, \tag{7.5}$$

with g's that are *independent of time*—that is, we have established a metric that is constant in time (as defined above). Solutions to Einstein's equations with this property are called **stationary**.

How do I know g_{00}, g_{0i}, and g_{ij} are independent of which τ-surface we are on? Suppose first that there is only a dt (i.e. we look at an infinitesimal line segment *along* an orbit). Because of the additivity of the parameterization (Eq. 7.4), a transformation $T(d\tau)$ just *adds* $d\tau$ to τ:

$$T(d\tau)\, T(\tau) = T(\tau + d\tau), \tag{7.6}$$

regardless of what τ is. Thus dt is unchanged by going from one τ-surface to another, and hence (since ds^2 is preserved by definition of an isometry), g_{00} must be the same for all τ-surfaces. Now suppose there is only a dx^1 (that is, $dx^2 = dx^3 = dt = 0$, on the "master" surface $\tau = 0$). Then by construction of the spatial coordinates there is *still* only a dx'^1 on the surface τ', and $dx'^1 = dx^1$. But

$$ds^2 = g_{11}(dx^1)^2 = ds'^2 = g'_{11}(dx'^1)^2, \tag{7.7}$$

so $g_{11} = g'_{11}$: it's independent of τ. And the same goes for the other components.

What we have done is to provide a coordinate-free way of defining time independence for a metric, so we can look for "stationary" solutions to Einstein's equation. How can we tell whether spacetime admits (in *some* coordinate system) a time-independent metric? The answer is that if the manifold possesses a continuous group $T(\tau)$ of isometries, such that (in a region $\mathcal{R}$) the tangent vectors to orbits are all time-like, then we can define time in such a way that the metric is time independent (in $\mathcal{R}$).

(2) Time-reversal invariance

If time reversal, T_r, is an isometry, it switches the sign of τ in the time translation isometries:

$$T_r\, T(\tau) = T(-\tau)\, T_r. \tag{7.8}$$

But there are many operations that change the sign of time—for example, (time reversal) $\times$ (parity). So we require in addition that

$$T_r(\text{entire orbit of } T(\tau)) = \text{same orbit}. \tag{7.9}$$

We take 7.8 and 7.9 together to define **time-reversal invariance**, and a stationary solution that is also time-reversal invariant is called **static**.

We can use time-reversal invariance to simplify the metric further. By definition of $T(\tau)$,

$$T(\tau)(t, x^i) = (t + \tau, x^i). \tag{7.10}$$

Now

$$T_r(t_0, x^i) = (t_0', x^i) \tag{7.11}$$

(it's on the same orbit, but the parameter could be different). But from 7.8,

$$T_r\, T(\tau)(t_0, x^i) = T(-\tau)\, T_r(t_0, x^i) = T(-\tau)(t_0', x^i), \tag{7.12}$$

so

$$T_r(t_0 + \tau, x^i) = (t_0' - \tau, x^i). \tag{7.13}$$

In particular, choosing $\tau = \frac{1}{2}(t_0' - t_0)$,

$$T_r\left(\frac{t_0 + t_0'}{2}, x^i\right) = \left(\frac{t_0 + t_0'}{2}, x^i\right). \tag{7.14}$$

Thus there is a fixed point (under T_r) on each orbit. Before, we picked the $t = \tau = 0$ surface arbitrarily, but now we choose $t = 0$ to be the surface given by these fixed points. Then

$$T_r(t, x^i) = (-t, x^i). \tag{7.15}$$

Invariance under T_r means the $dx\, dt$ cross terms in $g_{\mu\nu}$ must vanish:

$$ds^2 = g_{00}\,(dt)^2 + g_{ij}\, dx^i\, dx^j \tag{7.16}$$

for static solutions.

(3) Rotational invariance
We assume now an additional isometry: rotational invariance. Let R denote an element of the 3-dimensional rotation group SO(3); the corresponding isometry is $T(R)$, with

$$T(R_1)T(R_2) = T(R_1 R_2). \tag{7.17}$$

We assume it commutes with time translations and time reversal:

$$T(R)T(\tau) = T(\tau)T(R), \quad T_r T(R) = T(R)T_r. \tag{7.18}$$

The first of these implies that surfaces at constant time must be invariant under rotations. In particular, this is true of the $t = 0$ surface. We make two additional assumptions: (a) there exist only isolated one-point orbits under $T(R)$ (think the center of rotation) and (b) on any orbit $T(R_z(\alpha))$ there is at least one point P_N (the "north pole") such that

$$T(R_z(\alpha))P_N = P_N \tag{7.19}$$

(where α is the angle of rotation and z the axis).

I claim that we can put these isometries into a one-to-one correspondence with the 2-dimensional sphere in 3-dimensional Euclidean space. Let $R(\theta,\phi)$ be a rotation in the $\phi =$ constant plane, down from the north pole to θ:

$$T(R(\theta,\phi))P_N = P(\theta,\phi). \tag{7.20}$$

Now *any* rotation can be written as

$$R = R_z(\alpha)R(\theta,\phi) = R(\theta',\phi')R_z(\alpha'). \tag{7.21}$$

Thus we have divided up the $t = 0$ (hyper)surface into nested spheres; label them by the parameter r. For the static spherically symmetric case, then,[4]

$$ds^2 = f(r)\,dt^2 - g(r)\,dr^2 - h(r)\left(d\theta^2 + \sin^2\theta\,d\phi^2\right). \tag{7.22}$$

We can still choose the definition of r. What would be a convenient measure? The circumference of an equatorial great circle ($\theta = \pi/2, \phi: 0 \to 2\pi$) is $2\pi\sqrt{h(r)}$. Why not pick this circumference, divided by 2π, as our "radial" coordinate[5]—then $h(r) = r^2$, and

$$\boxed{ds^2 = f(r)\,dt^2 - g(r)\,dr^2 - r^2\,(d\theta^2 + \sin^2\theta\,d\phi^2).} \tag{7.23}$$

This is the line element for **spherically symmetric static solutions**.

7.2 The Exterior Solution

In this section we look for the gravitational field of a static point mass—the gravitational analog to Coulomb's law. Or (better) a stationary spherical sun at the center of the universe. We assume the field is static, and rotationally invariant:

$$ds^2 = e^\lambda\,dt^2 - e^\mu\,dr^2 - r^2\left(d\theta^2 + \sin^2\theta\,d\phi^2\right), \tag{7.24}$$

---

[4] What about cross terms of the form $dr\,d\theta$ and $dr\,d\phi$? They would transform (under rotations) as components of a 2-vector, and thus would not be rotationally invariant. Similarly, $d\theta\,d\phi$ would transform as a tensor.

[5] Are we sure the circumferences change size monotonically? For instance, on a cylinder the circumferences are all the same, and on a hyperboloid of revolution they decrease as you approach the neck, and then increase. But for now we shall assume that this is not a problem. Why not use the distance from the center as our radial coordinate? That distance involves a potentially nasty integral over $g(r)$, whereas the circumference is very simple.

where $\lambda(r)$ and $\mu(r)$ are functions to be determined from Einstein's equation. (Presumably the metric is Minkowskian at large distances, so λ and μ go to 0 as $r \to \infty$.)

Outside the sun, where $T_{\mu\nu} = 0$, Einstein's equation says

$$G_{\mu\nu} = 0. \tag{7.25}$$

But

$$G_{\mu\nu} = R_{\mu\nu} - \tfrac{1}{2} g_{\mu\nu} R \quad \Rightarrow \quad G \equiv G^{\mu}{}_{\mu} = R - \tfrac{1}{2} 4R = -R. \tag{7.26}$$

Thus, quite generally,

$$R_{\mu\nu} = G_{\mu\nu} - \tfrac{1}{2} g_{\mu\nu} G, \tag{7.27}$$

and 7.25 reduces to

$$R_{\mu\nu} = 0. \tag{7.28}$$

Number the coordinates in the obvious way:

$$x^0 = t, \; x^1 = r, \; x^2 = \theta, \; x^3 = \phi. \tag{7.29}$$

From 7.24, the nonzero elements in the metric and its inverse are

$$g_{00} = e^{\lambda}, \qquad g_{11} = -e^{\mu}, \qquad g_{22} = -r^2, \qquad g_{33} = -r^2 \sin^2 \theta, \tag{7.30}$$

$$g^{00} = e^{-\lambda}, \qquad g^{11} = -e^{-\mu}, \qquad g^{22} = -\frac{1}{r^2}, \qquad g^{33} = -\frac{1}{r^2 \sin^2 \theta}. \tag{7.31}$$

First we calculate the affine connections,

$$\Gamma_{\mu,\lambda\sigma} = \tfrac{1}{2} \left(-\partial_{\mu} g_{\lambda\sigma} + \partial_{\lambda} g_{\sigma\mu} + \partial_{\sigma} g_{\mu\lambda} \right), \qquad \Gamma^{\mu}_{\lambda\sigma} = g^{\mu\nu} \Gamma_{\nu,\lambda\sigma}. \tag{7.32}$$

Here $\Gamma_{\mu,\nu\lambda}$ is zero unless at least two indices are alike; the third index must be a 1 or a 2 (since $g_{\mu\nu}$ depends only on r and θ), and in the latter case the other two must be 3. Thus we can have permutations of $(1\mu\mu)$ and (233):

$$
\begin{array}{ll}
\Gamma_{1,00} = -\tfrac{1}{2} e^{\lambda} \lambda' & \Gamma^1_{00} = \tfrac{1}{2} e^{(\lambda-\mu)} \lambda' \\[4pt]
\Gamma_{0,10} = \Gamma_{0,01} = \tfrac{1}{2} e^{\lambda} \lambda' & \Gamma^0_{10} = \Gamma^0_{01} = \tfrac{1}{2} \lambda' \\[4pt]
\Gamma_{1,11} = -\tfrac{1}{2} e^{\mu} \mu' & \Gamma^1_{11} = \tfrac{1}{2} \mu' \\[4pt]
\Gamma_{1,22} = r & \Gamma^1_{22} = -r e^{-\mu} \\[4pt]
\Gamma_{2,12} = \Gamma_{2,21} = -r & \Gamma^2_{12} = \Gamma^2_{21} = \tfrac{1}{r} \\[4pt]
\Gamma_{1,33} = r \sin^2 \theta & \Gamma^1_{33} = -r e^{-\mu} \sin^2 \theta \\[4pt]
\Gamma_{3,13} = \Gamma_{3,31} = -r \sin^2 \theta & \Gamma^3_{13} = \Gamma^3_{31} = \tfrac{1}{r} \\[4pt]
\Gamma_{2,33} = r^2 \sin \theta \cos \theta & \Gamma^2_{33} = -\sin \theta \cos \theta \\[4pt]
\Gamma_{3,23} = \Gamma_{3,32} = -r^2 \sin \theta \cos \theta & \Gamma^3_{23} = \Gamma^3_{32} = \cot \theta
\end{array}
\tag{7.33}
$$

(the others are all zero). The (contracted) Riemann tensor is (6.77)

$$R_{\mu\nu} = \partial_\nu \Gamma^\lambda_{\lambda\mu} - \partial_\lambda \Gamma^\lambda_{\nu\mu} + \Gamma^\lambda_{\nu\rho} \Gamma^\rho_{\lambda\mu} - \Gamma^\lambda_{\lambda\rho} \Gamma^\rho_{\nu\mu}, \tag{7.34}$$

so

$$R_{00} = -\partial_1 \Gamma^1_{00} + \Gamma^1_{00} \Gamma^0_{10} + \Gamma^0_{01}\Gamma^1_{00} - \Gamma^1_{00}\left(\Gamma^0_{01} + \Gamma^1_{11} + \Gamma^2_{21} + \Gamma^3_{31}\right)$$

$$= -\frac{1}{2}e^{(\lambda-\mu)}\left[(\lambda' - \mu')\lambda' + \lambda''\right] + \frac{1}{2}e^{(\lambda-\mu)}\lambda'\left(\frac{1}{2}\lambda' - \frac{1}{2}\mu' - \frac{2}{r}\right)$$

$$= -\frac{1}{2}e^{(\lambda-\mu)}\left[\lambda'' + \frac{1}{2}\lambda'(\lambda' - \mu') + 2\frac{\lambda'}{r}\right] \tag{7.35}$$

and

$$R_{11} = \partial_1 \Gamma^0_{01} + \partial_1\Gamma^1_{11} + \partial_1\Gamma^2_{21} + \partial_1\Gamma^3_{31} - \partial_1\Gamma^1_{11} + \Gamma^0_{10}\Gamma^0_{01} + \Gamma^1_{11}\Gamma^1_{11}$$

$$+ \Gamma^2_{12}\Gamma^2_{21} + \Gamma^3_{13}\Gamma^3_{31} - \Gamma^1_{11}\left(\Gamma^0_{01} + \Gamma^1_{11} + \Gamma^2_{21} + \Gamma^3_{31}\right)$$

$$= \frac{1}{2}\lambda'' - \frac{2}{r^2} + \frac{1}{4}\lambda'^2 + \frac{2}{r^2} - \frac{1}{2}\mu'\left(\frac{\lambda'}{2} + \frac{2}{r}\right)$$

$$= \frac{1}{2}\left[\lambda'' + \frac{1}{2}\lambda'(\lambda' - \mu') - 2\frac{\mu'}{r}\right]. \tag{7.36}$$

Thus 7.28 already gives us two equations for λ and μ:

$$\lambda'' + \frac{1}{2}\lambda'(\lambda' - \mu') + 2\frac{\lambda'}{r} = 0, \tag{7.37}$$

$$\lambda'' + \frac{1}{2}\lambda'(\lambda' - \mu') - 2\frac{\mu'}{r} = 0. \tag{7.38}$$

Subtraction yields

$$(\lambda + \mu)' = 0, \tag{7.39}$$

so the sum, $(\lambda + \mu)$, is a constant. But both of them go to zero at infinity, so the sum is in fact 0:

$$\mu = -\lambda. \tag{7.40}$$

Putting this back into 7.37,

$$\lambda'' + \lambda'^2 + \frac{2}{r}\lambda' = 0. \tag{7.41}$$

Let

$$u \equiv e^\lambda, \tag{7.42}$$

so that

$$u' = \lambda' e^{\lambda} \quad \text{and} \quad u'' = (\lambda'^2 + \lambda'') e^{\lambda}. \tag{7.43}$$

Then

$$u'' + \frac{2}{r} u' = 0, \quad \text{or} \quad \frac{u''}{u'} = -\frac{2}{r}. \tag{7.44}$$

Integrating,

$$\ln u' = -2 \ln r + \ln a = \ln \left(\frac{a}{r^2} \right) \quad \Rightarrow \quad u' = \frac{a}{r^2} \quad \Rightarrow \quad u = -\frac{a}{r} + b \tag{7.45}$$

for some constants a and b. But $g_{00} = u \to 1$ as $r \to \infty$, so $b = 1$, and we conclude that

$$\boxed{g_{00} = \left(1 - \frac{a}{r} \right), \quad g_{11} = - \left(1 - \frac{a}{r} \right)^{-1}.} \tag{7.46}$$

The constant a can be determined either by matching boundary conditions at the surface of the "sun," or by fitting the Newtonian limit. Adopting the latter approach, we note that at very large r, where the gravity is weak, 6.135 and 6.141 tell us that

$$g_{00} = 1 + h_{00} = 1 - \frac{\alpha M}{4\pi r} = 1 - \frac{2MG}{r}, \tag{7.47}$$

so a, which is called the **Schwarzschild radius**, is given by[6]

$$\boxed{a = 2MG,} \tag{7.48}$$

where M is the mass of the "sun." For our sun, a is about 3 km. Notice that the metric is singular when $r = a$. At the moment we don't know whether this is a genuine singularity, or an indication that we have made a bad choice of coordinates. I'll address this question in due course, but for now we don't need to worry about it as long as the physical radius of the "sun" exceeds the Schwarzschild radius, because *inside* the sun a different metric prevails.

Problem 7.1

(a) Find R_{01}, R_{22}, and R_{33}, for a metric of the form 7.24.
(b) Evaluate these quantities for the Schwarzschild solution 7.46.

---

[6] Eds. In Section 6.3.2, r was the distance from the center, whereas in the present context r is the circumference of a great circle divided by 2π. In the former case there was no difference between the two "radii," but in the latter, at large distances, they differ by a constant. If this bothers you, assume we are far enough out that the offset is negligible, or wait for the matching boundary conditions to settle the matter more rigorously.

It is striking that we only needed two equations to determine $g_{\mu\nu}$. Given the structure of the metric (7.24) the off-diagonal elements of $R_{\mu\nu}$ *automatically* vanish, and once we have set R_{00} and R_{11} to zero, R_{22} and R_{33} are forced also to zero (see Problem 7.1(b)). Of course, since there were only two functions to be determined ($\lambda(r)$ and $\mu(r)$), this *had* to be the case, or there wouldn't have been a consistent solution at all. Still, it is of some interest to know in advance that $R_{00} = 0$ and $R_{11} = 0$ will suffice. Problem 7.2 will guide you through a proof.

Problem 7.2

For a metric of the form 7.24 the Ricci tensor is diagonal (that is, $R_{\mu\nu} = 0$ for $\mu \neq \nu$), regardless of the functions $\lambda(r)$ and $\mu(r)$. This suggests a potentially useful *theorem:* if the metric is diagonal (in some coordinate system), so too is the Ricci tensor (in those coordinates).

(a) Prove it for 2-dimensional spaces. (Use Problem 5.14(a).)

(b) Unfortunately, the "theorem" is *not* true (in general) for spaces of higher dimension.[7] Find R_{xy} for $ds^2 = dx^2 + x\,dy^2 + y\,dz^2$.

(c) From the definition (6.69), show that

$$\nabla^\mu G_{\mu\nu} = 0 \qquad (7.49)$$

always. [*Hint:* Use Problem 5.15(b).] From the case $\nu = 1$, show that if $R_{00} = R_{11} = 0$ then R_{22} and R_{33} are automatically zero, for a metric of the form 7.23 (use the result of Problem 7.1(a)).

7.3 Classic Tests of General Relativity

According to the Schwarzschild metric,

$$ds^2 = e^\lambda\,dt^2 - e^\mu\,dr^2 - r^2(d\theta^2 + \sin^2\theta\,d\phi^2), \qquad (7.50)$$

where

$$e^\lambda = e^{-\mu} = \left(1 - \frac{a}{r}\right) \quad \text{and} \quad a = 2GM. \qquad (7.51)$$

There are four classic experimental tests of this result:[8]

0. recovery of Newton's law of universal gravitation, in the appropriate limit;
1. precession of the perihelion of Mercury;
2. deflection of starlight passing near the sun;
3. the gravitational redshift.

---

[7] K. Z. Win (1996), arXiv:gr-qc/9602015v1. The "theorem" does hold in the case of rotational and time-reversal invariance (such as 7.24), and also when the metric is a function of only one coordinate.

[8] Eds. When Einstein first proposed these tests, the Schwarzschild solution was not yet known; he used the linearized approximation of Section 6.3.

We have already discussed the "zeroth" test in Section 6.3.2; here we consider the other three.

7.3.1 Precession of the Perihelion of Mercury

Recall that free particles move on geodesics (Eq. 6.14):

$$\delta \int \left(\frac{dx^\mu}{ds} \frac{dx^\nu}{ds} g_{\mu\nu} \right)^{1/2} ds = \int \delta \left(e^\lambda \dot{t}^2 - e^\mu \dot{r}^2 - r^2 \dot{\theta}^2 - r^2 \sin^2 \theta \, \dot{\phi}^2 \right)^{1/2} ds = 0,$$
(7.52)

where the dots denote differentiation with respect to s. First we vary θ:

$$\delta \left(\sqrt{()} \right) = \frac{1}{2} \frac{1}{\sqrt{()}} \left[-r^2 2\dot{\theta} \, \delta\dot{\theta} - r^2 2 \sin\theta \cos\theta \dot{\phi}^2 \, \delta\theta \right].$$
(7.53)

But

$$\delta\dot{\theta} = \frac{d}{ds} \delta\theta,$$
(7.54)

and we integrate the first term by parts:

$$\delta \left(\sqrt{()} \right) \rightarrow \left\{ \frac{d}{ds} \left(\frac{1}{\sqrt{()}} r^2 \dot{\theta} \right) - \frac{1}{\sqrt{()}} \left(r^2 \sin\theta \cos\theta \, \dot{\phi}^2 \right) \right\} \delta\theta = 0.$$
(7.55)

As in the Newtonian case, one solution is $\theta = \pi/2$, $\dot{\theta} = 0$; that is, the motion is confined to the equatorial plane. (This is simply a sensible choice of coordinates: if we orient the axes so that *initially* $\theta = \pi/2$ and $\dot{\theta} = 0$, then the motion remains in that plane.) With this,

$$\delta \int \left(g_{\mu\nu} \dot{x}^\mu \dot{x}^\nu \right)^{1/2} ds = \int \delta \left(e^\lambda \dot{t}^2 - e^\mu \dot{r}^2 - r^2 \dot{\phi}^2 \right)^{1/2} ds = 0.$$
(7.56)

Now we vary ϕ:

$$\frac{d}{ds} \left(\frac{1}{\sqrt{()}} r^2 \dot{\phi} \right) = 0.$$
(7.57)

Choose the parameter s such that

$$\sqrt{()} = 1$$
(7.58)

(that is, use proper time as the parameter). Then

$$r^2 \dot{\phi} = \ell$$
(7.59)

is a constant (think angular momentum, though for that we should really be including a factor of mass). Similarly, varying t,

$$\frac{d}{ds}\left(\frac{1}{\sqrt{()}}e^{\lambda}\dot{t}\right) = 0, \tag{7.60}$$

so

$$e^{\lambda}\dot{t} = E, \tag{7.61}$$

another constant of the motion (think energy).

Finally, we could vary r, but it's a little easier to invoke 7.58:

$$e^{\lambda}\dot{t}^2 - e^{\mu}\dot{r}^2 - r^2\dot{\phi}^2 = 1. \tag{7.62}$$

Using 7.59 and 7.61 to eliminate $\dot{\phi}$ and $\dot{t}$,

$$e^{-\lambda}E^2 - e^{\mu}\dot{r}^2 - r^2\frac{\ell^2}{r^4} = 1. \tag{7.63}$$

Putting in the Schwarzschild solution (7.51),

$$E^2 - \dot{r}^2 - \frac{\ell^2}{r^2}\left(1 - \frac{a}{r}\right) = \left(1 - \frac{a}{r}\right), \tag{7.64}$$

or

$$\frac{1}{2}\left(E^2 - 1\right) - \frac{\dot{r}^2}{2} - \frac{1}{2}\frac{\ell^2}{r^2}\left(1 - \frac{a}{r}\right) + \frac{a}{2r} = 0. \tag{7.65}$$

The various terms can be interpreted (roughly) as follows:

$$\frac{1}{2}\left(E^2 - 1\right): \quad \text{constant total energy,} \tag{7.66}$$

$$\frac{\dot{r}^2}{2}: \quad \text{radial kinetic energy,} \tag{7.67}$$

$$\frac{1}{2}\frac{\ell^2}{r^2}\left(1 - \frac{a}{r}\right): \quad \text{angular kinetic energy,} \tag{7.68}$$

$$-\frac{a}{2r}: \quad \text{gravitational potential energy.} \tag{7.69}$$

Thus the whole thing can be interpreted as conservation of energy. Note that the a/r term in the angular kinetic energy (or, if you prefer, the centrifugal potential) is the only one that would *not* be there in the Newtonian theory (also the derivative in the radial kinetic energy is with respect to *proper* time). In the case of the planet Mercury, $a/r \approx 10^{-8}$, so we're looking at a very small correction.

We're interested in the *orbit*, $r(\phi)$, not $r(t)$. From 7.59,

$$\left(\frac{ds}{d\phi}\right)^2 = \frac{r^4}{\ell^2}; \tag{7.70}$$

multiplying 7.64 by this quantity,

$$E^2 \frac{r^4}{\ell^2} - \left(\frac{dr}{d\phi}\right)^2 - r^2\left(1 - \frac{a}{r}\right) - \frac{r^4}{\ell^2}\left(1 - \frac{a}{r}\right) = 0. \tag{7.71}$$

As in the classical Kepler problem, we make the substitution

$$u \equiv \frac{1}{r} \quad \Rightarrow \quad \frac{dr}{d\phi} = -\frac{1}{u^2}\frac{du}{d\phi}. \tag{7.72}$$

Then

$$E^2 \frac{1}{\ell^2 u^4} - \frac{1}{u^4}\left(\frac{du}{d\phi}\right)^2 - \frac{1}{u^2}(1 - au) - \frac{1}{\ell^2 u^4}(1 - au) = 0, \tag{7.73}$$

or

$$\frac{du}{d\phi} = \pm (P(u))^{1/2}, \quad \text{where} \quad P(u) \equiv \frac{E^2}{\ell^2} - u^2(1 - au) - \frac{1}{\ell^2}(1 - au). \tag{7.74}$$

The motion is restricted to regions where $P(u) \geq 0$ (else $du/d\phi$ is imaginary), and, of course, $u \geq 0$. Here $P(u)$ is a cubic polynomial, with the generic form (a):

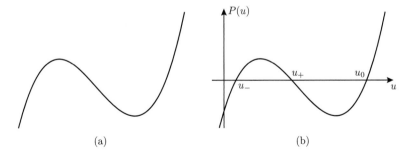

(a)   (b)

In particular, if the parameters are such that three zero crossings occur at positive u, as in (b), then we can have orbital motion confined between u_- (the **aphelion**: farthest from the sun) and u_+ (the **perihelion**: closest to the sun).

In the *classical* case, $P(u)$ was a *second*-order polynomial, $P_c(u)$; the au subtraction in the middle term is missing. The graph is then a parabola,

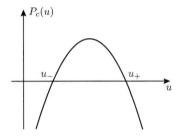

and

$$P_c(u) = (u - u_-)(u_+ - u). \tag{7.75}$$

The (classical) azimuthal angular displacement, $\Delta\phi_c$, as the planet goes from u_- to u_+, is

$$\Delta\phi_c = \int_{u_-}^{u_+} \frac{du}{\sqrt{P(u)_c}} = \int_{u_-}^{u_+} \frac{du}{\sqrt{(u - u_-)(u_+ - u)}} = 2\tan^{-1}\sqrt{\frac{u - u_-}{u_+ - u}}\bigg|_{u_-}^{u_+}$$

$$= 2\tan^{-1}(\infty) - 2\tan^{-1}(0) = \pi. \tag{7.76}$$

The planet sweeps out an angle π in going from perihelion to aphelion, and another π going from aphelion back to perihelion, so the entire circuit takes exactly 2π; the orbit is *closed*, and repeats itself indefinitely.

Now we want to correct that calculation by including the third root (u_0) in the polynomial:

$$P(u) = (u - u_-)(u_+ - u)(u_0 - u)a \tag{7.77}$$

(the factor of a is there to match the coefficient of u^3 in 7.74). Then

$$\Delta\phi = \frac{1}{\sqrt{au_0}} \int_{u_-}^{u_+} \frac{du}{\sqrt{(u - u_-)(u_+ - u)(1 - u/u_0)}}. \tag{7.78}$$

As we slowly turn on the au^3 term, the third root moves in from infinity; for u within the domain of the integral, $u \ll u_0$, and we can expand the last term:

$$\frac{1}{\sqrt{1 - u/u_0}} = 1 + \frac{1}{2}\frac{u}{u_0} + \frac{3}{8}\left(\frac{u}{u_0}\right)^2 + \cdots, \tag{7.79}$$

and keep only the first-order contribution:

$$\Delta\phi \approx \frac{1}{\sqrt{au_0}} \int_{u_-}^{u_+} \frac{du}{\sqrt{(u - u_-)(u_+ - u)}} \left(1 + \frac{1}{2}\frac{u}{u_0}\right)$$

$$= \frac{1}{\sqrt{au_0}}(\pi) + \frac{1}{2u_0\sqrt{au_0}} \int_{u_-}^{u_+} \frac{u\,du}{\sqrt{(u - u_-)(u_+ - u)}}. \tag{7.80}$$

Now,

$$\frac{d}{du}\sqrt{(u - u_-)(u_+ - u)} = \frac{1}{2}\frac{1}{\sqrt{(u - u_-)(u_+ - u)}}(-2u + u_+ + u_-), \tag{7.81}$$

so

$$\frac{u}{\sqrt{(u - u_-)(u_+ - u)}} = -\frac{d}{du}\sqrt{(u - u_-)(u_+ - u)}$$

$$+ \frac{(u_+ + u_-)}{2}\frac{1}{\sqrt{(u - u_-)(u_+ - u)}}. \tag{7.82}$$

Integrating,

$$\int_{u_-}^{u_+} \frac{u \, du}{\sqrt{(u-u_-)(u_+-u)}} = -\sqrt{(u-u_-)(u_+-u)}\bigg|_{u_-}^{u_+}$$

$$+ \frac{(u_+ + u_-)}{2} \int_{u_-}^{u_+} \frac{du}{\sqrt{(u-u_-)(u_+-u)}}$$

$$= \frac{\pi}{2}(u_+ + u_-). \tag{7.83}$$

Thus

$$\Delta\phi \approx \frac{\pi}{\sqrt{au_0}}\left(1 + \frac{u_+ + u_-}{4u_0}\right). \tag{7.84}$$

Equating the u^2 coefficients in the two expressions (7.74 and 7.77) for $P(u)$,

$$1 = a(u_+ + u_- + u_0) \quad \Rightarrow \quad au_0 = 1 - a(u_+ + u_-). \tag{7.85}$$

But remember, $u \ll u_0$ in the physical range $u_- \leq u \leq u_+$, so

$$a(u_+ + u_-) \ll 1 \quad \text{and} \quad au_0 \approx 1. \tag{7.86}$$

To first order, then,

$$\frac{1}{\sqrt{au_0}} \approx 1 + \frac{1}{2}a(u_+ + u_-) = 1 + \frac{1}{2}au_0\left(\frac{u_+ + u_-}{u_0}\right) \approx 1 + \frac{1}{2}\left(\frac{u_+ + u_-}{u_0}\right). \tag{7.87}$$

This leaves

$$\Delta\phi \approx \pi\left(1 + \frac{u_+ + u_-}{2u_0}\right)\left(1 + \frac{u_+ + u_-}{4u_0}\right) \approx \pi\left[1 + \frac{3}{4}\left(\frac{u_+ + u_-}{u_0}\right)\right]$$

$$\approx \pi\left[1 + \frac{3}{4}a(u_+ + u_-)\right] = \Delta\phi_{cl} + \Delta\phi_{gr}, \tag{7.88}$$

where $\Delta\phi_{cl}$ is the classical angular change, and $\Delta\phi_{gr}$ is that due to general relativity. Then the **precession of the perihelion** in one full orbit (out and back) is

$$2\Delta\phi_{gr} = \frac{3}{2}\pi a(u_+ + u_-) = 3MG\,\pi\left(\frac{1}{r_{min}} + \frac{1}{r_{max}}\right). \tag{7.89}$$

In the case of Mercury this turns out to be 43 seconds of arc per century.[9] Below, grossly exaggerated, is a figure of the precessing elliptical orbit; it starts at

---

[9] Of course, *anything* that adds a $1/r^3$ term to the potential will cause a precession in the perihelion of a planet—for instance, a nonspherical mass distribution within the sun, or the influence of other planets. Indeed, Einstein's correction is not the largest contribution. But the others were well understood, and the "missing" 43″ had been a notorious mystery since the middle of the 19th century. Eds. For Einstein's ecstatic reaction see Abraham Pais, *Subtle Is the Lord*, Oxford U. P., New York (1982), page 253.

perihelion P_1, goes out to aphelion A_1, and returns to perihelion P_2, which is displaced by an angle $2\Delta\phi_{gr}$ with respect to P_1.

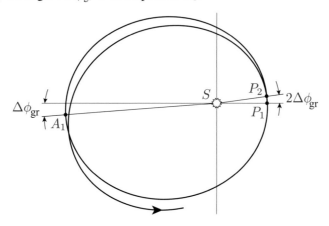

Problem 7.3

Look up the relevant quantities and confirm that Eq. 7.89 predicts a precession rate of 43″ per century.

7.3.2 Bending of Starlight

Next we consider the deflection of starlight as it passes by the sun. Light travels on *null* geodesics,

$$g_{\mu\nu}\frac{dx^\mu}{ds}\frac{dx^\nu}{ds} = 0, \tag{7.90}$$

for which the variational principle (7.52) is problematic, so we go back to the definition of a geodesic (6.13):

$$\frac{d^2x^\mu}{ds^2} + \Gamma^\mu_{\nu\lambda}\frac{dx^\nu}{ds}\frac{dx^\lambda}{ds} = 0. \tag{7.91}$$

For $\mu = 2$ this says

$$\frac{d^2\theta}{ds^2} + \frac{2}{r}\frac{dr}{ds}\frac{d\theta}{ds} - \sin\theta\,\cos\theta\frac{d\phi}{ds}\frac{d\phi}{ds} = 0, \tag{7.92}$$

which we again solve by picking $\theta = \pi/2$: the motion is confined to a plane. For $\mu = 3$ we have

$$\frac{d^2\phi}{ds^2} + \frac{2}{r}\frac{dr}{ds}\frac{d\phi}{ds} + 2\cot\theta\frac{d\theta}{ds}\frac{d\phi}{ds} = 0. \tag{7.93}$$

But $\cot \theta = 0$, so the last term drops out, and

$$\frac{d}{ds}\left(r^2\frac{d\phi}{ds}\right) = r^2\frac{d^2\phi}{ds^2} + 2r\frac{dr}{ds}\frac{d\phi}{ds} = r^2\left(\frac{d^2\phi}{ds^2} + \frac{2}{r}\frac{dr}{ds}\frac{d\phi}{ds}\right) = 0. \tag{7.94}$$

Once again (7.59),

$$r^2\dot{\phi} = \ell \tag{7.95}$$

is a constant. And from $\mu = 0$ we get

$$\frac{d^2t}{ds^2} + \lambda'\frac{dt}{ds}\frac{dr}{ds} = 0, \tag{7.96}$$

and therefore

$$\frac{d}{ds}\left(e^\lambda\frac{dt}{ds}\right) = e^\lambda\left(\frac{d^2t}{ds^2} + \lambda'\frac{dr}{ds}\frac{dt}{ds}\right) = 0, \tag{7.97}$$

so, as before (7.61),

$$e^\lambda\dot{t} = E \tag{7.98}$$

is a constant of the motion. We could now do $\mu = 1$, but it's simpler to work with 7.90; this is just like 7.63, only with zero on the right (and we're no longer using proper time as the parameter; proper time is a constant for photons—see Eq. 1.69):

$$e^{-\lambda}E^2 - e^\mu\dot{r}^2 - \frac{\ell^2}{r^2} = 0. \tag{7.99}$$

In place of 7.74 we now have

$$\left(\frac{du}{d\phi}\right)^2 = \frac{E^2}{\ell^2} - u^2(1 - au). \tag{7.100}$$

Einstein's contribution is the au term.

   Classically we would have

$$\left(\frac{du}{d\phi}\right)^2 = \frac{E^2}{\ell^2} - u^2 = (h - u)(h + u); \tag{7.101}$$

here $h \equiv |E/\ell|$ is 1 over the distance of closest approach (where $du/d\phi = 0$).

As the light ray comes in from infinity, the total angular deflection by the time it reaches the point nearest the sun is

$$\Delta\phi_c = \int_0^h \frac{du}{\sqrt{(h-u)(h+u)}} = \tan^{-1}\left(\frac{u}{\sqrt{h^2-u^2}}\right)\Big|_0^h$$

$$= \tan^{-1}(\infty) - \tan^{-1}(0) = \frac{\pi}{2}. \tag{7.102}$$

For the full trajectory (approaching and receding) it would be π—no deflection. (The classical trajectory is a straight line.)

In general relativity, however,[10]

$$\left(\frac{du}{d\phi}\right)^2 = h^2 - u^2(1-au) = h^2(1-\sigma^2), \quad \text{where} \quad \sigma \equiv \frac{u}{h}\sqrt{1-au}. \tag{7.103}$$

When $r = \infty$, $u = 0$, so $\sigma = 0$; at closest approach $du/d\phi = 0$, so $\sigma = 1$. Thus the total azimuthal angle swept out as the light ray comes in from infinity is

$$\Delta\phi_{gr} = \frac{1}{h}\int_{\sigma=0}^{\sigma=1} \frac{du}{\sqrt{1-\sigma^2}}. \tag{7.104}$$

Expanding σ as a power series in au,

$$\sigma = \frac{u}{h}\left[1 - \frac{1}{2}au - \frac{1}{8}(au)^2 - \cdots\right]. \tag{7.105}$$

By reversion of series,

$$u = h\sigma + \frac{1}{2}a(h\sigma)^2 + \frac{5}{8}a^2(h\sigma)^3 + \cdots, \tag{7.106}$$

so

$$du = h\left(1 + ah\sigma + \frac{15}{8}a^2h^2\sigma^2 + \cdots\right)d\sigma. \tag{7.107}$$

Putting this into 7.104,

$$\Delta\phi_{gr} = \int_0^1 \frac{d\sigma}{\sqrt{1-\sigma^2}}(1 + ah\sigma + \cdots) = \int_0^1 \frac{d\sigma}{\sqrt{1-\sigma^2}} + ah\int_0^1 \frac{\sigma\,d\sigma}{\sqrt{1-\sigma^2}} + \cdots$$

$$= \frac{\pi}{2} + ah + \cdots. \tag{7.108}$$

The total deflection (deviation from π for the whole trajectory) is $2ah$, where h (from the classical solution) is (approximately) 1 over the distance of closest approach. For a light beam grazing the surface that would be the sun's radius:

---

[10] Here h is no longer exactly 1 over the distance of closest approach, it's just a shorthand for $|E/\ell|$.

$$\frac{2a}{R_s} = \frac{4GM}{R_s} = 1.75 \text{ arc seconds.} \tag{7.109}$$

This prediction was confirmed by Eddington's famous measurement during the solar eclipse of 1919.

7.3.3 Gravitational Redshift

In special relativity a photon with wave vector k_μ moves along a null geodesic,

$$\frac{dx^\mu}{ds} = \alpha k^\mu, \quad k_\mu k^\mu = 0. \tag{7.110}$$

In flat space α is constant; we might as well choose the parameter s such that $\alpha = 1$:

$$\frac{d\alpha}{ds} = 0. \tag{7.111}$$

This is a covariant equation, and we shall adopt the same convention in general relativity. Now, the frequency (or energy) of the photon,[11]

$$\omega = k_0 = g_{00}k^0, \tag{7.112}$$

is constant along its path. Thus

$$\frac{d\omega}{ds} = \frac{d}{ds}\left(e^\lambda \frac{dx^0}{ds}\right) = 0, \tag{7.113}$$

for a metric of the form 7.24. We want the frequency in locally Minkowskian (elevator) coordinates; this requires a scaling of the time coordinate (since $g_{00} = 1$ in Minkowski coordinates):

$$dx^0_{\text{loc}} = e^{\lambda/2} dx^0, \tag{7.114}$$

and therefore also (since it transforms the same way)

$$k^0_{\text{loc}} = e^{\lambda/2}k^0 = e^{\lambda/2}e^{-\lambda}\omega = e^{-\lambda/2}\omega. \tag{7.115}$$

Evidently

$$e^{\lambda/2} k^0_{\text{loc}} = \omega \tag{7.116}$$

is constant along the trajectory, and hence

$$k^0_{\text{loc}}(r_1) = e^{-\lambda(r_1)/2} e^{\lambda(r_2)/2} k^0_{\text{loc}}(r_2). \tag{7.117}$$

---

[11] Note that the wave vector is naturally *covariant*, since the wave has the form $\exp(ik \cdot x)$, and the coordinates are naturally *contra*variant.

For example, suppose the photon (on a radial trajectory) is emitted at r_2, and detected at r_1; then 7.117 tells us the absorbed frequency in terms of the emitted frequency, each measured in the locally Minkowskian frame.[12] In particular, for the Schwarzschild solution,

$$
\begin{aligned}
e^{-\lambda(r_1)/2}\, e^{\lambda(r_2)/2} &= \left(1 - \frac{2MG}{r_1}\right)^{-1/2} \left(1 - \frac{2MG}{r_2}\right)^{1/2} \\
&\approx \left(1 + \frac{MG}{r_1}\right)\left(1 - \frac{MG}{r_2}\right) \approx \left(1 + \frac{GM}{r_1} - \frac{GM}{r_2}\right) \\
&= (1 + \Delta\varphi),
\end{aligned}
\tag{7.118}
$$

where

$$
\varphi(r) \equiv -\frac{GM}{r}
\tag{7.119}
$$

is the Newtonian gravitational potential energy and

$$
\Delta\varphi \equiv \varphi(r_2) - \varphi(r_1).
\tag{7.120}
$$

Thus

$$
\omega_{\text{loc}}(r_1) = (1 + \Delta\varphi)\,\omega_{\text{loc}}(r_2),
\tag{7.121}
$$

or

$$
\Delta\omega_{\text{loc}} \equiv \omega_{\text{loc}}(r_2) - \omega_{\text{loc}}(r_1) = -\Delta\varphi\,\omega_{\text{loc}}.
\tag{7.122}
$$

This is the **gravitational redshift**, which we encountered qualitatively in Example 4.1.

Einstein proposed measuring the frequency of spectral lines coming from atoms on the surface of the sun, which are emitted in a much stronger gravitational field than when they arrive on earth. Unfortunately, Doppler broadening and other confounding effects render the experiment unconvincing, and persuasive confirmation of the gravitational redshift had to await the Pound–Rebka experiment (R. V. Pound and G. A. Rebka Jr., *Phys. Rev. Lett.* **3**, 439 (1959)).

7.3.4 What Do They Really Test?

To what extent do these constitute definitive tests of general relativity?[13] First of all, they are at most testing the Schwarzschild solution, not Einstein's equation itself. But how well do the experiments check even the Schwarzschild metric?

---

[12] In Minkowski coordinates, of course, $k^0 = k_0$.

[13] Eds. There are now a number of more compelling experimental confirmations of Einstein's theory—most spectacularly the detection of gravitational waves by LIGO in 2015. See C. Will, *Theory and Experiment in Gravitational Physics*, 2nd ed., Cambridge U. P. (2018).

(The question is not how *accurate* the experiments are, but what they test *in principle*.) Let's assume the orbits are geodesics in *some* static, spherically symmetric metric, but not necessarily the Schwarzschild form 7.51.[14] Thus we take

$$e^{-\lambda} = 1 + \frac{a_1}{r} + \frac{a_2}{r^2} + \cdots ,$$

$$e^{-\mu} = 1 + \frac{b_1}{r} + \frac{b_2}{r^2} + \cdots . \tag{7.123}$$

Schwarzschild says

$$b_1 = -a = -2GM; \quad b_n = 0 \text{ for all } n > 1; \quad a_n = (a)^n. \tag{7.124}$$

Assume each term in the expansion is very small compared to the one before. How many of these coefficients have been probed by the four "classic" tests?

- **The third test** (gravitational redshift). This checks $a_1 = a$, and is actually superfluous, given that we assumed the principle of equivalence. How well does it check even the principle of equivalence? Borrowing the Planck formula ($\hbar\omega$) for the energy of a free photon, the effective mass is

$$m = \hbar\omega. \tag{7.125}$$

In a gravitational potential φ the *total* energy of the photon is

$$E = \hbar\omega + \varphi m = \hbar\omega + \varphi\hbar\omega, \tag{7.126}$$

and it is unchanging (as the photon moves from one place to another in the gravitational field):

$$\Delta E = \hbar\Delta\omega + (\Delta\varphi)\,\hbar\omega + \varphi\hbar\Delta\omega = 0 \implies (1+\varphi)\Delta\omega = -\omega\Delta\varphi, \tag{7.127}$$

or (dropping the much smaller potential term)

$$\Delta\omega = -\omega\,\Delta\varphi. \tag{7.128}$$

That just recapitulates 7.122; the point is that the gravitational redshift checks that gravity couples to the *total* energy of a photon, including its gravitational potential energy. But then, the Eötvös experiment had already shown that gravity couples to *all* forms of energy. Does the gravitational redshift do it *better*? Actually, it does not: Eötvös checked it to one part in 10^4, whereas the Pound–Rebka experiment tests it only to one part in 10^2. (Still, it *does* test the equivalence principle in the particular—and particularly clean—case of a photon in free fall.)

---

[14] This amounts to assuming the principle of equivalence for everything *except* gravity (application of the principle of equivalence to weak gravity in empty space is how we *got* Einstein's equation).

- **The second test** (bending of starlight). Returning to Eq. 7.99,

$$e^{-\mu}e^{-\lambda}E^2 - \dot{r}^2 - \frac{\ell^2}{r^2}e^{-\mu} = 0. \tag{7.129}$$

Using the generic forms 7.123 for $e^{-\mu}$ and $e^{-\lambda}$, and expanding to first order, we get

$$E^2\left(1 + \frac{a_1}{r} + \frac{b_1}{r}\right) - \dot{r}^2 - \frac{\ell^2}{r^2}\left(1 + \frac{b_1}{r}\right) = 0. \tag{7.130}$$

In the Schwarzschild case $(a_1 + b_1) = 0$, so (to lowest order) there will be a correction (to our earlier calculation) proportional to $(a_1 + b_1)$. Moreover, in the last term we now have b_1 in place of $-a$, so there will be another correction proportional to $(a + b_1)$:

$$\delta(\Delta\phi) = \alpha(h)(a_1 + b_1) + \beta(h)(a + b_1), \tag{7.131}$$

where α and β are some functions of the distance of closest approach. But the zeroth and third tests already check that $a_1 = a$, so the fact that the Eddington experiment confirmed Einstein's prediction is evidence that

$$b_1 = -a. \tag{7.132}$$

(To be certain, we should really check that $(\alpha(h) + \beta(h)) \neq 0$, which is in fact the case.)

- **The zeroth and first tests** (precession of the perihelion of Mercury). For a Newtonian planet of mass m_0, the conserved total energy (U) consists of the radial kinetic energy,

$$KE_r = \frac{1}{2}m_0\dot{r}^2, \tag{7.133}$$

plus the angular kinetic energy,

$$KE_\phi = \frac{1}{2}\frac{\ell^2}{m_0 r^2}, \tag{7.134}$$

plus the gravitational potential energy,

$$PE_g = -\frac{GMm_0}{r}. \tag{7.135}$$

Thus

$$2U = m_0\dot{r}^2 + \frac{\ell^2}{m_0 r^2} - \frac{2GMm_0}{r}. \tag{7.136}$$

Meanwhile, E is the total (special) relativistic energy of the planet:

$$E^2 - \mathbf{p}^2 = m^2, \qquad (7.137)$$

where, however, m is not just the rest mass m_0, but that diminished by the binding energy:

$$m = m_0 - \frac{GMm_0}{r} = m_0 \left(1 - \frac{GM}{r}\right). \qquad (7.138)$$

Taking GM/r to be small, we have

$$m^2 \approx m_0^2 \left(1 - \frac{2GM}{r}\right), \qquad (7.139)$$

so

$$E^2 \approx \mathbf{p}^2 + m_0^2 - \frac{2GMm_0^2}{r} = 2m_0 \left(\frac{\mathbf{p}^2}{2m_0} - \frac{GMm_0}{r}\right) + m_0^2, \qquad (7.140)$$

or

$$E^2 = m_0^2 + 2m_0 U. \qquad (7.141)$$

Now, to be consistent with our earlier notation, set $m_0 \to 1$; then $2U = E^2 - 1$, and the Newtonian equation of motion (7.136) reads

$$(E^2 - 1) - \dot{r}^2 - \frac{\ell^2}{r^2} + \frac{2GM}{r} = 0 \qquad \text{(Newton)}. \qquad (7.142)$$

By comparison, the Schwarzschild equation of motion (7.65) is

$$(E^2 - 1) - \dot{r}^2 - \frac{\ell^2}{r^2}\left(1 - \frac{a}{r}\right) + \frac{a}{r} = 0 \qquad \text{(Schwarzschild)}. \qquad (7.143)$$

For the generic spherically symmetrical metric (7.63),

$$e^{-\mu}e^{-\lambda}E^2 - \dot{r}^2 - \frac{\ell^2}{r^2}e^{-\mu} = e^{-\mu}, \qquad (7.144)$$

which becomes (for the expansion 7.123)

$$\left(1 + \frac{a_1}{r} + \frac{a_2}{r^2}\right)\left(1 + \frac{b_1}{r} + \frac{b_2}{r^2}\right)[1 + (E^2 - 1)] - \dot{r}^2 - \frac{\ell^2}{r^2}\left(1 + \frac{b_1}{r} + \frac{b_2}{r^2}\right)$$
$$= \left(1 + \frac{b_1}{r} + \frac{b_2}{r^2}\right). \qquad (7.145)$$

Now, for nonrelativistic velocities $E^2 \approx m_0^2 = 1$, so $(E^2 - 1)$ is already small, so we'll only keep the a_1 and b_1 terms in that factor. Then

$$(E^2 - 1)\left(1 + \frac{a_1 + b_1}{r}\right) - \dot{r}^2 - \frac{\ell^2}{r^2}\left(1 + \frac{b_1}{r} + \frac{b_2}{r^2}\right)$$

$$= \left(1 + \frac{b_1}{r} + \frac{b_2}{r^2}\right)\left(1 - 1 - \frac{a_1}{r} - \frac{a_2}{r^2}\right)$$

$$= -\frac{a_1}{r} - \frac{a_2}{r^2} - \frac{a_1 b_1}{r^2} - \frac{a_2 b_1 + a_1 b_2}{r^3} - \frac{a_2 b_2}{r^4}. \qquad (7.146)$$

Dropping the final terms in $1/r^3$ and $1/r^4$ (which after all would be contaminated by higher orders in the expansions for $e^{-\mu}$ and $e^{-\lambda}$), and also the $\ell^2 b_2/r^4$ term,

$$(E^2 - 1)\left(1 + \frac{a_1 + b_1}{r}\right) - \dot{r}^2 - \frac{\ell^2}{r^2}\left(1 + \frac{b_1}{r}\right) + \frac{a_1}{r}$$

$$= -\frac{a_1 b_1 + a_2}{r^2} \qquad \text{(Generic).} \qquad (7.147)$$

For Schwarzschild to match Newton at large r we must have $a = 2GM$ (of course; that's how we fixed a back in 7.48). For the generic case to match Newton (to lowest order), $a_1 = 2GM = a$. That's all the zeroth test tells us. If we grant (from the measured bending of sunlight) that $b_1 = -a$ (Eq. 7.132), then the generic case matches Schwarzschild provided

$$a_1 b_1 + a_2 = 0, \quad \text{which is to say} \quad a_2 = (a)^2. \qquad (7.148)$$

In short, the Newtonian limit determines a_1, the deflection of sunlight fixes b_1, and the precession of the perihelion of Mercury then sets a_2, but that's all the four "classic" tests can tell us. Any theory that reproduces these three parameters would predict the same results (to lowest order). To put it another way, any theory incorporating the principle of equivalence, with particles moving on geodesics, and a spherically symmetrical metric, that is consistent with the "classic" experiments, will match the Schwarzschild solution up through a_1, b_1, and a_2.

7.4 The Interior Solution

Inside the sun the stress tensor is *not* zero. For a perfect fluid at rest, in Minkowski coordinates,

$$T^{00} = \mu_0, \quad T^{0i} = 0, \quad T^{ij} = P\,\delta^{ij}, \qquad (7.149)$$

where μ_0 is the energy density and P is the pressure—both of them, presumably, functions of position (but not of time). Some **equation of state** relates μ_0 and

P to the temperature, chemical potential, etc. But this is too complicated for our purposes; we shall assume the sun is an *incompressible* fluid, so μ_0 is *uniform*.

In a *general* reference frame, with an element of fluid moving at proper velocity u^μ,

$$T^{\mu\nu} = (\mu_0 + P)u^\mu u^\nu - P\, g^{\mu\nu}. \tag{7.150}$$

In the Minkowski frame, a stationary object satisfies $u_0 = u^0 = 1$, but that's not the case in the static spherically symmetric coordinates 7.24. Rather,

$$g_{\mu\nu}u^\mu u^\nu = e^\lambda u^0 u^0 = 1 \quad \Rightarrow \quad u^0 = e^{-\lambda/2}. \tag{7.151}$$

But

$$u_0 = g_{00}\, u^0 = e^\lambda u^0 = e^{\lambda/2}, \tag{7.152}$$

so

$$T^0_{\ 0} = (\mu_0 + P) - P = \mu_0, \quad T^i_{\ 0} = T^0_{\ i} = 0, \quad T^i_{\ j} = -P\delta^i_j. \tag{7.153}$$

This is the stress tensor we will take to represent the interior of the sun, with μ_0 independent of r, and $P(r)$ a function yet to be determined.

Einstein's equation (6.74) says

$$G^\mu_{\ \nu} = R^\mu_{\ \nu} - \tfrac{1}{2}\delta^\mu_\nu R = -\alpha T^\mu_{\ \nu}. \tag{7.154}$$

We'll need three equations, for the three unknown functions $(\lambda(r),\ \mu(r),\ \text{and}\ P(r))$. I'll use

$$(1) \quad G^0_{\ 0} = -\alpha\mu_0, \tag{7.155}$$

$$(2) \quad G^1_{\ 1} = \alpha P, \tag{7.156}$$

$$(3) \quad \nabla_\mu T^\mu_{\ \nu} = 0. \tag{7.157}$$

The last of these follows from $\nabla_\mu G^\mu_{\ \nu} = 0$ (Eq. 7.49).[15] From 7.35 and Problem 7.1(a),

$$
\begin{aligned}
G_{00} &= R_{00} - \frac{1}{2}g_{00}\, R \\
&= -\frac{1}{2}e^{(\lambda-\mu)}\left[\lambda'' + \frac{1}{2}\lambda'(\lambda' - \mu') + \frac{2\lambda'}{r}\right] \\
&\quad - \frac{1}{2}e^\lambda\left\{e^{-\mu}\left[-\lambda'' - \frac{\lambda'}{2}(\lambda' - \mu') - \frac{2}{r}(\lambda' - \mu') - \frac{2}{r^2}\right] + \frac{2}{r^2}\right\} \\
&= \frac{e^{(\lambda-\mu)}}{r^2}\left(1 - r\mu' - e^\mu\right),
\end{aligned}
\tag{7.158}
$$

---

[15] You can use $G^2_{\ 2} = \alpha P$ or $G^3_{\ 3} = \alpha P$ for the third equation, if you prefer.

so

$$G^0{}_0 = g^{00} G_{00} = \frac{e^{-\mu}}{r^2} \left(1 - r\,\mu' - e^{\mu}\right). \qquad (7.159)$$

Similarly, using 7.36,

$$
\begin{aligned}
G_{11} &= R_{11} - \frac{1}{2} g_{11} R \\
&= \frac{1}{2} \left[\lambda'' + \frac{1}{2} \lambda'(\lambda' - \mu') - \frac{2\mu'}{r} \right] \\
&\quad + \frac{1}{2} e^{\mu} \left\{ e^{-\mu} \left[-\lambda'' - \frac{\lambda'}{2}(\lambda' - \mu') - \frac{2}{r}(\lambda' - \mu') - \frac{2}{r^2} \right] + \frac{2}{r^2} \right\} \\
&= -\frac{1}{r^2} \left(1 + r\lambda' - e^{\mu} \right),
\end{aligned}
\qquad (7.160)
$$

so

$$G^1{}_1 = g^{11} G_{11} = \frac{e^{-\mu}}{r^2} \left(1 + r\lambda' - e^{\mu}\right). \qquad (7.161)$$

The first equation (7.155) says

$$\frac{(re^{-\mu})'}{r^2} - \frac{1}{r^2} = -\alpha\mu_0, \quad \text{or} \quad (re^{-\mu})' = 1 - \alpha\mu_0 r^2. \qquad (7.162)$$

Thus

$$re^{-\mu} = r - \frac{\alpha}{3}\mu_0 r^3 + B \quad \Rightarrow \quad e^{-\mu} = 1 - \frac{\alpha}{3}\mu_0 r^2 + \frac{B}{r}, \qquad (7.163)$$

for some constant B. The third term introduces a spurious singularity at the center, so we assume $B = 0$, leaving

$$\boxed{g_{11} = -e^{\mu} = -\frac{1}{1 - (\alpha\mu_0 r^2/3)}.} \qquad (7.164)$$

Comments:

- This result invites an interesting geometrical interpretation. Consider 4-dimensional Euclidean space, with the Cartesian coordinates (x, y, z, w) and the metric

$$ds^2 = dx^2 + dy^2 + dz^2 + dw^2. \qquad (7.165)$$

In this space, imagine a 3-dimensional hypersphere of radius b:

$$x^2 + y^2 + z^2 + w^2 = b^2. \qquad (7.166)$$

Adopting polar coordinates, $(x, y, z) \rightarrow (r, \theta, \phi)$,

$$ds^2 = dr^2 + r^2 \, d\Omega^2 + dw^2, \tag{7.167}$$

and for points on the hypersphere $(r^2 + w^2 = b^2)$ we can eliminate dw:

$$w \, dw + r \, dr = 0 \implies dw = -\frac{r}{w} dr. \tag{7.168}$$

Thus

$$
\begin{aligned}
ds^2 &= \left(1 + \frac{r^2}{w^2}\right) dr^2 + r^2 \, d\Omega^2 = \left(\frac{b^2}{b^2 - r^2}\right) dr^2 + r^2 \, d\Omega^2 \\
&= \left(\frac{1}{1 - (r/b)^2}\right) dr^2 + r^2 \, d\Omega^2.
\end{aligned} \tag{7.169}
$$

This is the metric on a hypersphere of radius b embedded in Euclidean 4-space, and (with $b \equiv \sqrt{3/\alpha\mu_0}$) it is identical to the 3-dimensional geometry inside the sun. This suggests good coordinates to use in solving for λ and P, and it reveals that the singularity in Eq. 7.169 at $r = b$ is a geometrical artifact—a result of our choice of coordinates, not a physical singularity: r becomes a bad coordinate at the "equator" ($w = 0$), just as ϕ is a bad coordinate at $\theta = 0$, but there's nothing fishy in Cartesian coordinates. Moreover, b is a huge number—by the time $r = b$ we are already well outside the sun.

- Matching the exterior (7.46) and interior (7.164) Schwarzschild solutions at the surface of the sun (r_0),

$$e^{-\mu} = 1 - \frac{a}{r_0} = 1 - \frac{\alpha\mu_0 r_0^2}{3} \implies a = \frac{\alpha}{3}\mu_0 r_0^3. \tag{7.170}$$

But $a = 2GM$ (7.48) and $\alpha = 8\pi G$ (6.141), so

$$M = \left(\frac{4}{3}\pi r_0^3\right) \mu_0. \tag{7.171}$$

At first glance this looks perfect: the mass of the sun is its volume times the mass density. However, r_0 is not really the "radius" of the sun, and the energy density is not quite the mass density (there is also gravitational potential energy). Fortuitously, these two corrections cancel one another.

The second equation (7.156) says

$$\frac{e^{-\mu}}{r^2}\left(1 + r\lambda' - e^{\mu}\right) = \alpha P. \tag{7.172}$$

Using 7.163 (with $B = 0$),

$$\left(1 - \frac{\alpha\mu_0}{3}r^2\right)\left(\frac{1}{r^2} + \frac{\lambda'}{r}\right) - \frac{1}{r^2} = \alpha P, \tag{7.173}$$

or

$$\frac{\lambda'}{r}\left(1 - \frac{\alpha\mu_0}{3}r^2\right) - \frac{\alpha\mu_0}{3} = \alpha P. \tag{7.174}$$

But we can't solve this (yet), because we don't know P.

As for the third equation (7.157), referring to Eqs. 5.81 and 5.82,

$$\nabla_\mu T^\mu{}_\nu = \partial_\mu T^\mu{}_\nu + \Gamma^\mu_{\mu\lambda} T^\lambda{}_\nu - \Gamma^\lambda_{\mu\nu} T^\mu{}_\lambda = 0. \tag{7.175}$$

This *looks* like *four* equations (one for each ν). Let's take them one at a time, using the table of Christoffel symbols in Section 7.2, and Eq. 7.153:

$\nu = 0$:

$$\nabla_\mu T^\mu{}_0 = \partial_\mu T^\mu{}_0 + \Gamma^\mu_{\mu\lambda} T^\lambda{}_0 - \Gamma^\lambda_{\mu 0} T^\mu{}_\lambda = \partial_0 T^0{}_0 + \Gamma^\mu_{\mu 0} T^0{}_0 - \Gamma^\lambda_{\mu 0} T^\mu{}_\lambda$$

$$= \partial_0 T^0{}_0 - \Gamma^1_{00} T^{\cancel{1}}_{\cancel{1}} - \Gamma^0_{10} T^{\cancel{0}}_{\cancel{0}} = \frac{\partial\mu_0}{\partial t} = 0. \tag{7.176}$$

So μ_0 is constant—but we already knew that.

$\nu = 1$:[16]

$$\nabla_\mu T^\mu{}_1 = \partial_\mu T^\mu{}_1 + \Gamma^\mu_{\mu\lambda} T^\lambda{}_1 - \Gamma^\lambda_{\mu 1} T^\mu{}_\lambda = \partial_1 T^1{}_1 + \Gamma^\mu_{\mu 1} T^1{}_1 - \Gamma^\lambda_{\mu 1} T^\mu{}_\lambda$$

$$= \partial_1 T^1{}_1 + \left(\Gamma^0_{01} + \cancel{\Gamma^1_{11}} + \Gamma^2_{21} + \Gamma^3_{31}\right) T^1{}_1$$

$$\quad - \left(\Gamma^0_{01} T^0{}_0 + \cancel{\Gamma^1_{11} T^1{}_1} + \Gamma^2_{21} T^2{}_2 + \Gamma^3_{31} T^3{}_3\right)$$

$$= \partial_1 T^1{}_1 + \left(\frac{1}{2}\lambda' + \cancel{\frac{1}{r}} + \cancel{\frac{1}{r}}\right)(-P) - \frac{1}{2}\lambda'\mu_0 - \frac{1}{r}\cancel{(-P)} - \frac{1}{r}\cancel{(-P)}$$

$$= \frac{\partial}{\partial r}(-P) - \frac{1}{2}\lambda'(P + \mu_0) = 0. \tag{7.177}$$

Thus

$$\frac{P'}{P + \mu_0} = -\frac{1}{2}\lambda'. \tag{7.178}$$

This is the equation of hydrostatic equilibrium; it says that the pressure gradient balances the gravitational force.

---

[16] *Beware:* Although $T^0{}_1 = 0$ (for example), $\nabla_0 T^0{}_1 \neq 0$ (!).

$\nu = 2$:

$$\nabla_\mu T^\mu{}_2 = \partial_\mu T^\mu{}_2 + \Gamma^\mu_{\mu\lambda} T^\lambda{}_2 - \Gamma^\lambda_{\mu 2} T^\mu{}_\lambda = \partial_2 T^2{}_2 + \Gamma^\mu_{\mu 2} T^2{}_2 - \Gamma^\lambda_{\mu 2} T^\mu{}_\lambda$$

$$= \partial_2 T^2{}_2 + \Gamma^3_{32} T^2{}_2 - \Gamma^1_{22} T^1{}_1 - \Gamma^2_{12} T^2{}_2 - \Gamma^3_{32} T^3{}_3$$

$$= \partial_2 T^2{}_2 + \cot\theta(-P) - \cot\theta(-P) = -\frac{\partial P}{\partial\theta} = 0. \qquad (7.179)$$

So P is independent of θ—but we already knew that.

$\nu = 3$:

$$\nabla_\mu T^\mu{}_3 = \partial_\mu T^\mu{}_3 + \Gamma^\mu_{\mu\lambda} T^\lambda{}_3 - \Gamma^\lambda_{\mu 3} T^\mu{}_\lambda = \partial_3 T^3{}_3 + \Gamma^\mu_{\mu 3} T^3{}_3 - \Gamma^\lambda_{\mu 3} T^\mu{}_\lambda$$

$$= \partial_3 T^3{}_3 - \Gamma^1_{33} T^1{}_1 - \Gamma^3_{13} T^3{}_3 - \Gamma^2_{33} T^2{}_2 - \Gamma^3_{23} T^3{}_3$$

$$= -\frac{\partial P}{\partial\phi} = 0. \qquad (7.180)$$

So P is independent of ϕ, as we knew.

So $\nabla_\mu T^\mu{}_\nu = 0$ yields only one new condition (7.178)—together with the trivial results that P depends only on r and μ_0 is independent of time.

To solve for P we first switch to the notation suggested in Eqs. 7.165–7.169:

$$b^2 \equiv \frac{3}{\alpha\mu_0}; \quad w^2 \equiv b^2 - r^2 \Rightarrow r\,dr = -w\,dw, \quad \frac{d}{dr} = \frac{dw}{dr}\frac{d}{dw} = -\frac{r}{w}\frac{d}{dw}, \qquad (7.181)$$

so 7.174 becomes

$$\frac{1}{r}\left(-\frac{r}{w}\frac{d\lambda}{dw}\right)\left(1 - \frac{r^2}{b^2}\right) - \frac{1}{b^2} = \alpha P \Rightarrow w\frac{d\lambda}{dw} + 1 = -\alpha b^2 P. \qquad (7.182)$$

Meanwhile, 7.178 says

$$\frac{d\lambda}{dw} = -\frac{2}{P + \mu_0}\frac{dP}{dw}. \qquad (7.183)$$

Putting them together,

$$1 - \frac{2w}{P + \mu_0}\frac{dP}{dw} = -\frac{3}{\mu_0}P, \qquad (7.184)$$

or

$$\frac{dw}{w} = \frac{2\mu_0}{3}\frac{dP}{(P + \mu_0)(P + \mu_0/3)} = dP\left(\frac{1}{P + \mu_0/3} - \frac{1}{P + \mu_0}\right). \qquad (7.185)$$

Integrating,

$$\ln w = A + \ln \left(\frac{P + \mu_0/3}{P + \mu_0} \right) \implies w = e^A \left(\frac{P + \mu_0/3}{P + \mu_0} \right). \tag{7.186}$$

At the surface of the sun ($r = r_0$, or $w = w_0$) the pressure must vanish, so

$$\frac{w}{w_0} = \frac{3P + \mu_0}{P + \mu_0}, \tag{7.187}$$

or, solving for P,

$$\boxed{P = \frac{\mu_0(w - w_0)}{(3w_0 - w)}.} \tag{7.188}$$

As r decreases, w increases, so as you penetrate deeper into the sun the pressure increases monotonically. But for a large enough sphere you hit the pole at $w = 3w_0$, and the incompressible fluid model breaks down. The maximum value of w occurs at $r = 0$, where $w = b$, so to be safe we need $3w_0 > b$:

$$r_0^2 = b^2 - w_0^2 < \frac{8}{9}b^2, \quad \text{or} \quad \frac{r_0^2}{b^2} < \frac{8}{9}, \tag{7.189}$$

but

$$\frac{1}{b^2} = \frac{\alpha\mu_0}{3} = \frac{8\pi G\mu_0}{3}, \tag{7.190}$$

so the condition is

$$2GM < \frac{8}{9}r_0. \tag{7.191}$$

This tells us the minimum size an object of given mass can have, and yet remain an incompressible fluid; evidently you can never add enough mass to get the radius below the Schwarzschild radius. Actually, *all* equations of state have such singularities; hydrostatic forces prevent gravitational collapse, provided the mass is not too great (the problem is that mass contributes to gravity quadratically, but to pressure only linearly). But once collapse begins, it is no longer a static situation.

Problem 7.4

Plot the pressure as a function of r, for the interior of the sun, from the center ($r = 0$) to the surface ($r = r_0$); use the actual values for the mass and radius of the sun.

Problem 7.5

Show that

$$e^\lambda = e^{\lambda_0} \frac{(3w_0 - w)^2}{4w_0^2},$$ (7.192)

where λ_0 is the value at the surface of the sun.

Problem 7.6

For a fluid with proper energy density $\mathcal{E}$ and pressure P (7.150),

$$T^{\mu\nu} = (\mathcal{E} + P)v^\mu v^\nu - P g^{\mu\nu} \quad \text{and} \quad \nabla_\mu T^{\mu\nu} = 0,$$

show that

$$\nabla_\mu [(\mathcal{E} + P)v^\mu] - v^\mu \partial_\mu P = 0,$$ (7.193)

and therefore

$$(\mathcal{E} + P)v^\mu (\nabla_\mu v^\nu) - (g^{\mu\nu} - v^\mu v^\nu)\partial_\mu P = 0.$$ (7.194)

7.5 The Schwarzschild Singularity

We return now to the external Schwarzschild solution (assuming $a > r_0$):

$$ds^2 = \left(1 - \frac{a}{r}\right) dt^2 - \left(1 - \frac{a}{r}\right)^{-1} dr^2 - r^2 d\Omega^2.$$ (7.195)

Ostensibly, it has singularities at $r = a$ and $r = 0$. But are they *physical*, or just artifacts of bad coordinates? How can you tell when a singularity is genuine? One way is to examine the curvature scalar R—it's independent of the coordinate system, so if the singularity persists, you know it is real. This reveals that $r = a$ is a phony singularity, but $r = 0$ is genuine.

Another way is to study radial free fall. From 7.64 (with $\ell = 0$),

$$E^2 - 1 = \left(\frac{dr}{ds}\right)^2 - \frac{a}{r}.$$ (7.196)

The motion is just like Newtonian, in a $1/r$ potential—the falling object sails right through the "singularity" at $r = a$, with no problem at all! But what about its *velocity*, dr/dt? From 7.61 and 7.62,

$$\left(1 - \frac{a}{r}\right) - \frac{1}{(1 - a/r)} \frac{(dr/ds)^2}{(dt/ds)^2} = \frac{1}{(dt/ds)^2} = \frac{(1 - a/r)^2}{E^2},$$ (7.197)

or

$$\left(1 - \frac{a}{r}\right)^2 - \left(\frac{dr}{dt}\right)^2 = \frac{1}{E^2}\left(1 - \frac{a}{r}\right)^3, \tag{7.198}$$

so

$$\left(\frac{dr}{dt}\right)^2 = \left(1 - \frac{a}{r}\right)^2 - \frac{1}{E^2}\left(1 - \frac{a}{r}\right)^3. \tag{7.199}$$

In the immediate vicinity of the "singularity," $r \approx a$, and the cubic term is negligible:

$$\left(\frac{dr}{dt}\right)^2 \approx \left(1 - \frac{a}{r}\right)^2 \approx \left(\frac{r - a}{a}\right)^2. \tag{7.200}$$

As the object falls toward $r = a$,

$$\frac{dr}{r - a} \approx -\frac{dt}{a}, \quad \ln(r - a) = \frac{t_0 - t}{a}, \quad r = a + e^{(t_0 - t)/a}. \tag{7.201}$$

It doesn't actually *reach* a until $t = \infty$. It takes a finite amount of *proper* time to go through the Schwarzschild radius, but an *infinite* amount of coordinate time! If you're falling toward the singularity, you go faster and faster, so your clock runs slower and slower; to an external observer the process takes forever, but to you it's the blink of an eye.

Notice, finally, what happens to the *signs* when you pass through the Schwarzschild singularity ($r < a$):

$$ds^2 = \left(1 - \frac{a}{r}\right)dt^2 - \frac{1}{(1 - a/r)}dr^2. \tag{7.202}$$

The plus sign (which is the fingerprint of the temporal coordinate) now attaches to r, and the minus sign (which should indicate a spatial component) is now associated with t. It is as if the temporal and radial coordinates traded places.

Problem 7.7

Consider an object in radial free fall toward the Schwarzschild singularity (Eq. 7.199). When it's far away it accelerates toward the sun, obviously, but since it's going to take an infinite amount of coordinate time to reach the Schwarzschild radius, it must at some point slow down.[17] At what radius (r_0) does it reach its maximum speed ($|dr/dt|$)? What is that maximum velocity? Your answers will involve E, of course. Suppose now that it starts from rest at $r = \infty$. What then is E, where does the maximum velocity occur, and what is that maximum speed (as a multiple of c)?

---

[17] The *proper* acceleration—what the unfortunate astronaut's own accelerometer measures—does *not* go to zero here.

7.5.1 Kruskal Coordinates

If r and t are *bad* coordinates, in the sense that they create a false singularity at $r = a$, why not introduce *good* coordinates, that do not have this defect? What we want is a **maximal extension** of the (exterior) Schwarzschild solution: a manifold $\mathcal{M}$ such that

1. $R_{\mu\nu} = 0$ except at isolated points;
2. $\mathcal{M} \supset \mathcal{R}$, where on $\mathcal{R}$ we have the geometry of the Schwarzschild solution for $r > a$ and all time ($-\infty < t < \infty$), that is, it matches the Schwarzschild solution outside the Schwarzschild radius;
3. Every geodesic either (a) can be continued indefinitely to spatial infinity or (b) runs into a *true* singularity (*viz.* $r = 0$).[18]

Solution: For $r > a$, change coordinates from (r, t) to (u, v), defined as follows:[19]

$$u \equiv \sqrt{\frac{r}{a} - 1} \, \exp\left(\frac{r}{2a}\right) \cosh\left(\frac{t}{2a}\right), \tag{7.203}$$

$$v \equiv \sqrt{\frac{r}{a} - 1} \, \exp\left(\frac{r}{2a}\right) \sinh\left(\frac{t}{2a}\right). \tag{7.204}$$

We want to rewrite the metric in these coordinates:

$$du = \frac{1}{2a}\left[\frac{1}{(r/a - 1)} u \, dr + u \, dr + v \, dt\right]$$

$$= \frac{1}{2a}\left[\frac{1}{(1 - a/r)} u \, dr + v \, dt\right],$$

$$dv = \frac{1}{2a}\left[\frac{1}{(1 - a/r)} v \, dr + u \, dt\right],$$

$$dv^2 - du^2 = \frac{1}{4a^2}\left[\frac{1}{(1 - a/r)^2}(v^2 - u^2)\, dr^2 + (u^2 - v^2)\, dt^2\right]. \tag{7.205}$$

Thus

$$\frac{4a^2\,(1 - a/r)\,(dv^2 - du^2)}{(u^2 - v^2)} = -\frac{1}{(1 - a/r)}\, dr^2 + \left(1 - \frac{a}{r}\right) dt^2. \tag{7.206}$$

Now

$$u^2 - v^2 = \left(\frac{r}{a} - 1\right) e^{r/a}, \tag{7.207}$$

---

[18] This is the property *not* shared by the Schwarzschild solution; the latter admits geodesics such as radial free fall, which terminate (as $t \to \infty$) on the false singularity at $r = a$.

[19] Eds. M. D. Kruskal, *Phys. Rev.* **119**, 1743 (1960). The coordinates were discovered independently by G. Szekeres, *Publ. Mat. Debrecen* **7**, 285 (1960), and are now known as Kruskal–Szekeres coordinates.

so 7.206 becomes

$$\frac{4a^3}{r}e^{-r/a}(dv^2 - du^2) = ds^2 + r^2\,d\Omega^2. \tag{7.208}$$

Equation 7.207 determines r as some nasty function of u and v; t is a lot simpler: from 7.203 and 7.204,

$$t = 2a\tanh^{-1}(v/u). \tag{7.209}$$

In the new coordinates, then,

$$ds^2 = F(dv^2 - du^2) - r^2\,d\Omega^2, \tag{7.210}$$

where $F(u,v)$ is shorthand for $(4a^3/r)\exp(-r/a)$, expressed in terms of u and v. The singularity at $r = a$ has vanished, we are liberated from the restriction to $r > a$, and the three requirements for a maximal extension are met.

Let's explore the u, v space, using 7.207 and 7.209 for various r and t:

$$r = 0: \quad u^2 - v^2 = -1 \quad \text{(hyperbolas)}, \tag{7.211}$$

$$r = a: \quad u^2 - v^2 = 0 \Rightarrow v = \pm u, \tag{7.212}$$

$$r = 2a: \quad u^2 - v^2 = e^2 \quad \text{(hyperbolas)}, \tag{7.213}$$

$$t = 0: \quad v = 0 \quad \text{(the u-axis)}, \tag{7.214}$$

$$t = \pm\infty: \quad v = \pm u \quad \text{(same as $r = a$)}. \tag{7.215}$$

This divides the space into six regions: in I and II, $r > a$, and the Schwarzschild metric prevails; in III and IV, $0 < r < a$; in V and VI, r is negative (these two regions are presumably unphysical and inaccessible). Notice that the u-, v-coordinates are double valued, with two different points for a given r and t. Increasing v is the uniform definition of advancing time. In region I this means increasing t, in region II it means *decreasing* t, in region III it means decreasing r, and in region IV it means *in*creasing r.

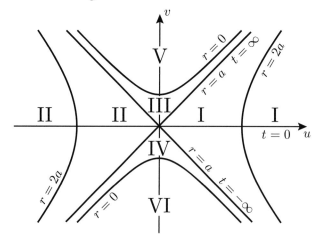

Radial null geodesics (photon lines) satisfy $dv^2 = du^2$; they are straight lines with slope ± 1 (dashed):

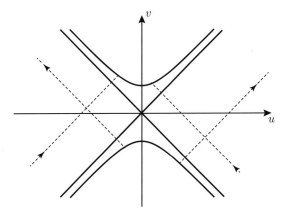

Causality: If $dv/ds > 0$, the trajectory is time-like; the direction of increasing v is the direction of increasing time. Thus, from region IV you can get to regions I and II, and from either of them you can get to region III. But once you are in region III you are bound eventually to hit the $r = 0$ singularity. It is impossible to find orbits that *stay* in IV or *leave* III, but there are orbits that remain in I or II.[20]

7.5.2 Geometry of the Equatorial Surface

At constant t and $\theta = \pi/2$ the spatial part of the Schwarzschild metric in regions I and II reduces to

$$d\ell^2 = \left(1 - \frac{a}{r}\right)^{-1} dr^2 + r^2\, d\phi^2. \tag{7.216}$$

It defines a 2-dimensional surface, which we can think of as embedded in a 3-dimensional Euclidean space (r, ϕ, z), according to the following:

Theorem: *An n-dimensional Riemannian manifold with all signs positive in the metric can (at least locally) be embedded in a larger Euclidean manifold.*

Proof: Let $(s_1, s_2, \ldots, s_n)$ be the coordinates in the Riemannian manifold. We want to construct a larger Euclidean space whose coordinates are

---

[20] Eds. In modern parlance region III is a **black hole**: nothing there can escape the singularity at $r = 0$. Region IV is its time-reversed twin: a **white hole**, to which nothing can return. Their boundaries, at $t = \pm\infty$ and $r = a$, are known as **event horizons**. But these terms, which were introduced in the early sixties, were not used by Coleman.

$$x^1 = x^1(s_1, \ldots, s_n),$$
$$x^2 = x^2(s_1, \ldots, s_n),$$

$$\vdots$$

$$x^N = x^N(s_1, \ldots, s_n) \quad (N \geq n), \tag{7.217}$$

such that when restricted to the Riemannian subspace the two measures agree. That is to say, when

$$dx^1 = \sum_{j=1}^n \left(\frac{\partial x^1}{\partial s^j}\right) ds^j, \quad dx^2 = \sum_{j=1}^n \left(\frac{\partial x^2}{\partial s^j}\right) ds^j, \quad \text{etc.,} \tag{7.218}$$

we have

$$(dx^1)^2 + (dx^2)^2 + \cdots + (dx^N)^2 = g_{ij}(s)\, ds^i\, ds^j. \tag{7.219}$$

This gives us $(n^2 + n)/2$ differential equations (one for each of the distinct components in g_{ij}), so if the Euclidean space has

$$N = \frac{n(n+1)}{2} \tag{7.220}$$

dimensions (or more), we can satisfy all the differential equations, at least locally.[21] (For instance, any 2-dimensional Riemannian manifold (with positive signature) can be thought of as a (2-dimensional) surface embedded in a 3-dimensional Euclidean space.)   QED

At $\phi = 0$, Eq. 7.216 defines a curve, with arc length ℓ:

$$d\ell = \frac{1}{\sqrt{1 - a/r}}\, dr, \tag{7.221}$$

which, in the embedding Euclidean space, becomes

$$d\ell = \sqrt{dr^2 + dz^2} = \sqrt{1 + \left(\frac{dz}{dr}\right)^2}\, dr. \tag{7.222}$$

Thus

$$1 + \left(\frac{dz}{dr}\right)^2 = \frac{1}{1 - a/r}, \tag{7.223}$$

or

$$dz = \pm\sqrt{\frac{a}{r-a}}\, dr \quad \Rightarrow \quad z = \pm 2\sqrt{a(r-a)}. \tag{7.224}$$

---

[21] If there are symmetries in the space, we may be able to get away with a smaller Euclidean space.

Now rotate this about the z-axis to generate the (r, ϕ) surface as embedded in the (r, ϕ, z) Euclidean space:

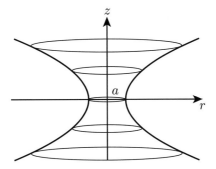

In regions III and IV, $0 < r < a$, so $(1 - a/r)$ is negative and the Schwarzschild metric (expressed in terms of positive quantities) becomes

$$ds^2 = -\left(\frac{a}{r} - 1\right) dt^2 + \left(\frac{a}{r} - 1\right)^{-1} dr^2 - r^2 \, d\Omega^2. \qquad (7.225)$$

The signature $(+, -, -, -)$ remains the same, but t has become a spatial coordinate (negative contribution to ds^2) and r the temporal coordinate (positive contribution):

$$d\ell^2 = \left(\frac{a}{r} - 1\right) dt^2 + r^2 \, d\Omega^2. \qquad (7.226)$$

In region III, advancing time (increasing v) means *decreasing* r; all objects are squeezed. At any *given* time (constant r) the spatial geometry is cylindrical:

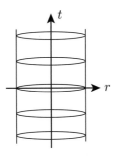

7.5.3 *Tidal Stress near* $r = 0$

Imagine a box in free fall as it approaches the (genuine) singularity at $r = 0$. In the box are two objects, a distance $\ell = r \, \Delta\phi$ apart (they could be two cells in the body of a falling astronaut). Ignoring nongravitational forces, they ride on adjacent radial geodesics, and because those are converging they will be pulled together:

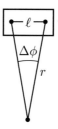

The "force" of attraction between them, as an observer in the box would experience it, is proportional to their (proper) acceleration

$$\frac{d^2\ell}{ds^2} = \frac{d^2r}{ds^2}\,\Delta\phi. \tag{7.227}$$

From Eq. 7.225,

$$\left(\frac{dr}{ds}\right)^2 = \frac{a}{r} - 1 \approx \frac{a}{r}, \quad \text{or} \quad \left(\frac{dr}{ds}\right) = \sqrt{\frac{a}{r}} \tag{7.228}$$

in the vicinity of $r = 0$:

$$\sqrt{r}\,dr = \sqrt{a}\,ds \quad \Rightarrow \quad \frac{2}{3}r^{3/2} = \sqrt{a}\,s \tag{7.229}$$

(setting $s = 0$ at $r = 0$). So $r \propto s^{2/3}$, or (differentiating twice)

$$\frac{d^2r}{ds^2} \propto s^{-4/3}, \tag{7.230}$$

and the "tidal" force goes like

$$\frac{d^2\ell}{ds^2} \propto s^{-4/3}, \tag{7.231}$$

which gets huge as $s \to 0$—the astronaut is crushed.

Problem 7.8

Given two metrics, $g_{\mu\nu}$ and $\bar{g}_{\mu\nu}$, such that

$$\bar{g}_{\mu\nu} = f(x)\,g_{\mu\nu}$$

(where x stands for the set of coordinates), show that the paths of photons in 4-space are the same for the two metrics. Use the following two different methods:
(a) Show that the free electromagnetic Lagrangian is the same in the two cases.
(b) Show that the photon equation of motion for one metric can be turned into the equation of motion in the other metric by an appropriate redefinition of the path parameter.

Problem 7.9

Consider a static metric of the form

$$ds^2 = f^2 dt^2 - h_{ij} dx^i dx^j, \tag{7.232}$$

where f and h are functions of position (but not of time). Show that the path of a photon in **x**-space (the conventional path of geometrical optics) is obtained by minimizing the integral

$$\int \frac{1}{f} \left(h_{ij} \frac{dx^i}{ds} \frac{dx^j}{ds} \right)^{1/2} ds, \tag{7.233}$$

with the endpoints of the path held fixed. [This is the gravitational analog of Fermat's principle of least time; it shows that an arbitrary static gravitational field acts like an anisotropic refractive medium.]

8

Conservation and Cosmology

8.1 Conservation Laws

8.1.1 Scalar Conservation Laws

In special relativity a divergenceless 4-vector current j^μ leads to a conserved scalar quantity Q:

$$\partial_\mu j^\mu = 0 \quad \Rightarrow \quad \int j^0(\mathbf{x},t)\, d^3\mathbf{x} \equiv Q; \tag{8.1}$$

Q is both independent of time (conserved) and invariant under Lorentz transformations (scalar). In general relativity the natural generalization of the continuity equation (the zero divergence of j^μ) would be

$$\nabla_\mu j^\mu = 0, \tag{8.2}$$

but this does not lead to a conserved quantity (you can't apply Gauss's theorem). However, if we first multiply the 4-vector j^μ by $\sqrt{-g}$, turning it into a vector *density*, the covariant derivative becomes an ordinary derivative, as we observed in footnote 13 in Chapter 6 (see also Problem 6.8):

$$\nabla_\mu \left(\sqrt{-g}\, j^\mu \right) = \partial_\mu \left(\sqrt{-g}\, j^\mu \right) = 0. \tag{8.3}$$

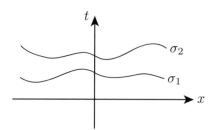

Then

$$Q(\sigma_1) \equiv \int_{\sigma_1} \sqrt{-g}\, j^0(\mathbf{x}, t)\, d^3\mathbf{x} = Q(\sigma_2), \tag{8.4}$$

where σ_1 and σ_2 are space-like surfaces. In this sense Q is a conserved quantity.

8.1.2 The Energy–Momentum Pseudotensor

The *matter* stress–energy tensor satisfies[1]

$$\nabla_\mu T^{\mu\nu} = 0 \tag{8.5}$$

(see Eq. 7.157). To extract a conservation law, though, we need the partial derivative, not the covariant derivative. We could again multiply by $\sqrt{-g}$, to convert the tensor into a tensor density:

$$\nabla_\mu \left(\sqrt{-g}\, T^{\mu\nu} \right) = 0. \tag{8.6}$$

If you like, this turns the *first* index into a vector density, but the second is still a vector, so (5.79 and 5.82)

$$\nabla_\mu \left(\sqrt{-g}\, T^{\mu\nu} \right) = \partial_\mu \left(\sqrt{-g}\, T^{\mu\nu} \right) + \Gamma^\nu_{\lambda\mu} T^{\mu\lambda} \sqrt{-g} = 0. \tag{8.7}$$

This tells us the rate at which energy/momentum is transferred from the gravitational field to matter (and back)—but $T^{\mu\nu}$ (or $\sqrt{-g}\, T^{\mu\nu}$) is not by itself conserved, in the presence of gravity (and this is reflected in the fact that its ordinary divergence is not zero). Could we construct a *total* stress tensor, combining a gravity contribution and a matter contribution, that *is* conserved in the sense that

$$\partial_\mu \left(\sqrt{-g}\, (T^{\mu\nu} + T^{\mu\nu}_{\text{grav}}) \right) = 0? \tag{8.8}$$

Well, whatever the gravitational part might be, it is certainly *not* a tensor, because (by the principle of equivalence) in a freely falling elevator gravity vanishes; if $T^{\mu\nu}_{\text{grav}}$ *were* tensorial it would be zero in *any* coordinates.[2]

What about *global* quantities? Could there exist a conserved energy–momentum 4-vector P^μ? How would it transform? Don't forget, vectors live in the tangent space at a particular point; there's no way to associate a vector or a tensor with the whole manifold. But we might plausibly assume that there exists an asymptotically Minkowskian manifold (defining a preferred global coordinate system),

---

[1] Coleman indicated that this section follows L. D. Landau and E. M. Lifshitz, *The Classical Theory of Fields*, Pergamon Press. In the third edition (1971) it is §101.

[2] Eds. This argument is not compelling. After all, the Ricci tensor is not zero in elevator coordinates, and this does not violate the equivalence principle. Coleman is evidently assuming something about the hypothetical gravitational stress tensor, but he does not elaborate.

and demand that P^μ transform as a 4-vector in this special (asymptotic) reference frame. Then, stepping backward, we might ask whether there could exist a local density that integrates to P^μ. Perhaps; but (as we just found) it cannot be a tensor. Maybe it's something different, like the Christoffel symbols. We will try an *ansatz* of the form[3]

$$\partial_\mu \left[-g(T^{\mu\nu} + t^{\mu\nu}) \right] = 0, \tag{8.9}$$

where $t^{\mu\nu}$ is this nontensor representing (in some sense) the local energy/momentum density of the gravitational field, such that the integral of the quantity in square brackets transforms as a 4-vector in the asymptotic Minkowski coordinates. That's the best we can hope for. What could this $t^{\mu\nu}$ be?

There are two methods for constructing $t^{\mu\nu}$: you can go back to the special relativity prescription, write Einstein's action in terms of the "field" $g_{\mu\nu}$, and deduce the conserved stress tensor (in Minkowski space) the same way we did in Section 2.4.2—or you can simply fiddle around. The latter approach is faster, and that's how I'll do it. Starting with Eq. 7.49 and Problem 6.2,

$$\nabla_\nu G^{\mu\nu} = 0 \quad \Rightarrow \quad \nabla_\nu(-gG^{\mu\nu}) = 0. \tag{8.10}$$

In locally geodesic coordinates (where the Γ's are zero)

$$\partial_\nu(-gG^{\mu\nu}) = 0. \tag{8.11}$$

Now, in electrodynamics, the continuity equation $\partial_\nu J^\nu = 0$ is enforced by

$$J^\nu = \partial_\mu F^{\mu\nu}, \quad \text{with} \quad F^{\mu\nu} = -F^{\nu\mu}. \tag{8.12}$$

This suggests, by analogy, that $\partial_\nu (-gG^{\mu\nu}) = 0$ can be enforced by

$$-gG^{\mu\nu} = \partial_\lambda \eta^{\mu\nu\lambda} \quad \text{with} \quad \eta^{\mu\nu\lambda} = \eta^{\nu\mu\lambda} = -\eta^{\mu\lambda\nu} \tag{8.13}$$

(symmetry in the first two indices ensures that of $G^{\mu\nu}$; antisymmetry in the last two mimics 8.12). In elevator coordinates,

$$\eta^{\mu\nu\lambda} = \tfrac{1}{2}\partial_\rho \left[g \left(g^{\mu\nu} g^{\lambda\rho} - g^{\mu\lambda} g^{\nu\rho} \right) \right], \tag{8.14}$$

as you can check for yourself (Problem 8.1). Now we *stipulate* that η has this *same form in every coordinate system* (it's not tensorial, obviously). In other coordinate systems 8.13 no longer holds, but Einstein's equation still does, and we define $t^{\mu\nu}$ as whatever is needed to make

$$(-g)G^{\mu\nu} = -\alpha(-g)T^{\mu\nu} = \partial_\lambda \eta^{\mu\nu\lambda} + (-g)\alpha t^{\mu\nu}. \tag{8.15}$$

---

[3] As it turns out, $\sqrt{-g}$ doesn't work, but $-g$ *does*.

It $(t^{\mu\nu})$ is also nontensorial (but we already knew it had to be—it's a "pseudotensor" that vanishes in locally geodesic coordinates, just like Γ).[4] Then

$$\partial_\lambda \eta^{\mu\nu\lambda} = -\alpha(-g)\,(T^{\mu\nu} + t^{\mu\nu}), \qquad (8.16)$$

from which it follows that

$$\partial_\nu \left[-g(T^{\mu\nu} + t^{\mu\nu}) \right] = 0. \qquad (8.17)$$

Using this we can construct a global energy–momentum 4-vector. Imagine two space-like surfaces that, in the asymptotic flat Minkowski space, go to a constant times t_1 and t_2:

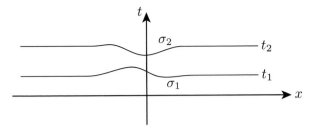

Then

$$\int_{\sigma 1} T_{\text{tot}}^{\mu 0}(\mathbf{x}, t)\, d^3\mathbf{x} = P^\mu(t_1) = P^\mu(t_2), \qquad (8.18)$$

where T_{tot} is the sum of the matter tensor and the gravitational pseudotensor. Now, we are free to choose flat surfaces that are *not* perpendicular to the t-axis in the asymptotic region, by ordinary Lorentz transformation (under which $t^{\mu\nu}$ *is* tensorial). Thus a *total* energy/momentum 4-vector,[5] including gravity, *is* possible in general relativity.

Problem 8.1

Check that 8.14 satisfies 8.13, in locally geodesic coordinates. (First show, using results from Problem 6.2, that $\partial_\nu(-g) = -2g\Gamma^\lambda_{\nu\lambda}$.)

8.2 The Universe at Large

We turn now to the structure of the universe as a whole. To begin with I will discuss some general principles that apply to *any* viable model. For the moment we will not

---

[4] Landau and Lifshitz give the (cumbersome) explicit formula for $t^{\mu\nu}$ in Eq. (101.6); it's a quadratic function of g's and Γ's.
[5] It's a 4-vector with respect to the asymptotic (Minkowskian) reference frames.

invoke Einstein's equation—these considerations apply just as well to (say) Milne's kinematic theory (Problem 1.3) or the Hoyle–Bondi steady-state model.

8.2.1 General Principles

We will adopt the perspective of a near-sighted God, and take the matter in the universe to be smoothly distributed, on a large enough scale. Stars and galaxies become the "atoms" in a continuous fluid—a "gas," whose small-scale lumpiness is presumably irrelevant. We make two assumptions:

1. **Weyl's postulate:** The universe is a 4-dimensional Riemannian manifold $\mathcal{M}$, whose metric $g_{\mu\nu}$ has signature $(+, -, -, -)$; at every point x in $\mathcal{M}$ there is a time-like curve passing through x (a **galactic world line**[6]). Galactic world lines do not intersect (except possibly at isolated points)—i.e. there is only one fluid velocity at each point.

2. **The cosmological principle:** It is an empirical fact that the density of galaxies, on a cosmic scale, is isotropic about the earth: we appear to be at the center of a spherically symmetrical universe. We will assume this is not a coincidence— that the universe appears isotropic *every*where, and at *any* time. An observer on another galaxy may see a matter distribution different from ours, but it will at least be spherically symmetric. Formally, for every galaxy $\mathcal{G}$ there exists a group of rotational isometries (with $\mathcal{G}$ as their center) that leave the world line of $\mathcal{G}$ invariant, and act nontrivially on all galaxies in the neighborhood of $\mathcal{G}$. These isometries take galactic world lines into other galactic world lines (this would *follow* from spherical symmetry if we assumed general relativity, because then galactic world lines would be geodesics).

In essence, Weyl's postulate says the universe is a gas, and the cosmological principle says it is isotropic (everywhere and always). An ostensibly stronger condition would state that there exists an isometry turning *any* galactic world line into *any other* galactic world line—but this can actually be *proved*, from the cosmological principle, so it's not an extra assumption.

Given these two assumptions, what would a convenient set of coordinates for the universe at large be?

---

[6] In general relativity, with galaxies moving only under gravity, galactic world lines would be geodesics, and Weyl himself assumed that they are. But for the moment a galactic world line is simply the trajectory of a "molecule" in the cosmic "gas." As we shall see, given the second postulate they must in fact be geodesics, but this is not an independent assumption.

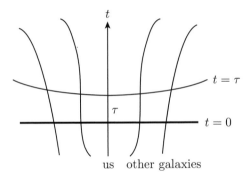

us   other galaxies

Pick some galaxy ("us") to be the "center," and construct the $t = 0$ surface so as to be invariant under the rotational isometry. On this surface, then,

$$d\ell^2 = h(r,0)\, dr^2 + k(r,0)\, d\Omega^2. \tag{8.19}$$

But we won't define the parameterization of the spheres as we did before ($k \to r^2$). Construct the $t = \tau$ surface by choosing points on each galactic world line at proper time τ later. Define the spatial coordinates on this surface by carrying them over (via the galactic world lines) from $t = 0$. The spatial coordinates of a galaxy will then be time independent (**comoving coordinates**). The full spacetime metric is

$$ds^2 = f(r,t)\, dt^2 + g(r,t)\, dr\, dt - h(r,t)\, dr^2 - k(r,t)\, d\Omega^2. \tag{8.20}$$

Along a given galactic world line, r, θ, and ϕ are constant, and $t = \tau$. For a galaxy, then, $ds^2 = f\, dt^2 = f\, d\tau^2 = f\, ds^2$, so in these coordinates $f = 1$.

Question: Is the world line of our central galaxy a geodesic? Consider the geodesic that starts with the same position and velocity as our galaxy. Could it subsequently deviate from our galactic world line? No, for to do so it would have to veer off in some direction, and that would violate isotropy. And the same applies to all other galaxies. *Conclusion: All galactic world lines are geodesics.* Note that this is a *conclusion*, not a postulate, and it does not depend on general relativity.

In these coordinates, then, the metric takes the form

$$ds^2 = dt^2 + g(r,t)\, dr\, dt - h(r,t)\, dr^2 - k(r,t)\, d\Omega^2, \tag{8.21}$$

and galactic world lines satisfy 6.13:

$$\frac{d^2 x^\mu}{ds^2} = -\Gamma^\mu_{\lambda\sigma} \frac{dx^\lambda}{ds} \frac{dx^\sigma}{ds} = 0. \tag{8.22}$$

Now, along a galactic world line $dx^0/ds = 1$, and the spatial coordinates don't change at all; therefore

$$\Gamma^\mu_{00} = 0, \quad \text{so} \quad \Gamma_{v,00} = g_{v\mu}\Gamma^\mu_{00} = 0 \tag{8.23}$$

(5.106), and it follows (from 5.110) that

$$\frac{\partial}{\partial t} g_{0v} = 0. \tag{8.24}$$

So

$$\frac{\partial g(r,t)}{\partial t} = 0; \quad \text{and hence} \quad g(r,t) = g(r). \tag{8.25}$$

We can therefore introduce

$$F(r) \equiv \int g(r)\,dr, \quad \text{so that} \quad \frac{dF}{dr} = g(r), \tag{8.26}$$

and redefine the initial time surface $(t = 0)$ by

$$t = t' - \frac{F(r)}{2} \Rightarrow dt = dt' - \frac{1}{2}g(r)\,dr. \tag{8.27}$$

Then

$$dt^2 + g\,dr\,dt \rightarrow (dt')^2 - g\,dr\,dt' + \tfrac{1}{4}g^2\,dr^2 + g\,dr\,dt' - \tfrac{1}{2}g^2\,dr^2$$
$$= (dt')^2 - \tfrac{1}{4}g^2\,dr^2. \tag{8.28}$$

This casts the metric in the following form:

$$ds^2 = (dt')^2 - H(r,t')\,dr^2 - K(r,t')\,d\Omega^2, \tag{8.29}$$

with just *two* undetermined functions. All this follows from the Weyl postulate and the cosmological principle.

Surfaces of constant time are perpendicular to galactic world lines, so they are unique (it doesn't matter what central galaxy you choose), and isotropic about any point on them. This raises Élie Cartan's classic problem in differential geometry: *What 3-dimensional spaces are rotationally isotropic about every point?* In 3-dimensional spherical coordinates,

$$d\ell^2 = e^{\mu(r)}\,dr^2 + r^2\,d\Omega^2. \tag{8.30}$$

Imagine embedding this as the $w = 0$ surface in a 4-dimensional manifold:

$$ds^2 = dw^2 + d\ell^2. \tag{8.31}$$

In this larger space the zero/zero component of the Einstein tensor, $G^0{}_0$, is a 3-scalar function, on the $w = 0$ surface, and $\partial_i G^0{}_0$ is a 3-vector. Suppose the latter is not zero; then there is a *preferred direction* on the $w = 0$ surface, which is forbidden. So $G^0{}_0$ is constant (independent of position), and we found before (7.169) that

$$G^0{}_0 = -\alpha\mu_0 \quad \Rightarrow \quad e^{-\mu} = \left(1 - \frac{r^2}{b^2}\right), \quad b^2 \equiv \frac{3}{\alpha\mu_0}. \tag{8.32}$$

In this context b^2 could be positive, negative, or zero. We might as well rescale lengths so that $b^2 = 1, 0,$ or -1:

$$e^{-\mu} = 1 + \epsilon r^2, \quad \text{where} \quad \epsilon = \begin{cases} -1, \\ 0, \\ +1. \end{cases} \tag{8.33}$$

Evidently there are three possible kinds of spaces. They are *necessary* for complete isotropy. They are also *sufficient*:

- $\epsilon = 0$ is a flat Euclidean space—it is trivially isotropic.
- $\epsilon = -1$ is a 3-dimensional hypersphere of radius 1, in a 4-dimensional Euclidean space:

$$w^2 + r^2 = 1 \implies 2w\,dw + 2r\,dr = 0 \implies dw = -\frac{r}{w}\,dr,$$

so

$$ds^2 = dw^2 + dr^2 + r^2\,d\Omega^2 = \frac{1}{1 - r^2}\,dr^2 + r^2\,d\Omega^2.$$

As a sphere it is manifestly isotropic about every point.

- $\epsilon = +1$ has the same metric as a hyperboloid in Minkowski space:

$$t^2 - r^2 = 1 \implies 2t\,dt - 2r\,dr = 0 \implies dt = \frac{r}{t}\,dr,$$

so

$$ds^2 = dt^2 - dr^2 - r^2\,d\Omega^2 = -\frac{1}{1 + r^2}\,dr^2 - r^2\,d\Omega^2.$$

You can bring any point on the hyperboloid to the center by a Lorentz transformation, and there the isotropy is manifest.

Conclusion: There are precisely three solutions to Cartan's isotropy problem.

8.2.2 The Robertson–Walker Metric

Returning to comoving coordinates (8.29), the metric of the universe takes the form

$$ds^2 = dt^2 - R(t)^2 \left[\frac{1}{(1 + \epsilon r^2)} dr^2 + r^2 d\Omega^2 \right]. \qquad (8.34)$$

There remains one function of time $R(t)$ (the "radius of the universe"—actually, a dimensionless scale factor) and a 3-valued discrete parameter ϵ. Depending on whether $R(t)$ is an increasing or a decreasing function, the universe is either expanding or contracting. Note that we have still made no use of Einstein's equation. This is the famous **Robertson–Walker metric**; it describes a **homogeneous** universe (something we did *not* assume—it emerges as a consequence of our two postulates): at any particular time t, all galaxies see the same thing. If the universe is expanding or contracting for one observer, it is for everyone.

Converting from spherical coordinates (r, θ, ϕ) to **isotropic coordinates** (ρ, θ, ϕ) (which show the isotropy explicitly),

$$\left[\frac{1}{(1 + \epsilon r^2)} dr^2 + r^2 d\Omega^2 \right] = F(\rho)^2 \left[d\rho^2 + \rho^2 d\Omega^2 \right]. \qquad (8.35)$$

Comparing the two expressions,

$$\frac{1}{(1 + \epsilon r^2)} dr^2 = F^2 d\rho^2 \quad \text{and} \quad r^2 = F^2 \rho^2 \quad \Rightarrow \quad \frac{dr^2}{r^2(1 + \epsilon r^2)} = \frac{d\rho^2}{\rho^2}. \qquad (8.36)$$

$$\int \frac{dr}{r\sqrt{1 + \epsilon r^2}} = \int \frac{d\rho}{\rho} \quad \Rightarrow \quad \ln\left(\frac{r}{1 + \sqrt{1 + \epsilon r^2}} \right) = \ln \rho - \ln \rho_0. \qquad (8.37)$$

Picking the constant of integration so that $r = \rho$ when $\epsilon = 0$ (which is to say, taking $\rho_0 = 2$),

$$\left(\frac{r}{1 + \sqrt{1 + \epsilon r^2}} \right) = \frac{\rho}{2}, \quad \text{or} \quad r = \frac{\rho}{1 - (\epsilon/4)\rho^2}. \qquad (8.38)$$

Then

$$F = \frac{r}{\rho} = \frac{1}{\left(1 - (\epsilon/4)\rho^2\right)}. \qquad (8.39)$$

Both coordinate systems have disadvantages: spherical coordinates have an artificial singularity at $r = 1$ if $\epsilon = -1$; isotropic coordinates have a phony singularity at $\rho = 2$ when $\epsilon = 1$.

8.2.3 Redshift and Luminosity

It is an empirical fact that the universe is expanding. Distant galaxies are moving away from us, with a speed proportional to their distance:

$$v = Hd. \tag{8.40}$$

This is **Hubble's law**, and H is **Hubble's constant**.[7] The speed is measured by the **redshift** of spectral lines; the distance is determined from the apparent luminosity of objects whose intrinsic luminosity is known (so-called **standard candles**). But how are these quantities related to elements of the Robertson–Walker metric? For a photon approaching the earth on a radial trajectory,

$$ds^2 = dt^2 - \frac{R(t)^2}{(1 + \epsilon r^2)} dr^2 = 0, \tag{8.41}$$

so

$$\frac{dt}{R(t)} = -\frac{dr}{\sqrt{1 + \epsilon r^2}}. \tag{8.42}$$

A time-independent measure of the distance to the galaxy from which the photon left is

$$\int_{t_1}^{t_2} \frac{dt}{R(t)} = \int_0^{r_1} \frac{dr}{\sqrt{1 + \epsilon r^2}} \equiv d, \tag{8.43}$$

where t_1 is the moment of emission (on the distant galaxy), t_2 is the moment of absorption (on earth), and r_1 is the radial coordinate of the galaxy. Since d is a constant (unlike the geometrical distance, $R(t)d$, which changes as the universe expands), the time between pulses (or maxima of the electromagnetic wave) is given by

$$\frac{dt_2}{R(t_2)} = \frac{dt_1}{R(t_1)}. \tag{8.44}$$

The frequency is inversely proportional to the time between peaks, so

$$\frac{\omega^{(\text{em})}}{\omega^{(\text{ab})}} = \frac{R(t_2)}{R(t_1)}. \tag{8.45}$$

This is the formula for the cosmic redshift, in terms of the ratio of the "radii of the universe" at the moment of emission and at the moment of absorption.

Meanwhile, we define the (intrinsic) emitted luminosity as

$$L \equiv \frac{dE^{(\text{em})}}{dt_1}, \tag{8.46}$$

---

[7] Eds. E. P. Hubble, *Proc. Nat. Acad. Sci.* **15**, 168 (1929).

where $dE^{(em)}$ is the energy emitted in time dt_1. In terms of the number of photons emitted, $dE = \hbar\omega\,dn$, so

$$\frac{dn}{dt_1} = \frac{L}{\hbar\omega^{(em)}}. \tag{8.47}$$

These photons spread out over a sphere of radius $R(t_2)d$, by the time they reach the earth, so

$$\frac{dn}{dA\,dt_1} = \frac{L}{4\pi[R(t_2)d]^2\hbar\omega^{(em)}}, \tag{8.48}$$

or, using 8.44 and 8.45,

$$\frac{dn}{dA\,dt_2} = \frac{L}{4\pi[R(t_2)d]^2\hbar\omega^{(ab)}}\left(\frac{R(t_1)}{R(t_2)}\right)^2. \tag{8.49}$$

In terms of the energy *absorbed* ($dE^{(ab)} = \hbar\omega^{(ab)}$), then,

$$\frac{dE^{(ab)}}{dA\,dt_2} = \frac{L}{4\pi[R(t_2)d]^2}\left(\frac{R(t_1)}{R(t_2)}\right)^2. \tag{8.50}$$

The **luminosity distance** $R(t_2)d$, which is the geometrical distance to the distant galaxy *at the time of absorption*, is the only way this result differs from what special relativity would give.

8.3 General Relativity and Cosmology

8.3.1 The Friedman[8] Universe

The Robertson–Walker metric,

$$ds^2 = dt^2 - R(t)^2 h_{ij}\,dx^i dx^j, \quad h_{ij}\,dx^i dx^j = \left(\frac{1}{1+\epsilon r^2}\right)dr^2 + r^2\,d\Omega^2, \tag{8.51}$$

holds for a broad class of cosmological models, including those of Milne and Bondi/Gold that are not consistent with general relativity. But what if we insist that it satisfy Einstein's equation? What does this tell us about $R(t)$ and/or ϵ? Let us assume that the universe on a cosmic scale is a perfect fluid (7.150):

$$T^{\mu\nu} = (\mu_0 + P)v^\mu v^\nu - P g^{\mu\nu}, \tag{8.52}$$

---

[8] Eds. Alexander Friedman's last name is often spelled with two n's. This is apparently due to a mistake by Einstein. "In his two notes, Einstein inadvertently introduced the following problem for future historians of cosmology: he christened Friedman with two n's in his last name whereas Friedman's 1922 paper was signed with one 'n.' It seems Friedman followed the 'advice' and submitted his 1924 paper with two n's." A. Belenkiy (6 Feb 2013), arXiv:1302.1498v1 [physics.hist-ph], footnote 15.

and in fact that the pressure is zero (dust).[9] Here v^μ is the (proper) velocity of the galaxies, which ride along with the coordinates, so

$$v^\mu = (1, 0), \quad \text{and hence} \quad T^{00} = \mu_0, \tag{8.53}$$

while all other components are zero. Equation (8.6) says

$$\nabla_\mu \left(\sqrt{-g}\, T^{\mu\nu} \right) = 0, \tag{8.54}$$

so

$$\partial_0 \left(\sqrt{-g}\, T^{00} \right) + \Gamma^0_{\lambda\sigma}\, T^{\lambda\sigma} \sqrt{-g} = 0. \tag{8.55}$$

But $\Gamma^0_{00} = 0$ for a metric of the form 8.51, so $\sqrt{-g}\, T^{00}$ is constant in time; for future convenience we'll write it as $3A/\alpha$, where A is time independent (and $\alpha = 8\pi G$, as always):

$$\sqrt{-g}\, T^{00} = \frac{3A}{\alpha}. \tag{8.56}$$

In locally Euclidean spatial coordinates,

$$g_{\mu\nu} = \begin{pmatrix} 1 & 0 & 0 & 0 \\ 0 & -R^2 & 0 & 0 \\ 0 & 0 & -R^2 & 0 \\ 0 & 0 & 0 & -R^2 \end{pmatrix}. \tag{8.57}$$

Its determinant is $-R^6$, so

$$\sqrt{-g} = R^3, \tag{8.58}$$

and therefore[10]

$$T^{00} = \frac{3A}{\alpha R^3}. \tag{8.59}$$

Now let's work out the Einstein tensor for the Robertson–Walker metric. The nonzero elements in the metric and its inverse are

$$g_{00} = 1, \quad g_{11} = -\frac{R^2}{(1 + \epsilon r^2)}, \quad g_{22} = -R^2 r^2, \quad g_{33} = -R^2 r^2 \sin^2\theta, \tag{8.60}$$

$$h_{11} = \frac{1}{(1 + \epsilon r^2)}, \quad h_{22} = r^2, \quad h_{33} = r^2 \sin^2\theta, \tag{8.61}$$

$$g^{00} = 1, \quad g^{11} = -\frac{(1 + \epsilon r^2)}{R^2}, \quad g^{22} = -\frac{1}{R^2 r^2}, \quad g^{33} = -\frac{1}{R^2 r^2 \sin^2\theta}. \tag{8.62}$$

---

[9] Unlike the interior Schwarzschild solution, μ_0 is presumably a function of time (as well as position).
[10] We should really include the transformation (from Euclidean coordinates back to spherical), but it doesn't matter: T^{00} doesn't change if the transformation is only on the *spatial* coordinates.

The nonzero Christoffel symbols are[11]

$$
\begin{array}{ll}
\Gamma_{0,ij} = R\dot{R}\,h_{ij} & \Gamma^0_{ij} = R\dot{R}\,h_{ij} \\[4pt]
\Gamma_{i,0j} = \Gamma_{i,j0} = -R\dot{R}\,h_{ij} & \Gamma^i_{0j} = \Gamma^i_{j0} = (\dot{R}/R)\delta^i_j \\[4pt]
\Gamma_{1,11} = \epsilon r R^2/(1+\epsilon r^2)^2 & \Gamma^1_{11} = -\epsilon r/(1+\epsilon r^2) \\[4pt]
\Gamma_{1,22} = R^2 r & \Gamma^1_{22} = -r(1+\epsilon r^2) \\[4pt]
\Gamma_{1,33} = R^2 r \sin^2\theta & \Gamma^1_{33} = -r(1+\epsilon r^2)\sin^2\theta \qquad\text{(8.63)} \\[4pt]
\Gamma_{2,33} = R^2 r^2 \sin\theta\cos\theta & \Gamma^2_{33} = -\sin\theta\cos\theta \\[4pt]
\Gamma_{2,21} = \Gamma_{2,12} = -R^2 r & \Gamma^2_{12} = \Gamma^2_{21} = 1/r \\[4pt]
\Gamma_{3,31} = \Gamma_{3,13} = -R^2 r \sin^2\theta & \Gamma^3_{13} = \Gamma^3_{31} = 1/r \\[4pt]
\Gamma_{3,32} = \Gamma_{3,23} = -R^2 r^2 \sin\theta\cos\theta & \Gamma^3_{23} = \Gamma^3_{32} = \cot\theta
\end{array}
$$

The Ricci tensor is

$$
R_{\mu\nu} = \partial_\nu \Gamma^\lambda_{\mu\lambda} - \partial_\lambda \Gamma^\lambda_{\mu\nu} + \Gamma^\rho_{\mu\lambda}\Gamma^\lambda_{\nu\rho} - \Gamma^\lambda_{\mu\nu}\Gamma^\rho_{\rho\lambda}. \tag{8.64}
$$

Its elements are

$$
\begin{aligned}
R_{00} &= \partial_0 \Gamma^\lambda_{0\lambda} - \partial_\lambda \Gamma^\lambda_{00} + \Gamma^\rho_{0\lambda}\Gamma^\lambda_{0\rho} - \Gamma^\lambda_{00}\Gamma^\rho_{\rho\lambda} \\
&= \partial_0 \left(\Gamma^1_{01} + \Gamma^2_{02} + \Gamma^3_{03} \right) + \left(\Gamma^1_{01}\Gamma^1_{01} + \Gamma^2_{02}\Gamma^2_{02} + \Gamma^3_{03}\Gamma^3_{03} \right) \\
&= \partial_0 \left(3\frac{\dot{R}}{R} \right) + 3\left(\frac{\dot{R}}{R} \right)^2 = 3\frac{\ddot{R}}{R} - 3\frac{\dot{R}^2}{R^2} + 3\frac{\dot{R}^2}{R^2} = 3\frac{\ddot{R}}{R},
\end{aligned} \tag{8.65}
$$

$$
R_{0i} = R_{i0} = 0, \tag{8.66}
$$

$$
R_{ij} = -\left(R\ddot{R} + 2\dot{R}^2 - 2\epsilon \right) h_{ij}. \tag{8.67}
$$

And the curvature scalar (written R_s now, to distinguish it from the "radius" $R(t)$) is

$$
R_s = g^{\mu\nu} R_{\mu\nu} = \frac{6}{R^2} \left(R\ddot{R} + \dot{R}^2 - \epsilon \right). \tag{8.68}
$$

Problem 8.2

(a) Using 8.64, find R_{11} and R_{01}.
(b) From 8.65–8.67, confirm 8.68.

---

[11] These are the same as we found in Section 7.2, with $r \to Rr$, $\lambda \to 0$, and $e^\mu \to 1/(1+\epsilon r^2)$, except that g_{ij} are now functions of t, so anything involving a time derivative is new—hence 8.63.

In particular,

$$G^0_{\ 0} = R^0_{\ 0} - \frac{1}{2} R_s = g^{00} R_{00} - \frac{1}{2} R_s = 3\frac{\ddot{R}}{R} - \frac{3}{R^2}\left(R\ddot{R} + \dot{R}^2 - \epsilon\right)$$

$$= -\frac{3}{R^2}\left(\dot{R}^2 - \epsilon\right), \tag{8.69}$$

and Einstein's equation says this is equal to

$$-\alpha T^0_{\ 0} = -\alpha\, g_{00}\, T^{00} = -\alpha\frac{3A}{\alpha R^3} = -3\frac{A}{R^3}, \tag{8.70}$$

so

$$\dot{R}^2 - \epsilon = \frac{A}{R}. \tag{8.71}$$

This is just like the formula for a classical particle in radial free fall,

$$E = \frac{1}{2}m\dot{r}^2 - \frac{GMm}{r} \tag{8.72}$$

with energy $E = \epsilon$, mass $m = 2$, and $2GM = A$. Remember that ϵ can be -1, 0, or $+1$. In the figure below, I plot the (constant) total energy and the potential energy:

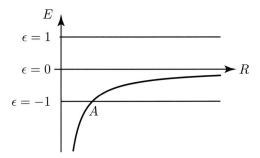

Suppose the particle starts out at some small r, with energy ϵ. If $\epsilon = 1$ it flies out to infinity, at an asymptotically constant speed (the universe expands forever); if $\epsilon = 0$ the expansion rate decreases, approaching zero (very slowly); if $\epsilon = -1$ it expands out to $R = A$, where it comes (momentarily) to rest, and falls back. These are the three **Friedman universes**.[12]

---

[12] Eds. For further discussion see A. Liddle, *An Introduction to Modern Cosmology*, 2nd ed., Wiley, Hoboken, NJ (2003).

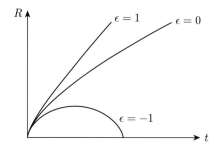

Problem 8.3

We used the 00 component of Einstein's equation to get Eq. 8.71, but what about the other components? Work out G_{11} and show that it automatically satisfies $G_{11} = -\alpha T_{11}$, given 8.71.

8.3.2 *The Cosmological Constant*

In the de Sitter theory (Section 6.2.2) Einstein's equation is replaced by 6.68:

$$G^{\mu}{}_{\nu} - \Lambda \delta^{\mu}_{\nu} = -\alpha T^{\mu}{}_{\nu} \tag{8.73}$$

(where Λ is the cosmological constant), so

$$-\frac{3}{R^2} \left(\dot{R}^2 - \epsilon \right) - \Lambda = -3 \frac{A}{R^3}, \tag{8.74}$$

and 8.71 becomes

$$\epsilon = \dot{R}^2 - \frac{A}{R} + \frac{\Lambda}{3} R^2. \tag{8.75}$$

This adds a harmonic oscillator potential to 8.72.

- **If $\Lambda > 0$.** The potential energy has no minimum; R goes from zero to the crossing point (dot, in the figure) and back to zero, in a time-symmetric manner: the universe expands and then collapses.

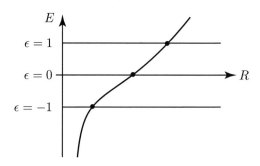

- **If Λ < 0.** The potential energy is always negative (the sum of two negative terms).

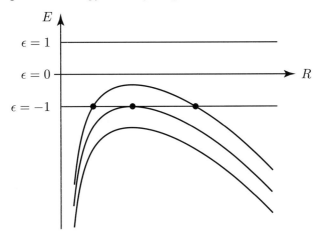

When $\epsilon = 1$ or $\epsilon = 0$, R expands from zero, slowing down, and then (after passing the maximum of the potential energy) speeds up again. The same goes for $\epsilon = -1$, if the potential maximum is less than -1, but if the potential maximum is greater than -1 then either R starts from zero, expands to the first crossing point, and then returns to zero, or else it comes in from infinity and reverses direction at the second crossing point, and returns to infinity. If the maximum is *exactly* -1, there are three possibilities: (1) it sits at the maximum, R is constant, and the universe is **static**, or (2) it expands from zero and comes to rest at the maximum, or (3) it starts from rest at the maximum, and expands out to infinity[13] (or the time-reversed contracting solution, which we ignore because of the observed expansion).

8.3.3 Singularities in the Robertson–Walker Metric

The Robertson–Walker metric (with $\Lambda = 0$) contains potential singularities (due to $R(t)$) at $t = 0$ and $t = \infty$. Are they genuine singularities (like $r = 0$ in the Schwarzschild metric), or mere coordinate singularities (like the one at $r = a$, which disappears when you use Kruskal coordinates)? Can a particle, for example, get to $t = \infty$ in a finite proper time (as it can, falling to the Schwarzschild radius)? Geodesics satisfy 6.13:

$$\frac{d^2 x^\mu}{ds^2} = -\Gamma^\mu_{\nu\lambda} \frac{dx^\nu}{ds} \frac{dx^\lambda}{ds}. \tag{8.76}$$

---

[13] This is the **Einstein–Lemaître solution**.

In the Robertson–Walker metric (8.34),

$$\frac{d^2 t}{ds^2} = -\Gamma^0_{ij} \frac{dx^i}{ds} \frac{dx^j}{ds} = -R\dot{R} \, h_{ij} \frac{dx^i}{ds} \frac{dx^j}{ds} < 0 \qquad (8.77)$$

(R is intrinsically positive, and $\dot{R} > 0$ in all models with an expanding universe). Therefore

$$\frac{dt}{ds} < \frac{dt}{ds}\Big|_{s=0} \qquad (8.78)$$

and you can only get to $t = \infty$ when $s = \infty$. We don't need to worry about any singularity at $t = \infty$, because we will never get to it. On the other hand, from 8.69 and 8.70,

$$G^0{}_0 = -3\frac{A}{R^3}, \quad \text{while} \quad G^i{}_j = -\alpha T^i{}_j = 0, \quad \text{so} \quad G^\mu{}_\mu = -3\frac{A}{R^3}, \qquad (8.79)$$

and therefore the singularity at $R = 0$ is real—it cannot be eliminated by a change of coordinates.

Comments:

- **What if $A = 0$?** Then the universe is empty of matter—or at any event the energy density of matter is zero (maybe there are very light galaxies and some kind of compensating negative energy). In this case 8.71 says $\epsilon = \dot{R}^2$, so $\epsilon = -1$ (the spherical case) is impossible; $\epsilon = 0$ would be OK—R would be constant (Minkowski space); and $\epsilon = 1$ is possible, with expansion at the speed of light. The latter is reminiscent of the Milne model (Problem 1.3), with galaxies on the forward light cone. Surfaces of constant cosmic time would be hyperbolas.

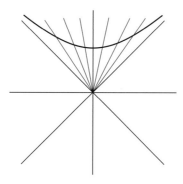

This model can be extended to Minkowski space with no galaxies in some regions, but because $T^{00} = 0$ it is not of much interest physically.
- **What about the genuine singularity at $R = 0$?** In the case $\epsilon = -1$ we found that R expands from 0 to a maximum of A, and then collapses back to 0 (the end of the universe). But this assumed zero pressure, which is clearly untrue when

the size is small. Maybe including nonzero pressure would push R out again, preventing it from reaching $R = 0$, and resulting in oscillatory behavior:

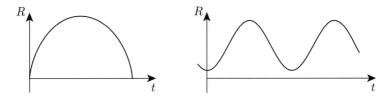

Unfortunately, this doesn't work. Einstein's equation says

$$R^{\mu}{}_{\nu} - \tfrac{1}{2}\delta^{\mu}_{\nu} R_s = -\alpha T^{\mu}{}_{\nu} . \tag{8.80}$$

Taking the trace of both sides,

$$-R_s = -\alpha T, \tag{8.81}$$

so

$$R^{\mu}{}_{\nu} = -\alpha \left(T^{\mu}{}_{\nu} - \tfrac{1}{2}T\delta^{\mu}_{\nu} \right). \tag{8.82}$$

In the case of spherical symmetry,

$$T^0{}_0 = \mu_0, \quad T^i{}_j = -P\delta^i_j, \quad \text{so} \quad T = \mu_0 - 3P. \tag{8.83}$$

Thus

$$R^0{}_0 = -\alpha[\mu_0 - \frac{1}{2}(\mu_0 - 3P)] = -\frac{\alpha}{2}(\mu_0 + 3P) = 3\frac{\ddot{R}}{R} \tag{8.84}$$

(8.65). Since μ_0 and P are intrinsically positive, it follows that $\ddot{R} < 0$, so even with pressure included there's no way to get a local minimum for R, and avoid the collapse. The pressure drives up the energy density more than it pushes outward.

- **The horizon.** Can we see the entire universe? Or could there exist galaxies that are simply "over the horizon," and (at present) invisible to us? Consider a light ray, coming to us from some distant object; suppose it left at time t_1, and arrives at t_2. The distance (on the sphere, hyperboloid, or whatever, depending on the value of ϵ) is (8.43)

$$d = \int_{t_1}^{t_2} \frac{dt}{R} = \int_0^{r_1} \frac{dr}{\sqrt{1 + \epsilon r^2}}. \tag{8.85}$$

If $R = 0$ at $t = 0$, then

$$d_h = \int_0^t \frac{dt}{R} \tag{8.86}$$

is the distance to the horizon at time t. Now, by 8.71,

$$\dot{R}^2 \sim \frac{1}{R} \;\Rightarrow\; R^{1/2}\dot{R} \sim 1 \;\Rightarrow\; R^{3/2} \sim t \;\Rightarrow\; R(t) \sim t^{2/3}, \qquad (8.87)$$

and therefore

$$d_h \sim t^{1/3}. \qquad (8.88)$$

As time goes on we can see farther and farther out (but photons reaching us from the horizon are redshifted out of existence).

In the case of a spherical universe ($\epsilon = -1$) t never gets to infinity. Can you see the back of your head before the big crunch occurs? From Eq. 8.71,

$$\dot{R}^2 = -1 + \frac{A}{R} \;\Rightarrow\; \frac{dR}{dt} = \sqrt{\frac{A}{R} - 1} \;\Rightarrow\; dt = \frac{dR}{\sqrt{(A/R) - 1}}. \qquad (8.89)$$

Define a new variable η, by $dt = R\,d\eta$, so that the Robertson–Walker metric (8.51) becomes

$$ds^2 = R^2(d\eta^2 - h_{ij}\,dx^i\,dx^j). \qquad (8.90)$$

For a radial light ray the distance traveled is equal to η, and

$$d\eta = \frac{dt}{R} = \frac{dR}{\sqrt{R}\sqrt{A-R}} \;\Rightarrow\; \eta = \cos^{-1}\left(\frac{A - 2R}{A}\right), \qquad (8.91)$$

so

$$R = \frac{A}{2}(1 - \cos \eta). \qquad (8.92)$$

Thus

$$dt = R\,d\eta = \frac{A}{2}(1 - \cos \eta)\,d\eta, \qquad (8.93)$$

and hence

$$t = \frac{A}{2}(\eta - \sin \eta), \qquad (8.94)$$

which is the equation for a cycloid. As R goes from 0 to maximum (A) and back to 0, η goes from 0 to π to 2π (8.92).

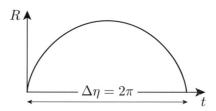

Thus $d = 2\pi$ is the maximum distance light can travel in the lifetime of the universe; at the halfway point you can see a distance $d = \pi$—halfway around the unit[14] sphere, in each direction. As η goes from 0 to π we see more and more stars, until at π we see *all* of them (some in front, and some in back). As η increases further we come to see each star *twice* (once redshifted and once blueshifted). But we never see the back of our own head. That would require $\eta = 2\pi$, when the universe has collapsed (and we have been crushed). Nor could you circumnavigate the universe, since you can't go faster than light.

---

[14] Remember, we rescaled distances back in Eq. 8.33.

Afterword

"You may have enjoyed this course and decided that you would like to do your thesis research in general relativity. DON'T. Einstein spent the last thirty years of his life working on general relativity, and it led to nothing. And he was smarter than you."[1]

Editors' Afterword

With the benefit of hindsight, Coleman's advice was clearly misguided. The years since Coleman gave his course have witnessed a rich bounty of experimental and theoretical results.

In the late 1950s, three leading schools of general relativity emerged: in Princeton under John Wheeler, in Cambridge (UK) under Dennis Sciama, and in Moscow under Yakov Zeldovich. Each of these leaders produced important successors: Wheeler taught Kip Thorne; Zeldovich taught Igor Novikov; and Sciama taught Stephen Hawking. There were of course many other schools where general relativity was taught and studied, but these three produced in the mid-1960s a renaissance in relativistic physics, which had been largely moribund since the 1930s.

Research in general relativity, or in cosmological physics based on relativity, has been recognized repeatedly by the Nobel Prize, and the pace is accelerating; no fewer than five and a half in the past three decades, and three and a half in the past decade (award date in parentheses):

- One half of the prize to A. Penzias and R. Wilson (1968), for their discovery of cosmic microwave background radiation.
- The discovery of pulsars, now understood as examples of gravitational collapse (neutron stars), by A. Hewish and M. Ryle (and though she was not included in the prize itself, J. Bell Burnell) (1974).

---

[1] End of last lecture, January 1970, as quoted by Erick Weinberg. (April meeting of the American Physical Society, 2016.) Coleman ended the 1966 course with similar advice.

- The 1974 discovery and subsequent study of the binary pulsar by R. Hulse and J. Taylor, providing the first indirect evidence for gravitational radiation (1993).
- The COBE satellite measurement of the black-body form and anisotropy of the cosmic microwave background radiation, led by J. Mather and G. Smoot (2006). This result supported earlier theoretical predictions of R. Penrose and S. Hawking, based on general relativity, that the universe began in an extremely dense (perhaps infinitely dense) state.
- The discovery of the accelerating expansion of the universe through observations of distant supernovae by S. Perlmutter, B. Schmidt, and A. Riess (2011) suggesting a nonzero value for the cosmological constant. The cause of this acceleration is often described by the term "dark energy."
- The design and implementation of LIGO, culminating in the direct detection of gravitational waves (2016) emitted during the coalescence of two black holes into one (B. Barish, K. S. Thorne, R. Weiss, 2017). The measurement of such tiny effects was pioneered by J. Weber in the late 1960s. Though that effort did not succeed, his audacity inspired the successful development of LIGO.
- For a lifetime of contributions to our understanding of the evolution of the universe and the earth's place in the cosmos, one half of the prize to J. Peebles (2019).
- For observational measurement of, and theoretical work predicting, black holes, to R. Genzel, A. Ghez, and R. Penrose (2020).

There has been much other notable work in general relativity. Observational discoveries include the following:

- I. Shapiro in the mid-1960s was able to establish the effect of gravity on the speed of light. Radar signals bounced off nearby planets in opposition to the earth, passing near the sun, were delayed by exactly the amount predicted by general relativity, thus providing a fourth classical test of the theory.
- V. Rubin, K. Ford, and their coworkers established (1970–1992) the existence of dark matter, hypothesized by F. Zwicky in 1933.
- We all unconsciously confirm a different gravitational effect (on time) whenever we invoke GPS on our smartphones to give us our location. The GPS system owes its accuracy in part to general relativistic corrections, including the gravitational redshift.
- In 2019, S. Doeleman and his colleagues produced the first image of a black hole, a massive collapsed object in the galaxy Messier 87, using data from radio telescopes all over the earth obtained through very long baseline interferometry.

Nor has theoretical work languished:

- Metrics describing rotating black holes, uncharged and charged, were found by R. Kerr and E. Newman (1965).

- Results by R. Feynman in the quantum theory of gravity (1963) were subsequently extended by B. DeWitt, L. Faddeev, and V. Popov (1966–1967), without which the quantum theory of non-Abelian gauge fields could not have been formulated. The quantum field theories of fundamental forces are based on this framework.
- Calculations by S. Hawking and J. Bekenstein (1973–1974) led to a first result in quantum gravity, that black holes would radiate, and that both temperature and entropy could be associated with them, thus linking thermodynamics with Einstein's theory.

Finally, there has been an explosion of books on relativity. Notable textbooks include (in order of publication)

- Y. B. Zeldovich and I. D. Novikov, *Relativistic Astrophysics*, 2 volumes, U. of Chicago P. (1971).
- P. J. E. Peebles, *Physical Cosmology*, Princeton U. P. (1971).
- S. Weinberg, *Gravitation and Cosmology: Principles and Applications of the General Theory of Relativity*, Wiley (1972).
- S. W. Hawking and G. F. R. Ellis, *The Large Scale Structure of Space-Time*, Cambridge U. P. (1973).
- C. W. Misner, K. S. Thorne, and J. A. Wheeler, *Gravitation*, W. H. Freeman (1973); reprinted with a new foreword by D. Kaiser and a new preface by C. W. Misner and K. S. Thorne, Princeton U. P. (2017).
- R. M. Wald, *General Relativity*, U. of Chicago P. (1984).
- S. M. Carroll, *Spacetime and Geometry: An Introduction to General Relativity*, Addison-Wesley (2004); reprinted by Cambridge U. P. (2019).
- L. Ryder, *Introduction to General Relativity*, Cambridge U. P. (2009).
- B. Schutz, *A First Course in General Relativity*, Cambridge U. P. (2009).
- A. Zee, *Einstein Gravity in a Nutshell*, Princeton U. P. (2013).

Notable popular expositions have also appeared, with the following among them:

- S. Weinberg, *The First Three Minutes*, Basic Books (1977).
- C. M. Will, *Was Einstein Right?*, Basic Books (1986).
- S. W. Hawking, *A Brief History of Time*, Bantam Dell (1988).
- K. S. Thorne, *Black Holes and Time Warps: Einstein's Outrageous Legacy*, W. W. Norton (1995).
- N. deG. Tyson, *Death by Black Hole*, W. W. Norton (2014).
- J. Levin, *Black Hole Blues and Other Songs from Outer Space*, Knopf (2016).

Deep questions remain to be answered, about dark energy and dark matter, the origin and fate of the universe, and the formulation of a mathematically consistent quantum theory of gravity. Were he alive today, Coleman might offer different advice to his students.

Appendix A

Compendium of Formulas

1. **Transformation laws**

   In passing from a set of coordinates $x(P)$ to a set of coordinates $y(P)$, a tensor density of weight w, with m contravariant indices and n covariant indices, transforms in the following way (5.35):[1]

   $$A^{\mu_1\cdots\mu_m}_{\nu_1\cdots\nu_n}(y(P)) = \frac{\partial y^{\mu_1}}{\partial x^{\alpha_1}} \cdots \frac{\partial y^{\mu_m}}{\partial x^{\alpha_m}} \frac{\partial x^{\beta_1}}{\partial y^{\nu_1}} \cdots \frac{\partial x^{\beta_n}}{\partial y^{\nu_n}} \left|\frac{\partial(x)}{\partial(y)}\right|^w A^{\alpha_1\cdots\alpha_m}_{\beta_1\cdots\beta_n}(x(P)),$$

   where $\partial(x)/\partial(y)$ is the Jacobian of the transformation. A tensor density of weight 0 is a tensor; the phrase "tensor density" (without specification of weight) usually means that the weight is 1.

2. **Special tensors**

   (a) The Kronecker delta is defined by

   $$\delta^\mu_\nu \equiv \begin{cases} 1 & \text{if } \mu = \nu, \\ 0 & \text{if } \mu \neq \nu. \end{cases}$$

   It is numerically invariant: its elements have the same numerical value (1 or 0) in any coordinate system.

   (b) The metric tensor is defined by (5.92)

   $$ds^2 \equiv g_{\mu\nu}\, dx^\mu\, dx^\nu,$$

   where, in general relativity, ds is the differential of proper time, and, in differential geometry, ds is the differential of arc length. The contravariant metric tensor is defined by (5.94)

   $$g^{\mu\lambda} g_{\lambda\nu} = \delta^\mu_\nu.$$

   (c) The determinant of $g_{\mu\nu}$ (written as a matrix) is denoted by g (5.97). It is a scalar density of weight 2 (5.99).

   (d) The metric tensors are used to raise and lower indices; thus, by definition (5.95),

   $$A^\mu = g^{\mu\nu} A_\nu, \quad A_\mu = g_{\mu\nu} A^\nu, \quad \text{etc.}$$

---

[1] For simplicity, the relative placement of the contravariant and covariant indices is not specified here.

(e) The Minkowski metric is the metric of special relativity; written as a matrix, (1.18),

$$\eta_{\mu\nu} = \begin{pmatrix} 1 & 0 & 0 & 0 \\ 0 & -1 & 0 & 0 \\ 0 & 0 & -1 & 0 \\ 0 & 0 & 0 & -1 \end{pmatrix}.$$

(f) The Levi-Civita symbol (ε: 5.31, ϵ: 5.36),

$$\epsilon^{\mu_1\cdots\mu_n} = \begin{cases} +1 & \text{if } \mu_1\cdots\mu_n \text{ is an even permutation of } 1, 2, \ldots, n, \\ -1 & \text{if } \mu_1\cdots\mu_n \text{ is an odd permutation of } 1, 2, \ldots, n, \\ 0 & \text{if any two indices repeat}, \end{cases}$$

is a numerically invariant tensor density of weight 1. Its covariant counterpart ($\epsilon_{\mu_1\cdots\mu_n}$) has the same elements, but transforms as a tensor density of weight -1.

3. **Affine spaces**

(a) Parallel transport is defined by (5.59)

$$dA^\mu = -\Gamma^\mu_{\nu\lambda} A^\nu \, dx^\lambda.$$

(b) The transformation rule for Γ is (5.67)

$$(\Gamma')^\mu_{\alpha\beta} = \left(\frac{\partial x'^\mu}{\partial x^\nu} \frac{\partial x^\sigma}{\partial x'^\alpha} \frac{\partial x^\lambda}{\partial x'^\beta} \right) \Gamma^\nu_{\sigma\lambda} - \frac{\partial^2 x'^\mu}{\partial x^\sigma \partial x^\lambda} \left(\frac{\partial x^\sigma}{\partial x'^\alpha} \frac{\partial x^\lambda}{\partial x'^\beta} \right).$$

4. **Elevator coordinates** (see footnote 7 in Chapter 4 and the sentence following Eq. 5.88.)

(a) In general relativity a set of coordinates is said to be "elevator" (or "locally geodesic") at a point P if, in these coordinates, at P,

$$g_{\mu\nu} = \eta_{\mu\nu} \quad \text{and} \quad \partial_\lambda g_{\mu\nu} = 0.$$

(b) In elevator coordinates the affine connections $\Gamma^\mu_{\nu\lambda}$ are zero (at P), but their spatial derivatives are *not*.

(c) One version of the equivalence principle is the statement that, for an arbitrary dynamical system in an arbitrary gravitational field, the equations of motion have the same form in elevator coordinates as they would have in the absence of the gravitational field.

5. **Covariant derivatives**

(a) The covariant derivative of a tensor (or density of weight w) is defined to be the ordinary derivative in elevator coordinates, and to transform like the appropriate tensor (or density). If we define the Christoffel symbols (5.106, 5.112)

$$\Gamma^\mu_{\nu\rho} \equiv \tfrac{1}{2} g^{\mu\lambda} (\partial_\rho g_{\nu\lambda} + \partial_\nu g_{\rho\lambda} - \partial_\lambda g_{\nu\rho}),$$

then the covariant derivative is given by (5.83)

$$\nabla_\lambda A^{\mu_1\cdots\mu_m}_{\nu_1\cdots\nu_n} = \partial_\lambda A^{\mu_1\cdots\mu_m}_{\nu_1\cdots\nu_n} + \Gamma^{\mu_1}_{\lambda\rho} A^{\rho\cdots\mu_m}_{\nu_1\cdots\nu_n} + \cdots + \Gamma^{\mu_m}_{\lambda\rho} A^{\mu_1\cdots\rho}_{\nu_1\cdots\nu_n}$$
$$- \Gamma^\rho_{\lambda\nu_1} A^{\mu_1\cdots\mu_m}_{\rho\cdots\nu_n} - \cdots - \Gamma^\rho_{\lambda\nu_n} A^{\mu_1\cdots\mu_n}_{\nu_1\cdots\rho} - w\Gamma^\rho_{\lambda\rho} A^{\mu_1\cdots\mu_m}_{\nu_1\cdots\nu_n}.$$

The final term can also be written using the identity (6.116)

$$\Gamma^\lambda_{\mu\lambda} = \partial_\mu(\ln(\sqrt{-g})).$$

(b) These expressions simplify considerably in certain special cases. For the metric tensor (5.104, Problem 6.2),

$$\nabla_\lambda g_{\mu\nu} = 0, \quad \nabla_\lambda g^{\mu\nu} = 0, \quad \nabla_\lambda g = 0.$$

For a scalar,

$$\nabla_\lambda A = \partial_\lambda A.$$

For a covariant vector (5.25, Example 5.1, Problem 5.9),

$$\nabla_\mu A_\nu - \nabla_\nu A_\mu = \partial_\mu A_\nu - \partial_\nu A_\mu$$

(the exterior derivative). For a contravariant vector density of weight 1 (Chapter 6, footnote 13),

$$\nabla_\mu A^\mu = \partial_\mu A^\mu.$$

(c) Given a path $x^\mu(s)$ parameterized by s, the covariant path derivative is defined by (6.11)

$$D_s \equiv \frac{dx^\mu}{ds}\nabla_\mu.$$

(d) Other notation:

$$\Gamma^\mu_{\nu\lambda} \equiv \left\{ {}^{\;\mu}_{\nu\;\lambda} \right\};$$

$$\partial_\mu A_\nu \equiv A_{\nu,\mu} = A_{\nu|\mu}, \quad \nabla_\mu A_\nu \equiv A_{\nu;\mu} = A_{\nu||\mu}.$$

6. **Covariant integration**

(a) If $A(x)$ is a scalar field,

$$\int d^4x \sqrt{-g}\, A(x)$$

is independent of the choice of coordinates (5.100).

(b) Within an integral of type (a), it is possible to "covariantly integrate by parts." That is,

$$\int d^4x \sqrt{-g}\,(A^{\cdots}\nabla_\mu B^{\cdots}) = -\int d^4x \sqrt{-g}\,(B^{\cdots}\nabla_\mu A^{\cdots})$$

(plus surface terms), provided that the suppressed indices are such that the quantity in parentheses is a scalar field.

7. **The Riemann tensor**

(a) The Riemann tensor is defined by (5.86)

$$\nabla_\mu \nabla_\nu A_\rho - \nabla_\nu \nabla_\mu A_\rho = R^\lambda{}_{\rho\mu\nu} A_\lambda.$$

The explicit expression is (5.87)

$$R^\lambda{}_{\rho\mu\nu} = \partial_\nu \Gamma^\lambda_{\mu\rho} - \partial_\mu \Gamma^\lambda_{\nu\rho} + \Gamma^\sigma_{\mu\rho}\Gamma^\lambda_{\sigma\nu} - \Gamma^\sigma_{\nu\rho}\Gamma^\lambda_{\sigma\mu}.$$

(b) The Riemann tensor has the following symmetry properties (5.89, 5.114):

$$R_{\lambda\rho\mu\nu} = -R_{\lambda\rho\nu\mu} = -R_{\rho\lambda\mu\nu} = R_{\mu\nu\lambda\rho}$$

and (5.90)

$$R_{\lambda\rho\mu\nu} + R_{\lambda\mu\nu\rho} + R_{\lambda\nu\rho\mu} = 0.$$

(c) From the Riemann tensor we define the Ricci tensor (5.119)

$$R_{\mu\nu} \equiv R^\rho{}_{\mu\rho\nu}$$

and the scalar curvature (5.120)

$$R \equiv R^\mu{}_\mu.$$

In a 2-dimensional manifold, the Riemann tensor can be written in terms of the scalar curvature (5.121):

$$R_{\mu\nu\lambda\sigma} = \frac{R}{2}(g_{\mu\lambda}g_{\nu\sigma} - g_{\nu\lambda}g_{\mu\sigma}).$$

In a 3-dimensional manifold, it can be written in terms of the Ricci tensor (Problem 5.14(b)),

$$R_{\mu\nu\lambda\sigma} = R_{\mu\lambda}\,g_{\nu\sigma} - R_{\nu\lambda}\,g_{\mu\sigma} - R_{\mu\sigma}\,g_{\nu\lambda} + R_{\nu\sigma}\,g_{\mu\lambda} - \tfrac{1}{2}R(g_{\mu\lambda}g_{\nu\sigma} - g_{\nu\lambda}g_{\mu\sigma}).$$

(d) The Riemann tensor obeys the Bianchi identities (5.122),

$$\nabla_\rho R^\sigma{}_{\lambda\mu\nu} + \nabla_\mu R^\sigma{}_{\lambda\nu\rho} + \nabla_\nu R^\sigma{}_{\lambda\rho\mu} = 0.$$

From this it follows that (Problem 5.15(b))

$$\nabla_\mu (R^{\mu\nu} - \tfrac{1}{2}g^{\mu\nu} R) = 0.$$

(e) If the Riemann tensor vanishes, the space is flat (5.131); that is to say, it is a union of open sets in each of which it is possible to choose coordinates such that $g^{\mu\nu} = \eta^{\mu\nu}$ (or, if it's a manifold embedded in a larger Euclidean space, $g^{\mu\nu} = \delta^{\mu\nu}$).

8. **Lagrangian dynamics**

   (a) The action for general relativity is (6.75, 6.99, 6.143)

   $$I = (16\pi G)^{-1} \int d^4x \sqrt{-g}\, R + I_m,$$

   where G is Newton's gravitational constant, and I_m is the action for matter (i.e. everything except gravity). Hamilton's principle leads to Einstein's field equation (6.69, 6.144)

   $$R_{\mu\nu} - \tfrac{1}{2} g_{\mu\nu} R = -8\pi G\, T_{\mu\nu},$$

   where $T_{\mu\nu}$, the energy–momentum tensor of matter, is (6.115)

   $$T_{\mu\nu} = \frac{2}{\sqrt{-g}} \frac{\partial \left(\mathcal{L}_m \sqrt{-g}\right)}{\partial g^{\mu\nu}}.$$

   It satisfies the conservation law (8.5)

   $$\nabla_\mu T^{\mu\nu} = 0.$$

   (b) An example of matter is the Klein–Gordon (scalar) field φ with no nongravitational interactions (6.104):

   $$I_m = \frac{1}{2} \int d^4x \sqrt{-g} \left(g^{\mu\nu} \partial_\mu \varphi\, \partial_\nu \varphi - \mu^2 \varphi^2\right),$$

   for which (6.108)

   $$T^{\mu\nu} = \partial^\mu \varphi\, \partial^\nu \varphi - \tfrac{1}{2} g^{\mu\nu} \left(g_{\lambda\rho} \partial^\lambda \varphi\, \partial^\rho \varphi - \mu^2 \varphi^2\right).$$

   The equation of motion is

   $$\frac{1}{\sqrt{-g}} \partial_\mu g^{\mu\nu} \partial_\nu \varphi = \nabla^\mu \partial_\mu \varphi = -\mu^2 \varphi.$$

   (c) Another example is the electromagnetic field $F^{\mu\nu}$ with no nongravitational interactions (6.109):

   $$I_m = -\frac{1}{4} \int d^4x \sqrt{-g}\, F_{\mu\nu} F^{\mu\nu},$$

   where (3.4)

   $$F_{\mu\nu} \equiv \partial_\mu A_\nu - \partial_\nu A_\mu.$$

   For this theory (6.114),

   $$T_{\mu\nu} = F_{\mu\rho} F^\rho{}_\nu + \tfrac{1}{4} g_{\mu\nu} F_{\lambda\rho} F^{\lambda\rho}.$$

   The equation of motion is

   $$\nabla_\mu F^{\mu\nu} = 0.$$

(d) A final example is a free-falling point particle (6.14):

$$I_m = -m \int ds \left[g_{\mu\nu} \frac{dy^\mu}{ds} \frac{dy^\nu}{ds} \right]^{1/2}.$$

In this case (2.165),

$$T^{\mu\nu} = m \int d\tau \, \delta^4(x - y(\tau)) \frac{dy^\mu}{d\tau} \frac{dy^\nu}{d\tau}.$$

The equation of motion is (6.12)

$$D_\tau^2 y^\mu = \frac{d^2 y^\mu}{d\tau^2} + \Gamma^\mu_{\nu\lambda} \frac{dy^\nu}{d\tau} \frac{dy^\lambda}{d\tau} = 0,$$

where τ is the proper time, defined by (6.19)

$$g_{\mu\nu} \frac{dy^\mu}{d\tau} \frac{dy^\nu}{d\tau} = 1.$$

(e) From the preceding formula it is possible to deduce an expression for tidal forces. Let two particles be separated by a small vector δ^μ, and let their velocities be equal (except for terms of order δ). Then (6.34)

$$D_\tau^2 \delta^\mu = R^\mu{}_{\lambda\nu\sigma} \frac{dy^\lambda}{d\tau} \frac{dy^\sigma}{d\tau} \delta^\nu.$$

9. Non-Lagrangian dynamics

(a) Sometimes it is possible to deduce the equations of motion for a dynamical system without resorting to a Lagrangian, either by using the conservation of the energy–momentum tensor or by appealing directly to the principle of equivalence.

(b) For example, for a fluid (7.150),

$$T^{\mu\nu} = (\mathcal{E} + P) v^\mu v^\nu - P g^{\mu\nu},$$

where v is the proper velocity (a time-like unit vector), $\mathcal{E}$ is the energy density in the rest frame of the fluid, and P is the pressure in the same frame. From the conservation of $T^{\mu\nu}$, it follows that (7.193)

$$\nabla_\mu [(\mathcal{E} + P) v^\mu] - v^\mu \, \partial_\mu P = 0$$

and (7.194)

$$(\mathcal{E} + P) v^\mu \nabla_\mu v^\nu - (g^{\nu\mu} - v^\nu v^\mu) \partial_\mu P = 0.$$

Together with the equation of state of the fluid, which gives $\mathcal{E}$ as a function of P, these form a complete set of equations of motion.

(c) Another example is an ideal gyroscope, a particle that has associated with it a space-like unit vector e^μ (the direction of the axis of the gyroscope), constrained to be orthogonal to the velocity vector of the particle. From the principle of equivalence, it follows that (6.26)

$$D_\tau e^\mu = \frac{de^\mu}{d\tau} + \Gamma^\mu_{\nu\lambda} \frac{dx^\nu}{d\tau} e^\lambda = 0.$$

10. Linearized gravity

(a) A weak gravitational field is one for which there exists a set of coordinates such that (6.35)

$$g_{\mu\nu} = \eta_{\mu\nu} + h_{\mu\nu},$$

where $h_{\mu\nu}$ is small. In what follows, we neglect terms of higher order in h than the first, and use the Minkowski-space metric tensor for raising and lowering indices.

(b) By making small coordinate transformations, we can alter $h_{\mu\nu}$ in the following way (6.49):

$$h_{\mu\nu} \rightarrow h_{\mu\nu} - \partial_\mu A_\nu - \partial_\nu A_\mu,$$

where A_μ is an arbitrary vector field. These are the "gauge transformations" of linearized gravitation theory.

(c) If we define (6.119)

$$\gamma_{\mu\nu} \equiv h_{\mu\nu} - \tfrac{1}{2}\eta_{\mu\nu}h^\lambda{}_\lambda,$$

it is possible to choose a gauge such that (6.129)

$$\partial^\mu \gamma_{\mu\nu} = 0.$$

In such a gauge, the equations of motion for the gravitational field are (6.130, 6.141)

$$\square^2 \gamma_{\mu\nu} = -16\pi G\, T_{\mu\nu}.$$

(d) The solutions to the homogeneous field equations are gravitational waves. A complete set of solutions are the plane waves (6.171)

$$\gamma_{\mu\nu} = E_{\mu\nu}e^{i\kappa\cdot x},$$

where κ is a null vector ($\kappa^\mu \kappa_\mu = 0$). It is still possible to choose a gauge such that (6.172)

$$E_{\mu 0} = \kappa^\nu E_{\mu\nu} = 0 \quad \text{and} \quad E^\lambda{}_\lambda = 0.$$

Thus, there are only two independent waves for each value of κ.

Appendix B

Final Exams

B.1 Final Exam, 1966

Problem 1. (20 points) Suppose there is an infinitely heavy point charge of charge Q at the center of the universe, and a charged particle with mass m, charge $-q$, moves in its field. In Newtonian dynamics, the forces in this problem are exactly the same as in the gravitational Kepler problem, and the bounded orbits are ellipses. In special-relativistic dynamics, this is no longer true: the ellipses precess. Find the magnitude and sign of this precession, in radians/revolution. How does this compare with the corresponding precession in the general-relativistic Kepler problem, for corresponding Newtonian orbits? (Neglect the radiation reaction—that is to say, consider the limit $q \to 0$ with q/m fixed.)

Problem 2. (40 points) Consider a manifold obeying the following restrictions:

(a) The manifold is completely covered by a set of four coordinates $(t, \mathbf{x})$ ranging from $-\infty$ to ∞.
(b) In terms of these coordinates the metric is static; that is to say,

$$ds^2 = \Phi^2(\mathbf{x})\, dt^2 - g_{ik}(\mathbf{x})\, dx^i\, dx^k.$$

(c) In terms of these coordinates, the metric is nonsingular: Φ is always positive and never infinite, g_{ik} is always positive definite and never infinite.
(d) The metric is asymptotically Minkowskian; that is to say,

$$\lim_{|\mathbf{x}| \to \infty} \Phi = 1, \qquad \lim_{|\mathbf{x}| \to \infty} g_{ik} = \delta_{ik},$$

where δ_{ik} is the Kronecker delta.
(e) Einstein's field equations for empty space are satisfied everywhere:

$$R_{\mu\nu} = 0.$$

The problem has two parts:

A. The surface $t = 0$ defines a 3-dimensional manifold, and Φ may be considered as a scalar field on that manifold. Rewrite Einstein's equations (e) as a set of equations involving the covariant derivatives of Φ on the 3-manifold and the curvature tensor for the 3-manifold.

230

B. Prove that the only manifold satisfying conditions (a) to (e) is flat space. (This proves that there is no solution to Einstein's equations that corresponds to a static gravitational field held together by the gravitational attraction of its own energy density.)

Problem 3. (40 points) We are going to explore an alternative gravitational theory, which is Lorentz invariant but not generally covariant. Gravity is described by a scalar field ϕ. The free action for gravity is

$$\frac{1}{2} \int d^4x \, \partial_\mu \phi \, \partial^\mu \phi.$$

The action for matter interacting with gravity is obtained from that of general relativity by the substitution

$$g_{\mu\nu} \rightarrow F(\phi) \, \eta_{\mu\nu},$$

where $\eta_{\mu\nu}$ is the Minkowski-space metric tensor and $F(\phi)$ is an analytic function of ϕ:

$$F(\phi) = 1 + \sum_{n=1}^{\infty} c_n \phi^n.$$

The coefficients c_n are to be adjusted so as to obtain agreement with experiment.
    General relativity satisfies four tests:

(a) It recreates Newtonian theory for planetary orbits, in the appropriate approximation.
(b) It predicts the correct gravitational redshift.
(c) It predicts the correct bending of light by the sun.
(d) It predicts the correct advance of the perihelion of Mercury.

Which of these four tests are satisfied by our theory, if the coefficients c_n are appropriately chosen? For those cases in which the answer is yes, what is the appropriate choice? (In your analysis, you may assume c_n is of the order of magnitude G^n, where G is Newton's gravitational constant.)

B.2 Final Exam, 1969

Problem 1. (25 points) For a spherically symmetric, but not necessarily static, metric, it is possible to use one's freedom in defining the time coordinate to put the metric in the form

$$ds^2 = e^\lambda \, dt^2 - e^\nu \, dr^2 - r^2 (\sin^2 \theta \, d\phi^2 + d\theta^2),$$

where λ and ν are functions of r and t. Assume that the metric is of the form shown, is Minkowskian at infinity, and that the energy–momentum tensor of matter vanishes for r greater than some radius r_0. Use the Einstein field equations

$$R_{\mu\nu} - \tfrac{1}{2} g_{\mu\nu} R = 0$$

in this region ($r > r_0$) to show that the metric is necessarily of Schwarzschild form in this region. (This is the generalization to the full field theory of the statement in the linearized theory that there is no spherically symmetric gravitational radiation.)

Problem 2. (25 points) Consider the Lorentz-invariant field theory of a scalar field θ in 2-dimensional spacetime, defined by the Lagrangian

$$\mathcal{L} = \tfrac{1}{2}\partial_\mu\theta\,\partial^\mu\theta + \lambda(\cos\theta - 1), \quad \mu = 0, 1,$$

where λ is positive. Let us call the two coordinates of spacetime t and x. We are interested in solutions of the field equations that are in steady motion, that is to say, solutions of the form

$$\theta(t, x) = F(x - vt),$$

with v some number, which also obey the boundary conditions

$$\theta(t, -\infty) = 0, \quad \theta(t, +\infty) = 2n\pi,$$

with n an integer.

(a) Show that the only solutions of the specified type are those for which n is ± 1. Show that for any v less than 1, there is only one solution of each type, up to a spacetime translation.

(b) We call these solutions "twists" and "antitwists," respectively. Calculate the energy and momentum of a twist as a function of v. Likewise for an antitwist.

(c) Let $F(x)$ be the solution of the field equations corresponding to a twist at rest ($v = 0$). Consider

$$\varphi(t, x) = F(x) + F(x + a),$$

for some number a. This configuration could be described as "two twists separated by an amount a"; it is not a solution of the field equations, but becomes an approximate solution as a becomes very large. Calculate the a-dependent part of the energy (chosen to go to zero as a goes to infinity) of this state, for large a. If it is positive, one would tend to believe that two distant twists repel; if negative, that they attract. Which is it? Answer the same question for one twist and one antitwist.

[*Remarks:* (1) This is a purely special-relativistic problem; there is no gravity. (2) In part (c), it suffices to represent the answer as an integral, provided you have it in a form where you can answer the question about signs.]

Problem 3. (50 points) You are to investigate a Minkowski-space theory of gravity, as a possible alternative to general relativity. In this theory, the gravitational field is a scalar field φ, and the action integral is of the form

$$I = \frac{1}{2}\int d^4x\,\partial_\mu\varphi\,\partial_\nu\varphi\,\eta^{\mu\nu} + I_m,$$

where I_m, the action integral for matter, is obtained from the corresponding object in general relativity by making the substitution

$$g_{\mu\nu} \to \eta_{\mu\nu}\left(1 + c_1\sqrt{G}\,\varphi + c_2\,G\,\varphi^2 + \cdots\right),$$

where the c's are coefficients to be adjusted to give agreement with experiment. [*Remarks:* (1) The substitution is to be made not only for $g_{\mu\nu}$, but for all quantities defined in terms of it, such as $g^{\mu\nu}$ and $\sqrt{-g}$. Thus, I_m is of the same form as in Einstein's theory, with a special choice for the metric tensor, and the principle of equivalence (in the form we used in class before we began discussing the gravitational field equations) is still valid. (2) $\eta_{\mu\nu}$ is, as always, the Minkowski-space metric tensor. (3) The factors of $\sqrt{G}$ are inserted to make the c's dimensionless. The power series can be continued indefinitely, of course, but if the c's are of reasonable magnitude, only the terms shown explicitly will have an observable effect.]

(a) Write down the field equations in the weak-field (linearized) approximation, and show that agreement with Newtonian theory in the appropriate approximation can be obtained if c_1 is properly chosen. What is the proper choice?

(b) Outside the sun (considered as a static, spherically symmetric source), φ must be of the form

$$\varphi = \frac{a}{r}.$$

Determine a in terms of M, the mass of the sun, by demanding agreement with Newtonian planetary theory in the appropriate approximation.

(c) A famous test of general relativity is the advance of the perihelion of Mercury. Does this theory also predict the right result, if c_2 is properly chosen? If so, what is the proper choice? (In this context, "the right result" means the prediction of general relativity, to lowest nonvanishing order in G.)

(d) The same for the gravitational redshift.

(e) The same for the bending of light by the sun.

Index

Page numbers for entries within a footnote are followed by an *n* and the footnote number; **bold** numbers indicate a term's definition. GR = General Relativity, SR = Special Relativity.

Abelian, 5
aberration of light, 32
Abraham, Max, 82
  Abraham-Lorentz–Dirac eq., 84
absorber, Feynman–Wheeler, 76, 83
acausal
  preacceleration, 88
  propagation, 71
acceleration
  as boost, 17n17
  Cavendish experiment, 10
  electron, 87
  energy conservation, 89
  field, 79, *see also* radiation field
  hyperbolic motion, 26, 90
  particle
    recovery of Newton's law, 151–153
    tensorial, 134
  proper, **25**, 34, 43, 45, 192n17
    tidal stress, 198
  twin paradox, 25
  uniform, 21, 25, *see also* proper acceleration
    equivalence principle, 92, 98–99
accelerometer, 192
action
  conservation laws, 52–55
  emphasized by Coleman, ixn1
  Euler–Lagrange equations, 52
  Feynman–Wheeler electrodynamics, 83
  general relativity, 227
    alternative, Lorentz-invariant, 231
    alternative, scalar field, 232
  gravitational field, 144
  group elements, 38
  matter as gravity source, 147
  nonuniqueness, 49

  particle mechanics, 48
    stress tensor, 61–62
  point charge in field, 68–70
  relativistic field, 51
  relativistic form, 123
  relativistic mechanics, 49
  weak, 141
action principle
  Einstein's equations, 144–147
  special relativistic fields, 133
action at a distance, 83
active transformation, 11–12
Adler, Ronald, x
affine
  group, **8**
  space, 100, **115**
    parallel transport in, 224
affine connection, 115, *see also* Christoffel symbol
  antisymmetric violates equivalence
    principle, 117
  coordinate transformation, 117, 125, 224
  determination for exterior Schwarzschild
    solution, 167–168
  elevator coordinates, 224
  Euclidean, polar coordinates, 117
  parallel transport, 116
  symmetric, 117, 121
  weak field, 141
alias, **11**, 101, *see also* passive transformation
  preferred by Einstein, 12
alibi, **11**, 101, *see also* active transformation
Alice and Bob, 21, 25
angular
  deflection
    bending of light, 178
  displacement
    Mercury perihelion, 174
  frequency, 42

angular momentum, 154
  conservation, 48
  conserved, 60
  density, **58**
  intrinsic (spin), **58**
  Mercury perihelion, 172
  relativistic, 47, 61
    conserved, 48
antisymmetric tensor
  angular momentum, 47
  any t. splits into symmetric t. and, 37
  canonical stress tensor, 58–62
  conserved, 47
  exterior derivative, 107–109
  $F^{\mu\nu}$, **63**
  $\Gamma^{\lambda}_{\mu\nu} - \Gamma^{\lambda}_{\nu\mu}$, 117
  group representations, 37–38
  infinitesimal Lorentz transformation, 56
  integration on manifold, 113–115
  Levi-Civita symbol, **109**
  Lorentz trans. do not mix symmetric and
    antisymmetric parts, 37, 39
  $R^{\mu}{}_{\nu\lambda\sigma}$, 121, 124
  scalars and vectors, 107
  Stokes's theorem in terms of, 107–109
  surface integral in terms of, 106
  tensor densities, 109–113
antitwist, 232
aphelion, **173**
appearance, relativistic cube, 29–31
arc length, 20, *see also* length
asymptotic conservation laws, 40–48
asymptotic field
  electromagnetic, 80
  gravitational, 201, *see also* Minkowski, space,
    asymptotic
axis, precession of gyroscope's, 136

bare mass, 68, 82, 89, 90
Barish, Barry C., 221
Bazin, Maurice, x
Bekenstein, Jacob David, 222
Belinfante, Frederik J., 58n26
Bianchi identities, 127, 162, 226
Birkhoff, Garrett, 10
black hole, 195n20, 220
Boas, Mary L., 31n20
Bondi–Gold cosmological model, 210
boost, 17–20
boundary conditions
  Green's function, electromagnetic wave
    equation, 70, 74, 76, 87
  Schwarzschild solution, 169
  twists and antitwists, 232
bounded orbits in Kepler problem, 230
bulldozer method, 91
Burnell, Dame Jocelyn Bell, 220

cake, baking by match, 44
candle, standard, 209
canonical stress tensor, 57, 60, *see also* stress
    tensor, canonical,
  free Maxwell field, 66
  Klein–Gordon field, 148
  not symmetric, 58
  symmetric for KG and Maxwell fields, 149
carrier space, 38
Carroll, Sean M., 222
Cartan, Élie, 206
  classic problem of, 206–207
Cartesian coordinates, 103, 106, 117, 121, 134
  Euclidean 4-space, 186
  nothing fishy in, 187
  overlapping grids, 100
Cauchy's integral formula, 72
causality near a singularity, 195
Cavendish experiment, 10
centrifugal force
  effective, Eötvös experiment, 96
charge
  action, 68–70
  analogous to mass, 132
  classical electron theory, 68n10
  classical radius, 90
  conservation, 39, 54
    continuity equation, 66
    current on or off, 75n14
    must be local, 55
  Coulomb potential, 78
  electromagnetic field, 78–79
  energy conservation in particle motion, 89
  hyperbolic motion, 90–92
    radiation with no reaction force, 92
  Liénard–Wiechert potential, 78
  Lorentz force, 70
  Minkowski force, 70n11
  nonrelativistic motion with radiation damping,
    84–88
  radiation from, 76–80
  relativistic Kepler problem, 230
  relativistic motion, massive, 82
    Larmor formula, 89
  self-force, 82
chemical potential, 185
Christoffel symbol $\Gamma^{\lambda}_{\mu\nu}$, 123, **224**, *see also* affine
    connection
  as gravitational "force", 134
  coordinate transformation, 117, 125, 224
  in terms of metric tensor, 124
  Robertson–Walker metric, 212
  table, Schwarzschild metric, 167
  vanishes in elevator frame, 137
Christoffel symbol $\Gamma_{\lambda,\mu\nu}$, 123
  table, Schwarzschild metric, 167
circumference, Schwarzschild spacetime, 169n6

classic test of GR
 deflection of light near sun, 170, 176–179
 gravitational redshift, 170, 179–180
 Mercury's perihelion shift, 170–176
 recovery of Newton's law, 153, 170
 tests Schwarzschild metric, 180–184
classical electron theory
 runaway as embarrassment, 85
classical physics, 3
 combined dilations and Lorentz
  transformations, 20
 geometric symmetries, 6–11
clock
 Doppler shift, 27
 gravitational time dilation, 97
 Lorentz contraction, 22
 near Schwarzschild radius, 192
 proper time, 21
 relativistic Doppler shift, 43
 synchronization, 21
 time dilation, 21
  paradox, 23
 twin paradox, 25
COBE, 221
Coleman, Sidney R., ix, 68n10
 no kinematic conservation laws in mechanics
  from discrete symmetries, 48
Coleman/standard terminology
 acceleration/boost, 17
 dilatations/dilations, 5
 elevator frame/locally inertial frame, 99
 induction field/generalized Coulomb field, 79
 Lorentz/Lorenz, 65n6
 parallel displacement/parallel transport, 115
comoving coordinates
 applied to galaxy, 205
 inertial frame, 25
 Robertson–Walker metric, 208
conformal transformation
 generators, 131
 Maxwell Lagrangian invariant, 198
connection, 115, *see also* affine connection
conservation
 global
  energy, in planetary orbits, 172
  energy–momentum, in GR, 201–203
  in general relativity, 200
  in special relativity, 200
 law, kinematic, 39, 48
 local energy–momentum
  Friedman universe, 211
  GR, 227
  requires $\nabla_\mu T^{\mu\nu} = 0$, 162
continuity equation, $\partial_\mu J^\mu = 0$, 39
 in electrodynamics, 202
 in general relativity, 200
 locally conserved charge, 55, 66

contraction
 lengths, 22, *see also* Lorentz–Fitzgerald
  contraction
 tensor indices, **36**
  invariant operation, 104
contravariant vector, **103**
convolution integral, **81**
coordinate transformation, **5**
 action, 61
 Christoffel symbol (2nd kind), 117, 125
 conformal group, 131
 contravariant vector, 102
 covariant derivative, 119
 covariant vector, 103
 gauge choice, 151
 gauge transformation, 156
 infinitesimal, 140
 Kruskal–Szekeres coordinates, 193
 light speed, 9
 matrix, **102**
 pseudotensor $t^{\mu\nu}$ not a tensor, 201
 scalar field, 5
 tensor, 104
 tensor density, 110, 223
 vector field, 34
coordinates
 active *vs.* passive transformation of, 11
 Cartesian, 100, *see also* Cartesian coordinates
 comoving, 25, *see also* comoving coordinates
 elevator, 121, *see also* elevator coordinates
 embedding, 195
 generalized, 48
 geometric transformation as change of, 5
 isotropic, **208**
 Kruskal–Szekeres, 193, *see also*
  Kruskal–Szekeres coordinates
 locally geodesic, 121, 125, 202, 203, **224**
 Minkowski, 202, *see also* Minkowski
  coordinates
 not vectors under coordinate transformations,
  102n2
 patches on manifolds, 100
 rotations and boosts, 17
 singularities, 208, 215, 216
cosmic background radiation, 221
cosmic redshift, 209, *see also* Hubble's law
cosmological
 constant Λ, 143, 214
 models
  Robertson–Walker metric, 210
 principle, 204, 206
Coulomb potential, 78
"covariant" meanings, various, 132n1
covariant
 integration, 225
  by parts, 225
 path derivative, **225**
covariant derivative
 contravariant vector, **119**

covariant vector, **120**
    does not commute with another, 120
    elevator frame, $\nabla \to \partial$, 137
    metric tensor, $\nabla_\lambda g_{\mu\nu} = 0$, 225
    obeys product rule, 120
    tensor, 120
    tensor density, **120**, 161, **224**
covariant vector, **103**
cube, appearance of, 29, *see also* appearance of
    relativistic cube
curl, 64
    4-dimensional analog, 130
    exterior derivative, 108, 115n9
current, **54**
    conserved scalar, 200
    density, 66
    divergence of, 55
    electromagnetic, 65, 66, 69, 77
    internal symmetry, 55
curvature
    light beam in freely falling elevator, 99
    measured by Riemann tensor, 121
    parallel transport around a loop, 127–129
    produced by matter and energy, 132
    scalar R, 126, 144, 191, 212, **226**,
        see also Ricci scalar
    zero if and only if $R^\mu{}_{\nu\lambda\sigma} = 0$, 131
cycloid, 218
cylinder
    intrinsically flat, 129
    its isometries, 163
    not simply connected, 108

d'Alembertian, **65**
dark energy, 221
dark matter, 221
de Broglie formula, 42, 156
de Sitter
    space
        locally Poincaré invariant, 142
    universe
        collapse of, 214, 216
deflection of light near sun, 170, 176–179
delta function, 33, 72, 73, 77, 80, 87
density
    energy, 184, 187, 191, 216, 217
    galaxies, 33, 204
    mass, 187
Derbes, David, x
derivative
    conserved quantity, 200, 201
    covariant, 119, *see also* covariant derivative
    exterior, 107, *see also* exterior derivative
    via affine connection, 116
determinant
    identity, 145
    Jacobian, 107, *see also* Jacobian
    Levi-Civita symbol, $\epsilon^{\mu\nu\lambda\rho}$, 114
    Lorentz group structure, 13

metric tensor, g, 122
    of general coordinate trans., 160
    proof of Stokes's theorem, 109
DeWitt, Bryce S., 222
Dicke, Robert H., 97
dilation, 10
    combined with Lorentz transformations, 19
    universe invariance, 10
dipole radiation
    electromagnetic, 159
    gravitational not allowed, 159
Dirac, Paul A. M., 82n23, 86n26
discrete symmetry
    no kinematic conservation law, 48n15
distance
    determined by metric tensor, 104
    extremized by geodesics, 134
    horizon, in Friedman universes, 217
    metric spaces, 100
    on a manifold, 121
distance–velocity relation, 209, *see also* Hubble's
    law
distribution
    as generalized function, 73
    Lorentz invariant, 33
    mass, 175
    matter, 204
    particles, Lorentz invariant, 33
    velocities, 32
divergence
    added to stress tensor, 58
    conserved quantity, 200
    covariant of a tensor density, 146
    current, 55
    electron mass renormalization, 80
    Lorenz gauge, 65
    stress tensor, 62, 67
    tensor density, 113
    theorems of Gauss, Stokes, 114
Doeleman, Sheperd S., 221
Doppler shift, 27, 43
    blurs gravitational redshift, 180
    relativistic, 44
downness, 9
dual space, 103
dual tensors, 112, 113
    integration on manifolds, 113
    tensor densities, 115, 118
dynamics, general relativistic
    Lagrangian, 227
    non-Lagrangian, 228

ϵ_{ijk}, Levi-Civita symbol, 64n2
Eddington's eclipse measurement, 179, 182
effective mass, 69
eigenstate, 5
eigenvalue, 122n21, 156, 158
Einstein tensor $G^{\mu\nu}$, **143**
    divergence $\nabla_\mu G^{\mu\nu} = 0$, 226

Einstein tensor $G^{\mu\nu}$ (cont.)
  electromagnetic field, 143
  pseudotensor $t^{\mu\nu}$, 202
  Robertson–Walker metric, 211
Einstein's equation, **227**
  action principle for, 144–147
  Bianchi identities, 162*n*2
  cosmological constant, Λ, 214
  Einstein–Lemaître solution, 215
  empty space, **142**
  gravitational analog to Maxwell's equations,
      132, 138
  in presence of matter, **144**
  isometries, 162
  Lagrangian, 153
  linearized, 150
    similar to Maxwell's equations, 151
  Newton's law from, 151–153
  pseudotensor, $t^{\mu\nu}$, 201–203
  Robertson–Walker metric, 210
  Schwarzschild solution
      exterior, 162–170
      interior, 185–190
  stationary solution, **164**
  stress tensor, $T^{\mu\nu}$, 147
  universe collapses if $\epsilon = -1$, 217
Einstein, Albert, ix
  bases of general relativity, 99
  equivalence principle, 98
    not coincidental, 95
  Lorentz invariance, 11*n*14
  preferred alias to alibi, 12
  summation convention, 8
  tests proposed by, 170
elastic collision, 42, 44
elastic modulus, 38
elasticity, 38
electric charge, 69, 132
electric field, 64
electrodynamics, 63, 67*n*9, 76, 77,
      see also Maxwell's equations
  Feynman–Wheeler, 76
  Lagrangian, source-free, 148
  logically similar to GR, 132
  Lorenz gauge, 151
  Poisson's equation, 152
  quantum, 80
  reduced to a wave equation, 75
  runaway
    classical limit, 86
    mode in QED, 88
  source fields, 75
  time asymmetry, 76
electromagnetic field tensor $F^{\mu\nu}$, **63**
  continuity equation, 202
  field equations, 154
    no source, 63
    sourced, 66

gauge invariance, 65, 154
  Lagrangian, 148
    gravity, 227
  Maxwell's equations, no source, 133
electromagnetism
  stress tensor, 70
  unlike gravity, 98
electron, 10
  classical radius, 90
  classical theory of, 68
  equation of motion, 82
  mass renormalization, 80
  radiates into future, 76
  scattering, 88
elevator
  coordinates, 99*n*7, **121**, 128, 137, 138, 179, 201,
      224
    locally geodesic, 121
  equivalence principle, 98
elliptical orbit, 4
  bounded in Kepler problem, 230
  precession, 175
Ellis, George F. R., 222
embedding space, 100, 121, 196, 206
embedding theorem, 195
energy
  analysis of perihelion shift, 172
  burning match, 44
  center of, 48
  conservation, 44
    charged particle, 89
  cosmological constant, Λ, 214–215
  couples to Einstein's gravity, 97
  density, 58, 62, 66, 187, 191
    fluid, 228
    internal Schwarzschild, 185
    universe collapse, 217
  electrostatic, 90
  field, 67, 68, 80
  Friedman universe, 211, 213–214
  gravitational
    field, 159
    global, 203
    Newtonian, 180
    potential, Newtonian, 182
    relativistic, global, 201–203
  kinetic
    angular, 172, 182
    asymptotic, 46*n*14
    charges, runaway, 89
    nonrelativistic, 41
    radial, 172, 182
  Lagrangian, 49
  levels, 5
  luminosity, 210
  mass due to electrostatic, 90
  negative, high deg. derivatives, 52
  photon, 42, 181

planet, relativistic, 183
potential, 45, *see also* potential energy
produces curvature, 132
radiation, 92
relativistic, 41, 47
runaway mode, 85
twist and antitwist, 232
energy–momentum tensor, 227, 228, 231,
    see also stress tensor
Eötvös, Loránd von, 96
experiment, 96, 97, 181
$\epsilon_{\lambda\mu\nu\rho}$, 224, *see also* Levi-Civita symbol
equations of motion, 4, 5, 49, 55, 63, 65, 88, 91,
    138, 224, 228
weak gravitational field, 229
equator, Schwarzschild surface, 195
equilibrium, hydrostatic, 188
equivalence of mass m_G and m_I, 95
Eötvös experiment, 96
Newton experiment, 96n2
equivalence principle, **92**, **95**, 224
as basis for GR, 99
coordinates, 121
elevator, 98
extended by Einstein, 98
free fall, 117
gravitational redshift, 181
gyroscope, 228
light beam, 99
non-Lagrangian dynamics, 228
pseudotensor $t^{\mu\nu}$, 201
scalar field gravity, 233
Schwarzschild metric, 184
source of gravity, 143
strong form, 99
symmetry of $\Gamma^\lambda_{\mu\nu}$, 117
very weak form, 99
weak form, 99
Euclidean manifold, 115, 163
embedding theorem, 195
Euclidean space, 117, 121
2 dimensions, Γ^i_{jk} in polar coordinates, 117
3 dimensions, 127, 131, 166
hypersurface embedded in, 195
integration, 134
4 dimensions, 186
$\epsilon = 0$ Friedman universe, 207
metric on embedded hypersphere, 187
n dimensions, 131
contravariant, covariant the same, 35n2
Γ's zero with Cartesian coordinates, 121
geodesics typically minimal in, 135n4
locally, Robertson–Walker metric in, 211
Euler angles, 13
Euler–Lagrange equation, 49, 52, 53, 55
event horizon, 195n20
event in spacetime, **3**

expansion of the universe
Einstein–Lemaître model, 215
Friedman model, 213, 216
experimental test, 170, *see also* classic test of GR
exponential growth, runaway, 85, 86
exterior calculus, 104–115
exterior derivative, **107**
antisymmetric tensor, dual, 113
$F^{\mu\nu}$, 148
vector, 225
exterior solution, Schwarzschild, 166,
    see also Schwarzschild metric
maximal extension, 193
extremum, Hamilton's principle, 48

$F^{\mu\nu}$, 133, *see also* electromagnetic field tensor
Faddeev, Ludvig Dmitriyevich, 222
faster than light, 32, *see also* tachyon
Fermat's principle
geodesic equation its apotheosis, 135
gravity and refraction, 199
Feyman–Wheeler electrodynamics, 76, 83
Feynman, Richard P., 76n16, 222
field
electric
in terms of F^{i0}, 63
transverse, 160
electromagnetic, 68, *see also* Maxwell field
bare mass, 69
covariant field equation, 133
covariant Lagrangian, 148
point charge, 79, 80
radiation, 79
gravitational
action, 141
Einstein equation, **141**
equivalence principle, 92, 98, 99, 117
Hamiltonian, 97
Lagrangian, 144
light beam, 98
matter and metric, 138
Newtonian from GR, 151
static point mass, 166, *see also* Schwarzschild
    metric
stress tensor, 202
sun, symmetry of, 4
tidal force, 137
weak, 138, **229**
"in" and "out", 75, 83
satisfy wave equation, 76
induction, 79, 89, 92
Klein–Gordon
covariant, 133
Lagrangian, 148
magnetic
in terms of F^{ij}, 64
Maxwell, A^μ, 63, *see also* Maxwell field
radiation, 92

field (cont.)
    scalar
        action, 51
        canonical stress tensor, 57
        conserved current, 55
        coordinate transformation, 104
        Klein–Gordon equation, 51
        symmetric stress tensor, 60
        transformation law, 5, 34
    vector, 5, 103
        covariant derivative, 119
        inner product, 105
        local, 123
        parallel transport, 130
        transformation law, 34
finite neighborhood, 129
Fitzgerald contraction, 22,
        see also Lorentz–Fitzgerald c.
fixed point theorem, 88
flashlight drive, 42, 45
flat space
    Euclidean, 117, 121
    Minkowski, 129, 132, 133, 142, 179, 231
flatness, 129
    measured by Riemann tensor, 131
fluid
    stress tensor, 184, 185, 191, 228
    sun as incompressible, 185
    universe as, 204, 210
Fock, Vladimir A., x
force
    delta function, 86
    gravitational, 10, 95, 151
        Eötvös experiment, 96
        hydrostatic equilibrium, 188
    Hooke's law, binding, 84
    Lorentz, **70**, 132
    Minkowski, 51
    tidal, 137, 228
        near singularity, 197
Ford, W. Kent, 221
Fourier transform, 71
free fall
    frame, 117, 138, *see also* elevator coordinates
        locally Minkowskian, 117
        same as elevator coordinates, 121
    Friedman, 213
    Galileo observation, 95
    geodesic world lines, 137
    photon, 181
    Schwarzschild radius
        finite proper time, infinite coordinate
            time, 192
    Schwarzschild singularity, 191, 192
    tidal force, 137, 197
frequency
    Doppler shift, 43
    gravitational redshift, 179, 180

Friedman universe, 210–214
Friedman, Alexander A., 210
    spelling confusion, 210n8
Friedmann, Alexander A., 210,
        see also Friedman, Alexander A.
Fulton, Thomas, 92n30
future, **13**

galaxies
    as atoms, 204
    beyond the horizon, 217
    Cartan's problem, 206
    cosmological principle, 204
    Hubble's law, 209
Galileo
    all objects fall at the same rate, 95
    relativity, 46
$\Gamma^{\lambda}_{\mu\nu}$, $\Gamma_{\lambda,\mu\nu}$, 123, *see also* Christoffel symbol
γ, Lorentz factor, **18**
Gamow, George, 31
gauge invariance
    electromagnetism, 65
        $F^{\mu\nu}$, 70
    gravitation, 140
        Bianchi identities, 162n2
        $R_{\mu\nu\lambda\sigma}$, 159
gauge transformation
    electromagnetism, 65, 154
    gravitation, 140, 156
        linearized theory, 229
Gauss's theorem, 59
    four dimensions, **53**
    rank 2 tensor, 115
general relativity
    alternative theories, 142
    analogs trivial if dim < 4, 127
    basic insights, 99
    classic tests, 170, *see also* classic test of GR
    classic tests of what?, 180
    Coleman's admonition, 220
    cosmological models, 210
    4-dimensional manifold, 112
    $\Gamma^{\mu}_{\nu\lambda}$ symmetric, 117
    global conservation of energy–momentum,
            200–203
    isotropy, 205
    logically akin to Maxwell's equations, 132
    no dipole radiation, 159
    strong equivalence, 121
    symmetric stress tensor, 58
    two parts of, 132
    Weyl's principle, 204
generator
    conformal group, 131
    translation need not commute, 142
Genzel, Reinhard, 221
geodesic, **134**
    equation, 134

Kruskal coordinates, 193
light moves along a null, 179, 195
local coordinates, 159
    Γ's zero, 121, *see also* elevator coordinates
orbits, 181
particle, 228
straight line, absent gravity, 134
straightest lines in curved space, 134
tidal forces, 137, 197
world line in free fall, 137
world lines of free particles, 135, 152
geometry
    affine spaces, 115–121
    classical physics and group
        theory, 6–12
    differential, 100–115
        Cartan problem, 206
    equal to gravity, 132
    interior Schwarzschild solution, 187
    Minkowski space and Lorentz group, 13–23
    Riemannian, 121–131
    Schwarzschild equator, 195
    Schwarzschild solution, 193
Ghez, Andrea, 221
global conservation, 55
    energy, in orbits, 172
    energy–momentum in GR, 201
    in general relativity, 200
    in special relativity, 200
    momentum, in GR, 201
global invariance
    Poincaré not certain, 142
God, near-sighted, 204
gradient
    archetype of covariant vector, 35
    curl-free vector as, 130
    exterior derivative of scalar field, 108
    of scalar field, transformation, 104
    pressure, 188
    vector in three dimensions, 34
gravitation
    equal to geometry, 132
    Newton's law of, 11n13, 95, *see also* Newton's
        law of g.
gravitational
    collapse, 190
    constant, G, 95, 227
    field, 4, *see also* field, gravitational
    mass, m_G, 95
    potential energy, 172
    redshift, 170, 179–180, *see also* Pound–Rebka
        experiment
        in scalar gravity, 231
        quantum mechanics, 97
    wave, x, *see also* wave, gravitational
graviton, **158**
gravity
    alters spatial geometry, 98
    Einstein, couples to energy and mass, 97

equivalence principle, 98
GR as a theory of, ix
Newton, couples to mass, 97
sourced by stress tensor, 143
stretches time, 97
tidal forces, 138
universal effects, 98
weak field Lagrangian, 139
Greeks, ancient
    symmetries, 4n4
Green's function
    advanced, D_A, 71n13, 75
    boundary conditions, 74
    D, not unique, 71
    retarded, D_R, 71, 72
        four potential A^μ, 75
        near $t = 0^+$, 72
        solves either wave equation, 73
Griffiths, David J., ix, x, 152n19
gun, jumping the, 88
gyroscope
    equation of motion, 228
    precession of axis, 136

Hall, Brian C., 145n12
Hamilton's principle, 50, 61, 227
Hamiltonian, 97
harmonic
    coordinates, 151n18, 154, 156, *see also* Lorenz
        gauge
    force, 84
    potential, 214
Harvard University, ix
Hawking, Stephen W., 220–222
Heaviside–Lorentz units, 66n8, 78
helicity
    graviton, 158
    photon, 156
Hewish, Anthony, 220
Hooke's law, 38
horizon, Friedman universe, 217
Hoyle–Bondi steady-state model, 204
Hubble, Edwin P. 209n7
    Hubble's constant, **209**
    Hubble's law, **209**
        in Milne's model, 33
Hulse, Russell, 221
Huygens' principle, 74
hydrostatic equilibrium, 188
hyperbola, 26
    constant cosmic time, 216
    Kruskal coordinates, 194
hyperbolic
    functions, 18
    motion, **26**, 91
        charge, 90–92
        lack of radiation reaction force, 92
        zero radiation reaction, 90

hyperboloid
  Friedman universe, 217
  Lorentz group, 14, 15
hypersphere, 186, 187

induction field, 79, 89, 92
inertial frame, 25, *see also* reference frame, inertial
inertial mass, m_I, 95
  $\propto m_G$, gravitational mass, 95
inhomogeneous Lorentz group, 11, *see also*
        Poincaré group
integration
  covariant, 225
  on a manifold, 113–115
internal symmetry, **54**
intrinsic geometry, 100, 129
invariance
  conformal, 198
  dilation, 10, *see also* dilation
  gauge, 70, *see also* gauge invariance
  geometric, 5
  group properties, 4
  inconsequential, 5
  internal symmetry, 54
  isometry, 163
  Lagrangian, 49
  laws of motion, 4
  Lorentz, 11, *see also* Lorentz group, Lorentz
        invariance
  Newton's first law, 8
  Poincaré, 6, *see also* Poincaré invariance
  rotational, 165
  time reversal, **165**
  time translation, 163
  translation, 46
irreducible representation, 38
  conservation laws, 40
  rotation group, 38
isometry, **163**
  rotational invariance, 165
  time reversal, 164
  Weyl's postulate, 204
isotropic coordinates, 208
isotropy
  all galaxies follow geodesics, 205
  Cartan's problem, 206
  cosmological principle, 204

Jackson, J. David, 65n6
Jacobian, 106, 107, 110, 111, 115,
        160, 223

Kaiser, David, ixn1, 222
Kepler problem, 173
  relativistic with Coulomb potential, 230
Kerr, Roy P., 221
kinematic quantities, 40, 46
  asymptotically conserved, 61

kinematic theory, Milne, 204
kinetic energy, 41, *see also* energy, kinetic
  Lagrangian, 49
Klein–Gordon field, 227
Kronecker delta, 36, 38, **223**, 230
Krotkov, Robert, 97n4
Kruskal, Martin David, 193n19
Kruskal–Szekeres coordinates, 193, 215

Lagrangian, **48**
  Einstein's equation, empty space, 144
  electromagnetism, 63
  not uniquely defined, 49
  particle mechanics, 48–51
    relativistic particle, 49–51
  relativistic fields, 51–58
    charged, 51
    geometric invariance, 55–58
    internal symmetry, 54–55
    Klein–Gordon, 51
Landau, Lev Davidovich, 201n1, 203n4
Laplace's equation, two dimensions
        conformal transformations, 131
Laplacian
  d'Alembertian as 4-dimensional extension
        of, 65
Larmor formula, **89**
length
  along a geodesic, 137
  contravariant vector A^μ, 122
  in embedded space, 121, 196
  in general relativity, **223**
  Lorentz invariance, 20
  on manifold, 100
  perpendicular to motion, 23
  preserved by isometry, 163
  preserved by parallel transport, 123
  proper time, 21
  proper velocity, $dx^\mu/d\tau$, 25
  spacetime interval, 122
lepton number, 39n9, 55
Levi-Civita symbol, 110
  in n dimensions, 224
  in four dimensions, $\epsilon_{\mu\nu\lambda\sigma}$, 36n3
  in three dimensions, ϵ_{ijk}, 64n2
Levin, David, ix
Levin, Janna J., 222
Liénard–Wiechert potential, point charge, 78
Lifshitz, Evgeny Mikhailovich, 201n1, 203n4
light cone, 23, 77, 91, 216
LIGO, 160n23, 180n13, 221
line integral, 107
line of simultaneity, 24
linearized Einstein equation, 150–153
local conservation, 55, *see also* conservation, local
Lorentz covariant, **132**
Lorentz gauge, 65, *see also* Lorenz gauge

Lorentz group, **11**
  hyperboloid
    1 sheet, 15
    2 sheets, 14
  proper *vs.* improper, 16, 115
  representations, 37
    conservation laws, 41
  structure, 13–19
  subgroup of Poincaré group, 11
  subgroups of, 17
Lorentz invariance, 11n14
  representations, 40
  tachyons consistent with, 32
Lorentz transformation, **11**, 20
  boost, 17
  infinitesimal, 56
  invariant quadratic form, 12
  Levi-Civita pseudotensor invariant, 36
  metric tensor invariant, 35
  Minkowski space isometry, 163
  parameters, 14
  proper, det = 1, 110
  symmetric, antisymmetric tensor pieces, 37
  tachyons, 32
  time and space transform, 34
  traceless, metric parts, 37
  with dilations, 19
Lorentz, Hendrik A., 65n6, 82, 86n25
  mass due to field energy, 90
Lorentz–Fitzgerald contraction, **22**, 28, 29
Lorenz gauge, **65**, 67, 70, 81, 151, 154
Lorenz, Ludvig V., 65n6
luminosity
  apparent, 209
  distance, **210**
  intrinsic, 209

MacLane, Saunders, 10
magnetic field, 64, *see also* field,
        magnetic
manifold, **101**
  affinely connected, 115
  coordinates differentiable, 101
  differentiable, **102**
  embedding theorem, 195
  integration on, 113–115
  Kruskal–Szekeres coordinates, 193
  Minkowski
    isometry, 163
  parallel transport, 115
  Riemannian, **122**
    isometry, 163
    metric, 132
  Stokes's theorem, 107
  structure, 101
  time-independent metric, 164
  Weyl's postulate, 204

mass
  bare, 68, 82
    negative, runaway mode, 89
  center of, rel. generalization, 48
  conserved $\neq$ in rel. collisions, 42
  couples to Newton's gravity, 97
  diminished by binding energy, 183
  effective, 69, 181
  electrostatic origin (Lorentz), 90
  equivalence principle, 98
  gravitational, m_G, 95
  incompatible with dilations, 10
  inertial, m_I, 95
  $m_I \propto m_G$, 95
    equivalence, 95
  photon rocket, 42
  photons, 42
  physical, 82
  renormalization, 80
  role in metric, 132
  Schwarzschild metric, 162
  unit of meters, 156n21
  with energy, couples to Einstein's gravity, 97
Mather, John, 221
matter
  curves spacetime, 132
  source of gravity, 147
Maxwell field
  field equations
    free, 64
    sourced, 66
  field tensor, $F^{\mu\nu}$, **63**
  gauge invariance, 65
  inhomogeneous wave equation, 75
  Lagrangian, 63
  Maxwell's equations, 67
  stress tensor, 66
Maxwell's equations, 43
  analogous to Einstein's equation, 138
  charge in, similar to role of mass in metric, 132
  curved space, 133
  incompatible with materialized charge, 75
  like linear Einstein equation, 151
  Lorenz gauge, 81
  source-free, **65**
  sourced, **67**
  time-reversal invariant, 76
mechanics
  classical, 38
    discrete symmetries, 48
  nonrelativistic
    Lagrangian, 49
  quantum, 42
    gravitational redshift, 97
    runaway solution, 86
  relativistic particle, 34
    Lagrangian, 49–51
  statistical, 76

Mercury's perihelion shift, 170–176
  as classic test, 182–184
  in scalar gravity, 231, 233
metric
  altered by gravity, 98
  Cartan's problem, 206
  comoving coordinates, 205
  conformal transformations, 198
  embedding, 186
  Euclidean
    isometry, 163
  gauge invariance, 162n2
  hyperboloid and Friedman universe, 207
  hypersphere embedded, 187
  isometry, 164
  Kruskal–Szekeres coordinates, 193
  Minkowski $\eta_{\mu\nu}$, 9, 138, **224**
    elevator frame, 99
    Lorentz tensor, 35
  point mass, 162, *see also* Schwarzschild m.
  Robertson–Walker, **208**
    Christoffel symbols (Γ's), 212
    cosmology, 210–215
    Einstein equation, 211
    singularities, 215
  Schwarzschild, x
  signature, 204
  spherically symmetric, 162, 183, *see also*
      Schwarzschild metric
  stationary, **164**
  time-independent, 164
metric space, 100
metric tensor $g_{\mu\nu}$, **9, 223**
  $\approx$ Minkowski if gravity weak, 138
  conformal trans., 131
  constant $\Rightarrow$ flat space, 129
  determines distance, 104
  Γ's expressed in terms of, 124
  geometry encoded in, 132
  $g^{\mu\nu} \neq g_{\mu\nu}$ in general, 122
  gravitational potential, 134
  invariant under Lorentz trans., 36
  obtained from $G^{\mu\nu}$, 162
  raise/lower indices, 36, 111, 122
  Schwarzschild, 162–170
  spherically symmetric, 166, *see also*
      Schwarzschild metric
  symmetric on Riemann manifold, 122
  variation of its determinant, 146
Milne model, 32, 204, 210, 216
Minkowski
  coordinates, 179, 180, 184
    asymptotic, 202
  force
    as relativistic form of Newton's
        second law, 51
  metric $\eta_{\mu\nu}$, **224**
  space, 202

$\epsilon = 0$ Friedman model, 216
  asymptotic, 201–203, 230
  hyperboloid in, 207
  indistinguishable from empty space, 117, 121
Misner, Charles W., 222
model
  Bondi–Gold, universe
    Robertson–Walker metric, 210
  Einstein–Lemaître, universe, 215
  Friedman, universe, 213
    $\epsilon = -1$, sphere $\neq$ empty, 216
    $\epsilon = 0$, flat, 216
    $\epsilon = 1$, expands, 216
  Hoyle–Bondi steady state, 204
  Lorentz, electron, 86n25
  Milne, universe, 32, 216
    kinematics, 204
    Robertson–Walker metric, 210
  sun as incompressible fluid, 190
Møller, Christian, x
momentum
  conservation, global
    divergence, 201
  conservation, local
    covariant divergence, 162
    Friedman universe, 211
  conserved, globally
    pseudotensor $t^{\mu\nu}$, 202–203
  electromagnetic density, 66
  field, 67, 80
    radiated, 89
  nonrelativistic
    conservation, 41
  photon, 42
    de Broglie wavelength, 156
  relativistic, **41**
    canonical stress tensor $T^{\text{CAN}}_{\mu\nu}$, 57
    conservation, 42, 46, 47
    non-Lagrangian dynamics, 228
    particle, 61
    stress tensor, 227
    symmetric stress tensor $T^{\text{SYM}}_{\mu\nu}$, 60
  stress tensor
    source of Einstein tensor $G_{\mu\nu}$, 143
  twist and antitwist, 232
momentum space, 62
Moniz, Ernest J., 86n25
monopole, 159
Mr. Tompkins in Wonderland, 31n20

neutron star, 220
Newman, Ezra T., 221
Newton's law of gravitation, **95**
  Kepler problem (Mercury), 171, 172
  kinetic energy
    angular, 182
    radial, 182

potential energy, 180, 182
recovery from Einstein's theory, 151–153, 170
weak-field limit of scalar field theory, 233
Newton's laws
  first, 6, 8, 11
    affine group, 8
    geodesic equation as generalization of, 135
    Poincaré group, 10
  second
    equivalence of m_I and m_G, 95
    from Euler–Lagrange equation, 49
    relativistic form, 51
Newton, Isaac, 4*n*4
  pendulum test of m_I and m_G equivalence, 96*n*2
Newtonian limit
  determination of constant in Einstein's equation, 144
  Schwarzschild metric, 169, 184
Newtonian line element, **154**
Noether, Amalie Emmy, 55*n*24
  Noether's theorem, 55*n*24
nonorthochronous, **15**, 16
Novikov, Igor Dmitriyevich, 220, 222

observation, **23**
observer, 8, 21–23, 32, 92, 192, 198, 204, 208
Okun, Lev B., 65*n*6
open set, 102, 226
orbit
  Einstein's equation, 164
  isometry, 163–166
  Mercury, 172–176
  range of transformation, **163**
orthochronous, **15**
oscillatory motion
  cosmological constant, Λ, 214
  radiation, 159
  reaction, 84

Pais, Abraham, 175*n*9
paradox
  bandits and train, 27–28
  prisoner's escape, 28–29
  twin, 25–26, 90
parallel displacement, **103**
parallel transport, **115**
  about a loop, 127
  contravariant vector, **224**
  density, 118
  distant, 130
  metric tensor, 123
  tensor, 118
parity, 16
  in classical mechanics, 48*n*15
passive transformation, 11–12
  $\not\Rightarrow$ conservation laws, 101
past, **13**

patch, coordinate, 100, 101, 129
Pauli, Wolfgang, x, 91
Peebles, James, 221, 222
pendulum
  Eötvös experiment, 96
  Newton's experiment, 96*n*2
Penrose, Roger, 31*n*20, 221
Penzias, Arno A., 220
perihelion, **173**
  precession of Mercury's, 171–176
    as classic test, 182–184
    in scalar gravity, 231
Perlmutter, Saul, 221
photon
  energy and effective mass, 181
  gravitational redshift, 181
  helicity, 156
  Hubble's law, 209
  mass, 80
  momentum, 156
  null geodesic, 195
  rocket, 42
  spin, 159
  zero mass, 156
Pisa, Leaning Tower of, 96
Planck's formula, 42
plane wave
  electromagnetic
    polarization, 155
    potential A^μ, 43
    solution, 155
  gravitational solution, 157, 160
Poincaré invariance, 6–11, 46, 142
  conservation laws
    fields, 55–58
    particles, 61–62
  group, **11**
  local, 142*n*9
    translations $\not\Rightarrow$ commute, 142
  transformation, **11**
point charge
  electromagnetic interaction, 68
  field of moving, 79
  Kepler problem, 230
  Liénard–Wiechert potential, 78
point mass
  metric produced by, 162, *see also* Schwarzschild metric
  Newton's law of gravitation, 10
Poisson's equation, 152
polarization
  electromagnetic waves, 155–156
  gravitational waves, 157–159
pole singularity
  interior Schwarzschild solution, 190
Politzer, H. David, ix
Popov, Victor Nikolaievich, 222
positron, 88

potential
  Coulomb, 78
  electromagnetic A^μ, 43, 63
    gauge invariance, 70
    $\not\exists$ runaways $\Leftarrow$ bounded, 88
    wave equation with source, 75
  gravitational $g_{\mu\nu}$, 134
  gravitational gh, 97
potential energy, 45
  de Sitter universe, 214
  Friedman universe, 213, 215
  gravitational, 172, 182, 187
  Lagrangian, 49
  mass units, 156n21
  Newtonian, 180
  photon, 181
Pound, Robert V., 180
  Pound–Rebka experiment, 180, 181
Poynting vector, **66**
preacceleration, 88, 90
precession
  gyroscope, 136
  Mercury perihelion, 171, *see also* perihelion,
    precession
principle of equivalence, *see* equivalence principle
prisoner's escape, 28, *see also* paradox, prisoner's
    escape
propagation, gravitational wave, 154
proper time τ, 20, **228**
  acceleration, 34
  extremized on geodesics, 135
  parameter, 50
  velocity, 32, 34
pseudotensor, gravitational $t^{\mu\nu}$, 201–203
pulsar, 220

quadrupole oscillation, 159

radiation
  electromagnetic, 76
    damping, 84–88
    point charge, 76–80
  field, 79
  gravitational, 159
    linearized theory, 231
  reaction
    finite, 82
    hyperbolic motion, 90–92
    oscillatory motion, 84
radius
  Schwarzschild, 169, *see also* Schwarzschild
    radius,
  of universe, 208
rank, tensor, 35
rapidity, **18**
Rebka, Glen A., 180
redshift
  cosmic
    formula, 209

Hubble's law, 209
  gravitational, 97, 170, 179–180,
    see also gravitational redshift,
    see also Pound–Rebka expt.
reference frame, 44
  appearance of moving object, 29
  asymptotic Minkowski, 202
  comoving, 25
  elevator coordinates, 99, 121
  inertial, 25, 36
  local Minkowski, 180, 185
  prisoner paradox, 28
  radiation measurement, 92
  SR *vs.* GR, ix
regularization, 80–82
relativity
  general, viii, *see also* general relativity
  special, viii, *see also* special relativity
renormalization, mass, **80**
reparametrization, **45**
representation
  group, **37**
    irreducible, **38**
  Lorentz group, 37, 40
rest frame, 42, 86, 228
  radiative absorption, 44
Ricci scalar R, **226**
  coordinate independent, 191
  gravitational Lagrangian, 144
  trace of Ricci tensor, **126**
Ricci tensor $R_{\mu\nu}$, **126**, 144, **226**
  symmetric, 127
Riemann tensor $R^\mu{}_{\nu\lambda\sigma}$, **121**, **226**
  Bianchi identities, 226
  if zero, space is flat, 226
  in three spatial dimensions, 127, 226
  in two spatial dimensions, 127, 226
  symmetry, 226
Riess, Adam G., 221
Robertson–Walker metric, **208**
  Einstein tensor in, 211
  in cosmological models, 210
  in locally Euclidean coordinates, 211
  singularities in, 215
rocket
  photon, 42
  twin paradox, 25
Rohrlich, Fritz, 92n30
Roll, Peter G., 97n4
rotation
  group representations, 38
  with time translation, 165
Rubin, Vera, 221
runaway mode, 85–89
Ryder, Lewis H., 222
Ryle, Sir Martin, 220

scalar
  conserved quantity, 55, 200

curvature, 126, *see also* curvature, s.
field, 51, *see also* field, scalar
product, relativistic, 103
Ricci, 126, *see also* Ricci scalar
scattering, relativistic, 39–41, 88
Schiffer, Menaham, x
Schmidt, Brian P., 221
Schrödinger, Erwin, x, 117n14, 120n17
Schutz, Bernard F., 222
Schwarzschild metric, x
    Christoffel symbol table, 167
    classic tests of, 180–184
    derivation, 162–170
    equatorial geometry, 195
    gravitational analog to Coulomb's law, 162
    in Kruskal–Szekeres coordinates, 193–195
    interior solution, 184–190
    singularity, 191–192
        compared with Robertson–Walker metric, 215
        real *vs.* phony, 191
        tidal stress near genuine, 197–198
        time and space trade places, 192
Schwarzschild radius, **169**, 192
    fall time, 192
Schwinger, Julian, 58n26
Sciama, Dennis, 220
Shapiro, Irwin I., 221
Sharp, David H., 86n25
shear modulus, 38
signature, 9, 196, 197, 204
simultaneity, lines of, 23
singularity
    Schwarzschild metric, 169, 191,
        see also Schwarzschild metric, s.
    geometrical artifact, 187
    pole, 190
Smoot, George, 221
Sohn, Richard, x
source
    electromagnetic, 66, *see also* current
    gravitational, 143, *see also* stress tensor
space-like
    separation, 77
    surface, **52**
special relativity, x
    c constant, 8
    conservation law, 200
    holds absent gravity, 117
    Lorentz covariance, 132
    mc^2 as energy, 97
    Minkowski metric, 224
    Minkowski space, 13
    $T^{\mu\nu}$ defined in, 148
    transformations are linear, 109
spectral lines, 180, 209
sphere
    intrinsically curved, 129

spherical symmetry
    Friedman universe, $\epsilon = -1$, 207
    generic metric, 183
    no gravitational radiation, 231
    Schwarzschild metric, 184
    static coordinates, 185
    static metric, 166, 181
    stress tensor, 217
    universe, 204
spin, **58**
    graviton, 158, 159
    photon, 156, 159
standard candle, 209
star
    aberration of starlight, 32
    Doppler shift, 43
state, equation of, 184, 190, 228
stochastic electrodynamics, 75n15
Stokes's theorem, **108**
    dual to Gauss's theorem, 114
    rank-1 tensor, 115
strain tensor S_{ij}, 38
stress tensor T_{ij}
    elastic, **38**
stress tensor $T^{\mu\nu}$
    canonical, **57**
    divergence-free, 153
    electromagnetic, 70
    fluid, 191, 228
    gravitational, 143
        recovering Einstein's equation, 147
        recovering Newton's law, 151
        with pseudotensor $t^{\mu\nu}$, 201
    interior of sun, 185
    Klein–Gordon field, 148
    Minkowski space, 202
    particle, 228
    perfect fluid
        moving, 185
        rest, 184
    spherically symmetric, 217
    symmetric, **60**
SU(3), 38n7
summation convention, Einstein, **8**, 64
Sylvester's law of inertia, 10
symmetric tensor
    sum of traceless, and multiple of metric, tensors, 37
symmetry, **4**
    coordinate-free definition, **163**
    cyclic, 124
    discrete, 48n15
    group, 38
    internal, **54**
        conservation law, 55
        relativistic field, 55
    O(3, 2), 143
    O(4, 1), 143
    spherical, 166, *see also* spherical symmetry

synchronization
    clocks, 21
Synge, John L., x
system
    coordinate, 16, 106, 110, 117, 121, 129, 164,
        170, 191, 201, 202, 208, 223
    inertial, 36
    physical, **3**
    reference frame, 28
Szekeres, George, 193*n*19

tachyon, 32
tangent space, **101**, 102, 103, 116, 123, 201
tangent vector, 101
    isometry orbit, 163
Taylor, Edwin F., 23*n*19
Taylor, Joseph, 221
tensor
    antisymmetric, **37**
        Stokes's theorem, 107
    contraction, 36
    contravariant, 35
    coordinate transformation, 104
    covariant derivative, **119**
    curvature, 138, *see also* Riemann tensor $R^{\mu}{}_{\nu\lambda\sigma}$
    $\delta\Gamma^{\lambda}_{\mu\nu}$, 145
    density, **110**, 201, 223
        $g^{\mu\nu}$, **146**
        coordinate transformation, 110, 223
        covariant derivative, 146, 224
        Levi-Civita pseudotensor, 111
        parallel transport, 118
    duality, 113
    Einstein, 143, *see also* Einstein tensor $G^{\mu\nu}$
    electromagnetic field, 133, *see also*
        electromagnetic field tensor $F^{\mu\nu}$
    energy–momentum, 57, *see also* stress tensor
    gravitational plane wave, 157
    metric, 9, 35
    mixed, 36, 104
    parallel transport, 118
    raise/lower index, 122
    rank, 35
    rank 0, 36
    Ricci, 126, *see also* Ricci tensor $R_{\mu\nu}$
    Riemann, 120, *see also* Riemann tensor $R^{\mu}{}_{\nu\lambda\sigma}$
    stress, 57, *see also* stress tensor
    symmetric, **37**
    tensor density, 120
    trace, **37**
Teplitz, Doris R., 68*n*10
Terrell, James, 31*n*20
test
    classic, 153, *see also* classic test of GR
    function, 73
    zeroth, 171, 182, *see also* classic test of GR,
        recovery of Newton's law

thermodynamics, second law, 76
theta function, 72, 77, 78
Thorne, Kip S., 220–222
tidal force, 137, **159**, 228
    near singularity, 197, 198
    Riemann tensor, 138
time
    asymmetry, 76
    dilation
        gravitational, 97
        in SR, **21**
    paradox, 23
    translation
        commutes with rotations, 165
        invariance, 163
Tompkins, Mr., 31*n*20
torsion pendulum, 96
torus, 108
trace of tensor, **37**
trajectory, 3, *see also* world line
transformation
    coordinate, 102, *see also* coordinate
        transformation
    Lorentz, 11, *see also* Lorentz transformation
    Poincaré, 11, *see also* Poincaré invariance
translation
    generator, 142
    Poincaré group, 11, 46
    local, **57**
twist, 232
Tyson, Neil deGrasse, 222

uniform acceleration, 21, *see also* acceleration,
        uniform
    equivalence principle, 92, 98
    hyperbolic motion, 26, 90
    Lorentz invariant, 25
    twin paradox, 25
uniform gravitational field
    equivalence principle, 98
    gravitational redshift, 97
units
    acceleration, 26
    Heaviside–Lorentz, 66*n*8, 78
    mass, 156*n*21
universe
    as a fluid, 204, 210
    cosmological constant, Λ, 214
    cosmological principle, 204
    expanding or contracting, 208
    Friedman models, 213
    homogeneous, 208
    horizon, 217
    Hubble's law, 209
    model, 32, *see also* model
    structure as a whole, 203
    Weyl's postulate, 204
upness, 9

vacuum-cleaner drive, 45
vector
  contravariant, **35**
    coordinate transformation, 102
  covariant, **35**
    coordinate transformation, 103
  field, 5, *see also* field, vector
  light-like, **12**
  space-like, **12**
  time-like, **12**
velocity, proper, **25**, 32, 34, 44, 70, 134, 228
  fluid, 185
  $\perp$ to proper acceleration, 25

Wald, Robert M., 222
watch time as proper time, 20
wave, electromagnetic, 154–156
  two polarizations only, 155
  wave equation
    Green's function, 70–74
    homogeneous, 65
    inhomogeneous, 70
  wave number, 42
  wave vector, 179
    Doppler shift, 44
wave, gravitational, 154, 156–159
  LIGO, 160*n*23
  linearized Einstein equation, 151
  plane-wave solutions, 156, 229
    two independent, 229
wavelength, de Broglie, 156
Weber, Joseph, 221

weight
  gravitational force, 98
  of tensor density, 110
Weinberg, Erick, 219*n*15
Weinberg, Steven, 222
Weiss, Rainer, 221
Weisskopf, Victor F., 31*n*20
Weyl, Hermann, x
Weyl's postulate, **204**, 206
Wheeler, John A., 23*n*19, 76*n*16, 220, 222
white hole, 195*n*20
Wichmann, Eyvind H., x
Will, Clifford M., 180*n*13, 222
Wilson, Robert W., 220
work
  external force, 89
  fields on charges, 68
world line, **3**, 20
  as parameter, 49, 51
  curved, 21
  everywhere time-like, 20
  free particle
    geodesic, 135, 137
  galaxy, 204
    geodesic, 205
    $\perp$ to t = const. surfaces, 206
  hyperbolic motion, 90
  proper time τ, 21, 68, 137
  tachyon, 32
  trajectory, 3

Young's modulus, 38

Zee, Anthony, 222
Zeldovich, Yakov Borisovich, 220, 222
Zwicky, Fritz, 221